The Conduct of Inquiry

An Introduction to Logic and Scientific Method

Richard F. Kitchener

University Press of America,® Inc.
Lanham • New York • Oxford

University Press of America,® Inc.
4720 Boston Way
Lanham, Maryland 20706

12 Hid's Copse Rd.
Cumnor Hill, Oxford OX2 9JJ

Printed in the United States of America
British Library Cataloging in Publication Information Available

Library of Congress Cataloging-in-Publication Data

Kitchener, Richard F.
The conduct of inquiry : an introduction to logic and scientific method / Richard F. Kitchener.
p. cm.
Includes bibliographical references and index.
1. Science—Methodology. 2. Logic. I. Title.
Q175.K532 1998 001.4'2—dc21 98-33435 CIP

ISBN 0-7618-1307-1 (pbk: alk. ppr.)

™ The paper used in this publication meets the minimum requirements of American National Standard for Information Sciences—Permanence of Paper for Printed Library Materials, ANSI Z39.48—1984

To Willard O. Eddy

who started me on this path

Contents

Analytical Contents

Preface

Over a decade ago, I started teaching a course entitled, *Logic and Scientific Method.* The course had been created during the post-World War II era by Willard Eddy, who taught it for over 20 years. It had been one of the first courses in philosophy taught at our university, the orientation of which was toward the applied and practical sciences. As a result, any course in philosophy had to be one that students in these applied areas would find relevant to their studies, but also one with philosophical substance. Willard used the standard text, Cohen and Nagel's *Introduction to Logic and Scientific Method*, and the course proved to be a phenomenal success.

When Willard's teaching and administrative duties became too burdensome, he asked me to take it over. Naturally, I continued in the same vein as Willard had; but the Cohen and Nagel textbook was out of print, and for all of its admirable qualities, I felt there were certain features of it that needed to be either updated or supplemented with more current topics. After trying a number of other textbooks, I finally decided that what was needed was a textbook that was a revision and updating of the Cohen and Nagel book. As I started working on such a project, I soon realized that it was taking me in a somewhat different direction. Still, I persevered over the years as I taught the course and wrote chapters when I could.

As I pursued this project, a single theme guided my work: the manuscript should be a *textbook.* Textbooks should be written for students and not, as they often are, for one's professional colleagues. Hence, not only must the material be presented at an elementary level, certain topics and issues that attract the major attention of philosophers are not necessarily the issues that should be the focus of a textbook. In addition, a textbook on logic and the scientific method should, in my opinion, be a practical textbook, i.e., it should be useful to the student. This means, first of all, that students should have considerable practice in developing certain reasoning skills necessary for thinking about scientific issues and evaluating the arguments of scientists. This best comes, in my opinion, by having the students work exercises. Hence, whenever possible, I have included exercises and suggested answers. I have also been driven by the conviction that philosophical textbooks should be relevant to the scientist (and not just to the philosophy student). That has dictated a certain style of presentation and selection of topics. But I also believe that philosophy students need to be conversant with science; hence I have tried whenever possible to include examples and case studies from the history of science.

According to a widespread view, the greatest achievement of mankind has been science—scientific knowledge. But what is unique about scientific knowledge and what gives it a claim to superiority over other forms of knowledge, is its method of acquiring knowledge. Hence, a knowledge of the scientific method is essential for a proper appreciation and understanding of the nature and place of science in the modern world. Even though I have some misgivings about the above view, its conclusion seems to me to be correct (as I indicate in Part I). But one cannot understand the scientific method, what we can call the *logic* of the scientific method—understanding by that term the rationale and justification of the scientific method—without understanding the rudiments of logic. Hence I have presented those topics in logic that seem absolutely essential for understanding the nature of scientific method; this includes an introduction to deductive and inductive logic, a summary of the propositional calculus, a discussion of the various types of inductive inferences together with a brief presentation of the problem of induction, and a chapter on probability. These topics on the formal aspects of scientific reasoning form the second part of the book. Finally, the bulk of the book—part 3—is taken up with aspects of scientific method per se and this has meant the *procedures* of science: observation, qualitative classification, definition, quantitative classification, measurement, analogy, eliminative induction, hypothesis testing, experimentation, and statistical inference. Some of these topics, e.g., the logic of experimental design and statistical hypothesis testing, are not usually presented in textbooks on logic and scientific method, but they seem to be absolutely essential to modern science. There are numerous other scientific procedures one could have included, but this particular selection of topics resulted in a book longer than I had intended. In a one-semester courses, for example, the instructor will need to select the material he or she wishes to stress; in my experience, one can cover about a chapter a week. However, most of the topics in scientific method have been included, so that there should be enough material for several different type of courses, e.g., instructors can easily include readings from original sources to supplement the course material.

I wish to thank Rod Adams, Jeff Gliner, and Phil Wood for their helpful reading and comments on several chapters, and the Endowment for Applied Philosophy of the Department of Philosophy, Colorado State University, for its financial assistance during several summers. I also wish to thank all of my past students on whom I have tried out various chapters and who have given me many suggestions for improvement.

PART ONE

Introduction

1

Logic and Scientific Method

General Introduction

Introduction

Although there are many things characteristic of the period from the Renaissance to the 20th century, perhaps the most important feature of this period is the rise of modern science, a change so radical that this period is often known as **the scientific revolution**.[1]

First in astronomy and cosmology and then in physics, chemistry, and later biology, there was a twofold radical change. First, there was a revolutionary change in the kind of theories of the world advanced. Prior to this period, the dominant view of the world was the medieval world view, a combination of theoretical, philosophical and theological views characterized by an allegiance to Aristotelian physics and cosmology, Ptolemaic astronomy, and Christian theology. Beginning with Copernicus and Vesalius, and then including Brahe, Bruno, Galileo, Kepler, Harvey, Gilbert, Borelli, Boyle, Hooke, Descartes, Leibniz, and Newton (to mention just some of the most important figures), a series of new scientific theories were advanced that would prove to be the foundations of our modern scientific view of the world. In this new picture of the world, Aristotelian physics was replaced by Galilean and Newtonian physics, Ptolemaic astronomy was replaced by Copernican and Keplerian astronomy, and a **teleological** cosmology and metaphysics were replaced by a mechanistic conception of the world.

Secondly, these same individuals began thinking about the world in a different way. What appeared as problems to be solved and questions answered, what was taken to be acceptable answers to these questions and solutions to these problems, in fact, the very methods employed to solve these problems and to answer them radically changed. In their style of thinking, in the type of inferences drawn, in the pattern of reasoning employed, individuals began to think about the world in different way. According to J. F. West (1965), the intellectual revolution of the 17th century

> took place principally as a result of scientific discovery. Before 1600, men searching for truth would look to the writings of the ancients; after 1700, reason and experiment were held to be the crucial test of truth. There was, moreover, an increasing use of mathematics; and of the inductive method, which derives generalizations from a large number of instances, rather than the deductive method, which derives laws of nature from supposedly unquestionable axioms. (p. 2)

For example, prior to the 17th century, a teleological and qualitative way of thinking was often employed in reasoning about problems of nature. During the scientific revolution, however, individuals began to think about these issues in a non-teleological (mechanistic) way, involving more quantitative modes of thought;[2] experimentation increasingly replaced mere naturalistic observation and a naive trust in one's senses; an egocentric perspective gave way to a de-centered perspective; etc.

Individuals of the 17th century were aware that the "moderns" (as they were later called) thought differently from the "ancients." In his monumental work of 1620, *The New Organon,* **Sir Francis Bacon** attempted to describe the "new" logic (or organon) then being employed by scientists in conscious opposition to the older logic of Aristotle. What Sir Francis Bacon was after was a description of "the scientific method," that way of reasoning characteristic of the new science. Although individuals such as **René Descartes** largely disagreed with Bacon's description of the scientific method and proposed a different type of "logic," Descartes agreed with him over the importance of describing the essential features of the scientific method. Hence, many historians have characterized this period as "the search for method."

Clearly, therefore, a key feature of this modern world view is the scientific way of thinking, the essential feature of which is the use of the scientific method. This book is about the nature of this scientific method: it is concerned with characterizing the underlying logic or rationale of this method. As we will see, it is important to understand the rudiments of logic, since logic plays an essential role in understanding the scientific method. But before we can understand the nature of scientific method we must first understand what a method is. We will then briefly discuss the scientific method and finally the connection between logic and scientific method.

What is a Method?

What then is a method? Most generally, we can say that a method involves two things: a set of procedures and a goal such that the procedures are supposed to result in the goal:

Procedures ⇒ Goal

By a *procedure* is understood anything in a broad category of things including rules, strategies, steps, techniques, means, or actions undertaken. By the *goal* we mean the aim, outcome, effect or (in general) what one hopes to achieve in following these procedures. For example, if one has a method for making a German chocolate cake; one has, first of all, a set of procedures one follows--a set of steps specified in the recipe for making the cake: beating the eggs, mixing flour and salt, adding chocolate, etc.--and a goal, which is the production of the cake. The recipe for making such a cake is evaluated in terms of its ability to produce an excellent German chocolate cake and likewise a scientific method is evaluated in terms of its ability to produce the respective goal in question. In employing scientific method, therefore, it is understood that there will be a set of procedures which should result in a certain goal. But what is the goal of using the scientific method, and what are these procedures?

The Goal of Science

What is the goal of using the scientific method? What do scientists hope to attain when they follow such a method? The most obvious answer here is *knowledge*, scientific knowledge. After all, the word "scientia" is just the Latin word for knowledge. Hence, when we speak of "scientific knowledge," we are really saying the same thing twice. But since many individuals distinguish the knowledge science obtains from the knowledge religion, philosophy or literature gives, it is still perhaps useful to speak of "scientific knowledge."

Scientific method is a set of procedures that, if followed, will (hopefully) result in the goal of scientific knowledge. In short, scientific method is a method for obtaining scientific knowledge. The study of scientific method is therefore part of the study called **epistemology**, the theory of knowledge: *Episteme* = the Greek term for knowledge, *logos* = the Greek word for 'theory of.' In particular, the **philosophy of science** is the specialty concerned with questions regarding the nature of scientific knowledge, scientific method, etc.[3] Because the philosophy of science (the epistemology of science) is such a large field, we will have little to say about the nature of scientific knowledge, at least explicitly. Our general perspective is that one should first understand the procedures (methods) used in the acquisition of scientific knowledge and, as a result, an account of the nature of scientific knowledge will emerge.

Procedures

The aspect of scientific method that will constitute the bulk of this book is concerned with the appropriate means to employ in order to reach this goal of science. These means are, we have said, the appropriate procedures to follow such that scientific knowledge will likely result. For example, many individuals have suggested that the steps of the scientific method can best be characterized as *the hypothetico-deductive method* in which one first constructs a hypothesis, then deduces a testable consequence from this hypothesis, then tests to determine whether the consequence is true or not, and finally draws a conclusion the truth or falsity of the initial hypothesis. Other individuals have argued that *the inductive method* is a better characterization of the steps involved in scientific method: first, one makes an initial observation, then forms a generalization (or law) from these observations, and finally constructs (or "induces") a more general theory. As we will see, both of these two views are (partially) correct since science uses both methods.

Methodology

Before proceeding, we must briefly discuss an important (but often confused) distinction—that between the *methods* involved in practicing the scientific method, and **methodology**, the theory of the methods involved. Although this distinction is not always easy to draw in a sharp way, it is the distinction between the actual *practice* of following the methods—the actual doing of science, which consists of a set of actions performed—and a descriptive account of what these methods are. This distinction concerns different levels of discourse in which the higher level of discourse (the meta-level) is about the lower level of discourse. For example, when an automobile mechanic fixes one's car, he is engaging in the practice of automobile repair. We might ask him to explain this practice to us, to describe it in a rationally coherent way. To do this, however, the mechanic must temporarily cease repairing an automobile, reflect upon his activities, and then he comes up with some kind of account of what these practices are and why he engages in them. When he provides us with such account, he can be said to be giving us a "theory" of automobile repair, and such theorizing is going on at a level different from the actual practice of repairing automobiles. This account of automobile repairing is what the Greeks meant by a "theory" and the Greek term 'logos' is a rough translation of "account." Hence, methodology is an account of scientific method (from 'methodos' + 'logos': a rational account of a set of methods). We can represent the difference between scientific method and scientific methodology, therefore, in the following way:

Second Level: methodology--theory of methods

First Level: methods = practices

This book deals with scientific methodology--the theory of scientific method. It is important to keep the distinction between practice and theory in mind. If the student wishes to learn how to engage in the actual practice of science, he or she is advised to take a science course in which he or she will receive practice in the actual doing of science. But if one desires an account or theory of scientific method, one must forsake actual scientific practice for reflection upon such practice.

The Theory of Scientific Method

If one is concerned with scientific methodology, how should one proceed? How does one determine what these scientific methods are and how should they be described? What constitutes an adequate theory of scientific method?

To begin with, we must distinguish *descriptive methodology* from *prescriptive methodology*, an example of the age-old **fact vs. norm (is vs. ought)** distinction. A **descriptive** methodology is a theory of scientific method in which one describes what scientists actually do when they are doing science; it is a description of the methods actually used by scientists. As ordinarily conceived, descriptive methodology is an empirical science and is pursued by psychologists, sociologists, and historians of science. Such endeavors are called 'descriptive' as opposed to 'prescriptive' because these attempt to describe what actually happens when scientists engage in science. What scientists actually do when they are doing science belongs to the factual realm and, hence, descriptive methodologies are factual methodologies.

By contrast, a **prescriptive methodology** is not so much concerned with describing the actual things done by scientists but with what they ought to do in order to obtain this knowledge. A prescriptive methodology thus prescribes, it doesn't describe. In contrast to a descriptive methodology, which might conclude that scientists do x, y and z in the hope of obtaining knowledge, a prescriptive methodology would conclude that in order to obtain knowledge, scientists ought to do x, y and z. As usually understood, therefore, a prescriptive methodology is a normative enterprise. Such an endeavor would be called philosophy of science by many people.

These same people would proceed to distinguish the philosophy of science, which is prescriptive, from the history of science, the psychology of science, and the sociology of science, which are descriptive. Although it is very difficult to distinguish these three areas, something (however brief) must be said about each of these areas.

The *psychology of science* studies the behavior and cognitive processes of individual scientists. According to the dominant view, it is concerned with constructing general theories about such behavior and cognition. As such, the psychology of science can be concerned either about the thought processes of particular scientists (e.g., what went on in Charles Darwin's mind when he came up with his theory of evolution) or general thought processes (e.g., what generally goes on in scientists' minds when they create a theory). The psychology of science is a very new field and, in fact, is now just being established.

The *sociology of science* is a much older field than the psychology of science and is concerned with the social processes involved in science. It studies the relation between science and society (the influence of social factors on science, the influence of science on society), the social institution of science, etc. For example, it might study the question of how frequently scientists cheat and what social and institutional factors encourage such cheating. Or it might study the question of how the governmental funding of science influences the course of science. As ordinarily conceived, the psychology of science and the sociology of science are differentiated in terms of whether one is studying an individual scientist or scientists in groups.

The oldest discipline is the *history of science.* It is difficult to distinguish the history of science from the psychology and sociology of science. For example, the history of science studies the behavior and cognitive processes of past scientists such as Darwin, just as the psychology of science does. Likewise, the history of science might study the influence of certain economic factors on the growth of science in 17th century England. But two things seem to be distinctive about the history of science. First, it is concerned with the past, whereas the psychology of science and the sociology of science are concerned with the present and the future. Secondly, according to the standard view, history of science is concerned with describing (and perhaps explaining) the unique and unrepeatable events of the past, whereas the psychology and sociology of science are concerned with general and repeatable events subsumable under generalizations.

Although no sharp line can be drawn between these three areas, they have one thing in common: they are all concerned with the facts of scientific activity. They are descriptive (empirical). But a quite different question concerns the normative or prescriptive study of science.

Suppose, for example, we discover that a certain percentage of scientists actually cheat: they fake their data, they "doctor" it up, they take credit for the work of others, etc. These are all factual matters. However, we could proceed and ask whether they ought to do this? And if not, what should they do? These questions concern what normative standards ought to be followed by scientists, whether they actually follow them or not, and they appear to belong to what is prescriptive methodology, which traditionally has belonged to the disciplines of logic and philosophy of science.

Prescriptive vs. Descriptive Methodology

In determining a methodology for science, which approach should be used--a descriptive one or a prescriptive one? The point of view adopted in this book is that both approaches are needed and that neither one is adequate by itself.

Can a purely descriptive approach to methodology be sufficient? Why can't we just observe what scientists actually do and leave it at that?

We have a problem here, for what should we observe and record? Is everything the scientist does relevant to scientific method? Suppose, for example, we observe a scientist coming into the laboratory, putting on a white coat, pouring a cup of coffee, walking over to her desk, inspecting a computer printout, etc. Should we include all of this material in our account of the procedures used by scientists? Should we, for example, include the drinking of coffee? Clearly this is not part of scientific method. But why not? The obvious answer is that it is not relevant to obtaining scientific knowledge. But how do we determine what is relevant? We need some theoretical account for deciding what is and what is not appropriate; simply recording everything the scientist does will not do. If this is what a purely descriptive methodology involves, it clearly is inadequate.

Furthermore, suppose we happen to observe a very bad example of a scientist, e.g., someone who is inefficient or a "hack" as an experimenter. Clearly, we don't want bad examples of scientific method, only good ones. So, we don't want merely to describe what actually happens everywhere; we only want to describe the good methods used by good scientists. But the notion of an ideal method or a good one is a normative concept and not a simple descriptive one.

On the other hand, we can't construct an adequate theory of scientific method in a purely prescriptive way, without regard to how science is actually practiced. If one advanced a purely ideal set of procedures, one every scientist ought to follow, this would be a purely prescriptive endeavor. But by itself this would not be adequate since it might be too idealistic, too far removed from the real world of actual scientists to be adequate. As many philosophers of science now claim, a prescriptive methodology must be tested and evaluated against actual science in order to be adequate. There are two reasons for this.

First, we are concerned with discovering an adequate methodology for science--science as it is--and not ideal, fictitious science. If our methodology were such that no actual scientist ever followed it, one could reasonably wonder what value such a methodology could have. Hence, a methodology must be a theory of the methods of actual science.

Second, it is a well-established principle of ethics that "ought implies can": to say that one ought to do something implies one can (is able) to do it. For example, there would be no point to my saying, "you ought to drive a car", if you are crippled and incapable of moving. Now, if one prescribes some norm that scientists ought to follow and it turns out that it is biologically, psychologically ,or sociologically impossible for them to follow it, then (according to the 'ought implies can' principle) such a norm cannot be legitimately prescribed. But what scientists can do is a question for the empirical sciences to decide, and such an answer is relevant to evaluating what scientists should do. Once again, our methodology must be both prescriptive and descriptive, both normative and factual.[4]

Scientific Method--One Method or Many?

Throughout most of the history of discussions of scientific method, individuals have written as if there were a single method called the scientific method. Why individuals have believed this is not clear. Perhaps they thought that there had to be something called the essence of doing science and such an essence required a single set of necessary and sufficient properties. Perhaps they thought that if there were several scientific methods, then confusion would follow since one wouldn't be able to tell if any alleged example was a true example of scientific method or not. Or perhaps they thought that such a situation would open the floodgates so that everything would eventually count as 'scientific,' much as one now reads about an individual being a "sanitation scientist," when what is really meant is that the individual is a janitor. In any case, it has become something of a tradition to accept the notion that there is just one thing called the scientific method.

But partly as the result of the work of **Ludwig Wittgenstein** and several contemporary cognitive scientists, the common assumption that every concept must have a set of essential, defining features has less plausibility than it once did. This opens up the possibility that there might be several different things called scientific method. For example, it might be the case that both the hypothetico-deductive method and the inductive method are examples of scientific method, even though they have importantly different properties. Indeed, it may very well be the case that there is a rather large list of very diverse activities that can be called 'scientific method,' including making an observation, performing an experiment, deducing mathematical proofs, constructing a mathematical model, applying such a model to concrete problem situations, reasoning about the plausibility of a certain theory, attempting to clarify an important concept, etc. Wittgenstein (1953) suggested that rather than looking for a set of necessary and sufficient conditions which might function as the essential definition of a concept (scientific method), it is more appropriate to think of examples of that thing as constituting what he called *family resemblances.* Members of a family may resemble each other in a variety of different and overlapping ways: brothers may resemble each other by way of their smiles, brother and sister may resemble each other by way of their noses, sister and father may resemble each other perhaps in an overall, nondescript way, etc., and yet between all of these members there is a family resemblance. Similarly, several things might count as 'scientific method' because there is a family resemblance between them.

Following up on the notion of a family resemblance, some cognitive scientists (Rosch & Mervis, 1975) have proposed a non-traditional account of the nature of a concept and conceptual categorization. An ordinary concept, such as bird, is a cognitive mental representation containing a paradigm-case of a bird. A robin, for example, is a prototype of bird, whereas a penguin is not. When we encounter examples of birds (e.g., owls), we classify these examples as birds because of their similarities to the prototype. The greater the number of similarities or family resemblances, the closer the example is to the prototype and the more certain we are that it is a bird. On this account, there is no single set of features making up a definition of 'bird'; instead, we categorize objects as birds according to their degree of resemblance to the prototype.

Suppose we employ such an account and apply it to the question, Is there a single thing called the scientific method? That seems to presuppose there is a set of necessary and sufficient conditions making up the essence or definition of scientific method, and it is notoriously difficult to come up with such a list. In some ways, doing philosophy is like doing science; in some ways, doing art is like doing science; in some ways engineering is like doing science, etc. Furthermore, we clearly have a *graded* conception of the sciences: biology, for example, is considered to be more scientific than psychology is, which, in turn, is more scientific than history is.

There seems to be a wealth of different activities that fall under the rubric of science: constructing a proof, measuring chemical elements, analyzing scientific concepts, observing stars through a telescope, etc. Are all of these things part of the essence of the scientific method? Must one do all of them to be scientific? How can one classify as science the disciplines of mathematics, ecology, nuclear physics, and social psychology? These questions have seemed perplexing to any one attempting to answer the question, What is the scientific method?, and it has led some individuals (e.g., Feyerabend, 1993) to argue there is nothing called the scientific method.

But using the concept of family resemblances and prototypes, we can reply that the concept of scientific method can be analyzed the way in which some cognitive concepts have been analyzed. True, there may not be a certain set of necessary and sufficient conditions (procedures) for something x to belong to the category *scientific method*. There are several activities and procedures classified as 'scientific' because they have a certain kind of similarity or resemblance to our prototypical case of scientific method, namely the set of procedures found in our best example of science--physics. Two activities in particular are characteristic of the new method found in the work of Galileo and

Newton, present also in the work of other great physicists, namely, the mathematical formulation of the underlying theoretical principles together with their experimental testing under controlled laboratory conditions.

Other disciplines are classified as scientific to the degree that they resemble our paradigm science and to the degree to which their procedures resemble the procedures of physics. Mathematics, computer science, even logic are taken to be scientific not because they do everything that physics does, but because these activities, in particular the activities of formalization and derivation, resemble the mathematical activities of physicists.

Likewise, some procedures are thought to be scientific because they resemble the experimental activities of physicists; the more experimental the discipline, the more scientific they are judged to be (e.g., chemistry, experimental biology). Disciplines that engage in activities closely resembling experimentation, e.g., naturalistic observation, are also judged to be scientific: observational biology, psychology, geology, and sociology but to a lesser degree.

The further away these sciences are from this experimental activity, the less they resemble physics and the less scientific they are judged to be. Some sciences, e.g., astronomy, are very mathematical but non-experimental; some sciences are very experimental but not very mathematical, etc. In short, there is a family resemblance between the family of the sciences, whose activities bear a resemblance to the scientific method found in our prototypical science. So, although there is no single set of procedures all and every science follows it is useful to group them together as 'scientific' and as following 'the scientific method' understanding that they are related by virtue of their family resemblances. It must be remembered, therefore, that if we write as if there were something called the scientific method, this is to be meant in the above sense.

The Logic of Scientific Inference

Up to now we have characterized scientific methodology as being concerned with a set of procedures to be used in obtaining scientific knowledge. Throughout most of this book we will be concerned primarily with this set of procedures, e.g., the steps involved in scientific observation, hypothesis testing, measurement, etc. As a result of following these procedures, scientists often obtain certain results and draw conclusions from them. For example, as a result of performing certain measuring operations, a scientist may conclude: "This object is x meters long." As a result of performing several laboratory procedures involved in experimentation, a scientist may conclude: "78% of our subjects improved their test scores" or "the latest scientific results indicate that large amounts of sodium chloride are harmful to one's health." Scientists typically spend a great deal of time drawing numerous conclusions such as: "Einstein's theory of general relativity is well-supported," "The theory of evolution is the only theory we currently have," etc.

In these judgments, scientists are making inferences, they are (1) drawing a conclusion, (2) which is presumably based upon some evidence. (Much of scientific method is concerned with the production of such evidence through appropriate procedures.) In encountering such scientific inferences, several questions naturally arise: Did the conclusion follow from the evidence?, Are other conclusions equally compatible with such evidence?, etc. These questions concern the nature of what we can call "good inference." Such questions belong to the realm of logic, which is concerned with determining the principles or rules involved in good reasoning. A substantial part of this book will be concerned with determining whether certain examples are examples of good scientific reasoning or not, with determining whether the scientific conclusion actually drawn is a good conclusion to be drawn from the evidence. In matters of scientific method, questions of logic are unavoidable, since scientific method is largely concerned with questions of good reasoning in science. This is why this book is entitled, *The Conduct of Inquiry: An Introduction to Logic and Scientific Method,* In summary, therefore, scientific methodology--the theory of scientific method--overlaps both with epistemology and with logic. These issues will be explicitly discussed in Part 3.

For Further Reading

The classical but somewhat outdated textbook on scientific method is Cohen & Nagel (1934). For historical overview of scientific methodology see Losee (1972). Burtt (1932), Randall (1940), and Butterfield (1957) discuss the rise of modern science. A recent textbook in the psychology of science is Gholson, et al. (1989); an introductory account of the sociology of science is given in Mulkay (1979); and a good introduction to the discipline of the history of science can be found in Kuhn (1988).

On fraud in science see Broad & Wade (1987), which makes for delightful reading.

Notes

[1] Whenever the reader encounters bold-face words, these are defined in the glossary accompanying each chapter.
[2] This is something of a caricature of the earlier period since one can find examples of medieval thinkers, e.g., the Merton School at Oxford, who also reasoned in this quantitative way.
[3] Scientific methodology is just one branch of the philosophy of science. Most of the recent work in philosophy of science has been concerned with questions about the (formal) structure of scientific theories, the nature of scientific explanation, the rationality of scientific change, etc. In this book, we will be concerned with these latter topics.
[4] The issue of facts versus norms, the question of the relation between the factual and the empirical and how (if at all) the empirical realm impinges on the normative realm, are extremely controversial topics in contemporary ethics, epistemology, and the philosophy of science. Clearly, it would be impossible to discuss these issues here in any systematic way.

2

Scientific Knowledge

Introduction

Throughout the history of western thought, many individuals have taken science to be the paradigm of knowledge. Although ordinary common sense, religion, literature, etc. may claim to know a variety of things, the judgement of history has been that it is scientific knowledge that sets the standard of what constitutes real knowledge. The dethronement of these other authorities began with the Greeks, was interrupted during the Middle Ages, and was finalized with the Scientific Revolution in the 17th Century. Some contemporary intellectuals (e.g., Feyerabend, 1993) have challenged this picture of the hegemony of scientific knowledge, but these challenges have, by and large, been unsuccessful challenges to the authority of scientific knowledge.

In any case, it is of crucial importance to inquire into the nature of scientific knowledge and how it is different from other ways of knowing. These questions arise and are dealt with in two major fields of philosophy—epistemology and the philosophy of science; they have been the subject of numerous articles and books. Needless to say, only the briefest survey of these fields can be given here.

Skepticism

When individuals first had the leisure to reflect upon the nature of man and his place in the universe, one of the initial questions asked concerned the nature and possibility of knowledge. This inquiry began with the Greek philosophers, and one of their most important schools was **Skepticism**.

There are several different types of skepticism. According to one (epistemic skepticism), *no knowledge is possible,* whereas another type (justification skepticism) maintains *no justification of one's beliefs is possible*. With respect to scientific knowledge, these positions would maintain that *there is no scientific knowledge (just a system of beliefs); the claims of science can't be rationally justified or shown to be objective* (hence, the claims of science are not epistemically better than any other system of discourse such as religion, astrology, poetry or crystal-ball reading).

Although there are several arguments advanced by the skeptic in favor of this position, one of the most popular ones is the following: no knowledge is possible because in order to have knowledge, one must have *certainty* and no such certainty is possible. To be certain about something it must be impossible to doubt it. But consider our alleged knowledge of the physical world. Is this knowledge absolutely certain and indubitable? If one looks at the history of science the answer is obviously "no,"since what was accepted as knowledge at one point in time was later rejected. The same, no doubt, will happen in the future: our current scientific theories will be radically revised and/or rejected by subsequent scientists who will have a different theoretical account of the world. But if so, then none of our scientific knowledge can be counted as certain, infallible, and indubitable.

Another way of expressing this doubt is the following: For something to count as knowledge, it must be true. But if it is true, then you can't be wrong about it, for if you were wrong about it, then obviously it wasn't true. But if scientific theories are changing, this means that scientists were wrong about what they earlier believed to be the case; but if they were wrong about these beliefs and theories, then these things weren't true and they were not examples of knowledge.

This argument of the skeptic is a powerful one. Throughout the centuries, philosophers (e.g., Descartes) have attempted to answer the skeptic and to show that we do have knowledge—*absolutely certain knowledge*. But few people have agreed that these arguments have been effective against the skeptic. In fact, many contemporary thinkers have come to believe that *it is impossible to answer skepticism.* This is because the skeptic has created the rules of the game in such a way that it is impossible to win. How can one *show* with absolutely certainty that there is such a rarefied kind of knowledge—knowledge that is absolutely certain and impossible to doubt since what it is logically possible to doubt seems unbounded? History is strewn with examples of such failure; hence this seems impossible. But, on the other hand, it certainly seems to be the case that we do have genuine examples of knowledge. Hence there seems to be something radically wrong with the demands of the skeptic. Perhaps, then, one shouldn't try to play the skeptic's game: perhaps one shouldn't try to answer the skeptic and to show that there is infallible knowledge. Recently a version of this position has been advanced as **fallibilism,** and it is one that many scientists hold.

According to the fallibilist, knowledge does not have to be infallible to be knowledge; for there is such a thing as *fallible knowledge*. To say knowledge (e.g., a scientific theory), is an example of fallible knowledge is to say that this theory is a case of genuine scientific knowledge but that it is not absolutely certain; furthermore, it is possible that this theory might be shown to be wrong at any point in the future, say, by new evidence. All of our scientific knowledge is *revisable*—revisable as we learn more and more about the world and about the nature of science. During this historical process, some of our more cherished theories will have to be rejected as inadequate and replaced by other theories; for example, as Newton's theory was replaced by Einstein's theory.

Although the skeptic will not be pleased with this line of argument, it is one that has been defended by several contemporary epistemologists and philosophers of science. It is also one that most scientists seem to adopt.

The Sources of Knowledge

If there is something that is genuine (scientific) knowledge, how can it be obtained? This is sometimes put as: *what are the sources of knowledge?*

By a source of knowledge, one understands something that not only produces knowledge—something that not only generates the knowledge—but also something that validates it by turning mere belief (or a hypothesis) into true knowledge. A source of knowledge accomplishes this by virtue of a property inherent in its very nature.

Throughout the history of thought, various sources of knowledge have been suggested:

1. sacred scripture
2. revelation
3. faith
4. authority
5. tradition
6. intuition
7. reason
8. experience
9. practice

A little reflection reveals that most of these sources of knowledge are not really sources of knowledge at all. Although some traditions may accept these as sources of knowledge, they are not really sources of (scientific) knowledge.

It is clear, first of all, that sacred scripture (e.g., the Bible, the Koran, etc.) cannot be a source of knowledge. The mere fact that a belief appears in these famous texts, or that an alleged event occurs, does not guarantee that the belief is really an item of knowledge. There are at least two problems with this approach. (1) Nothing about these texts in and of themselves constitutes the knowledge-making mechanism any more than finding a belief written down in a notebook makes this an item of knowledge. Perhaps these other texts might be sources of knowledge if we knew they were really sacred, but we don't. Someone might claim that they are sacred—they are the "word of God—and hence what they proclaim is true. But then the question is: How do you *know* they are the Word of God?; and to answer this question, one would have to cite a source of knowledge different from the alleged sacred text, for otherwise one would be giving a circular argument. Hence, it is this other source of knowledge that is really the validating principle.

(2) We have an additional problem, however, in that there are several different sacred texts, and they often disagree about the advice they give or the commandment they make. But if there are different claims being made by these sacred scriptures, how can we decide which one to believe, or which one really constitutes knowledge? An answer to this question could not appeal to the same sacred text—since that would be begging the question (arguing in a circle). Instead, one must consult a different source of knowledge and so the process would continue.

We have similar problems with faith and revelation. Can faith be a source of knowledge? If by "faith" one means "belief" then the answer is "no." If I believe, for example, that Darwin's theory of evolution is a good theory, and if something asks me for the source of knowledge underlying this belief, I cannot answer "faith" (i.e., belief), for then I would be saying: "I believe Darwin's theory because I believe it." What is needed, therefore, is an account of faith that makes it something different than mere belief.

Revelation is no better off. In fact, it seems to be equivalent to either sacred scripture or faith. But in so far as it is different from these, it seems to be claiming: "it has been revealed to an individual during a certain kind of religious experience". But, if so, then it is this religious experience that is the primary source of knowledge.

If faith, revelation, and sacred scripture are not sources of knowledge, what about the other candidates? What about *tradition* or *authority*? Here, we have candidates that are somewhat better (although as we shall see, ultimately they, too, are not successful).

Can one know something because of tradition? Much of our ordinary and scientific knowledge seems to be warranted by tradition, by an appeal to "that is the way we (individuals in a certain social group, our culture) have always done things," or "this is the way we have always believed." For example, we do say: "this is the way scientists have always (or for the most part) proceeded," or "scientists have traditionally believed . . . ." But a little reflection will show that authority per se is not a source of knowledge (even though it may be a secondary source of knowledge). For in virtue of what properties of tradition does it confer knowledge on a particular belief or way of acting? To say "in our tradition we believe (do) so and so. ." is just to say that we have always believed it (or done it). But habitual action and conformity of belief are not, in themselves, sources productive of knowledge; there is nothing about habit or custom per se that makes them a source of knowledge (although perhaps habit in addition to other epistemic principles might provide some initial epistemic warrant. For example, some epistemologists have suggested that such a supplementary epistemic assumption is the following: "any traditional belief or course of action has some initial epistemic presumption in its favor (unless shown otherwise)." Now, this principle in conjunction with a certain claim about what is a traditional belief or course of action would provide some warrant to the traditional belief or action. But here it is this secondary principle that is bearing most of the epistemic burden, and it is this principle that would need to be examined for its epistemic credentials. Even if such a conservative epistemic principle were impeccable, it would provide, at best, only some initial (prima facie) warrant, with most of the epistemological arguments centering on what additional matters would or would not defeat such a presumption.

If tradition is not a source of knowledge, neither is authority. Now clearly in ordinary life and in science we frequently rely upon authority for what to believe. Scientists are widely recognized as authorities in certain fields of specialty, e.g., quantum mechanics, ecology, medicine. When a question arises about the particular subject matter in question (e.g., Does Bell's Theorem cast doubt upon the EPR thought experiment?, does Vitamin C cure the common cold?), we tend to rely upon their answer to these questions. But authority, like tradition, although it may be a secondary source of knowledge, is not a primary source of knowledge. There is nothing about authority per se that would validate a knowledge claim; after all, authorities can be wrong about something just as we can. Furthermore, a particular authority (e.g., the Dali Lama) does not turn a belief into knowledge just by virtue of his believing it, pronouncing it, or commanding it. As children unfortunately know all too well, something is not true just because their parents say it is true and of any authority who proclaims something to be true, one can ask, "why? In virtue of what source of knowledge do they know this thing to be true?" If this is so, then clearly it is this source of knowledge that is really validating the claim to knowledge and not the individual's pronouncement.

Ordinarily, authorities are authorities in their respective fields because they have knowledge about these fields. But how do they know these fields? Not, in turn, by authority—here we would be embarking on an infinite regress—but by other means, e.g., because they have studied and reflected upon the relevant scientific theories, read the experimental studies themselves, actually performed the experiments, etc. It is these things that constitute whatever validity an authority can claim to have.

Furthermore, it must be remembered that an individual is an authority in a particular area of specialty, one in which he or she has special competence and training. But this expertise is not necessarily true of other areas of intellectual endeavor. Just because an individual is a leading expert in, say, cosmology, this does not necessarily make him or her an expert in another field, say, religion. Authority is always circumscribed authority!

Throughout the history of philosophy, individuals have often claimed to know something "by intuition". Is **intuition** a source of knowledge?

What is meant by 'intuition' has never been clearly stated, nor has it been free from conceptual puzzles. Sometimes, one hears talk about "woman's intuition" or "mystical intuition," where all that seems to be meant is that someone knows something, but doesn't know how they know it; e.g., My mother says: "I knew something was wrong with you and it was." I ask: “How did you know it, mom?” She answers: “I just felt it. " Here, my mother (who is often correct about these things) knows something but there is no recognizable way she could know it. Unless there is some account or theory about what intuition is and how it works, an appeal to it is of little help. In short, when an appeal to intuition is just a label for an unknown way of knowing, we can dispense with it.

There is, however, another use of the term intuition that can be found running throughout the history of epistemology and the history of science. In this traditional sense, "intuition" means "a direct and immediate cognitive grasp of something that is directly present to the mind or to the senses." For example, one knows something is red by intuition, which means: one is directly and immediately aware of something red. Likewise, one can know that A=A by intuition, meaning that one directly and immediately is aware of its truth (it couldn't be false). When we come to discuss Aristotle's theory of induction in a subsequent chapter, we will encounter intuition again; and one can also find great scientists (e.g., Galileo, Newton) using intuition to grasp the basic ingredients of the world—the fundamental variables or quantities of motion. So, we must (temporarily at least) retain intuition as a candidate for a source of knowledge. It will turn out, however, that it is reducible to other sources of knowledge, namely, sensory experience or reason.

The remaining candidates—reason, experience and practice—have all been advanced as candidates for sources of knowledge and, upon reflection, they do seem to be sources of knowledge. As a matter of fact, these three candidates have been closely associated with the three great epistemologies of rationalism, empiricism, and pragmatism of which we will have much to say throughout this book. It will be useful, therefore, to turn our attention to a classical debate in the history of epistemology, that between rationalism and empiricism.

Rationalism vs. Empiricism

According to the standard interpretation, the epistemological debate concerning the sources of knowledge pits rationalism against empiricism. This debate began in the 17th century in the aftermath of the Scientific Revolution and continued into the 18th century. These two respective schools seized upon two different aspects of the scientific method as practiced during the scientific revolution—the role of mathematics and mathematical proof versus the crucial place of experimentation and observation in the acquisition of scientific knowledge.

Rationalism

The school of rationalism is represented by the views of René Descartes (1596-1650) and Gottfried Leibniz (1646-1716), both of whom made important contributions to the history of science. The basic thesis of rationalism is the following:

All knowledge comes from and is reducible to the rational operations of the mind.

The paradigm-case of knowledge for the rationalist is a mathematical proof. In such a proof, the mind is engaged in reasoning, in inferring a proposition from other propositions. The inference in question is a deductive inference (see Chapter 4), which, if valid, is absolutely certain. Here, then the challenge of the skeptic to show that some knowledge is absolutely certain is partially met, since deductive inference is certain. But what about the original propositions, the premises, from which the conclusion is deduced? How does one know this proposition?

This issue has been discussed since ancient times, in particular, by Aristotle, and continues to be challenging today. There seem to be only four possibilities (see Figure 2.1): (1) one knows the original premises by inferring them deductively from still other propositions, and these would be deduced from still other propositions, which, in turn, would be deduced from other propositions and so on *ad infinitum*—an example of an *infinite regress*; (2) one knows the original premises by deducing them from other premises, and these premises would be deduced from other premises until finally we found ourselves back with our conclusion; here we do not have an infinite regress, we have a circle—most individuals would say a *vicious circle;* (3) one could arbitrarily stop anywhere one wanted and simply *assume* these to be true; this would seem to be an example of a kind of *conventionalism* where the first premises are not known but are arbitrary starting point—beliefs that are not warranted; (4) finally, these initial premises must be known, but not known in a deductive way, i.e., the initial premises are used as premises, warranting subsequent conclusions, but these premises are not themselves deductively warranted by other premises: they must be warranted in some other way. These principles are, what Aristotle called first principles—the foundations for everything else (hence this is a version of version of **foundationalism**). But how can one know these first

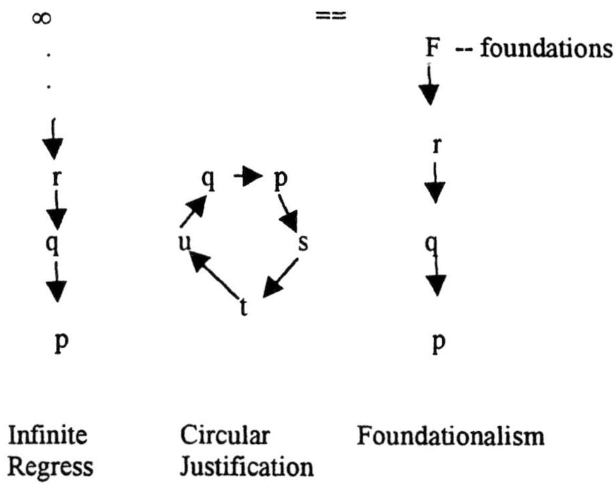

Figure 2.1. Varieties of Justification

principles?

Rationalists have replied that these first principles (or axioms) are known by an operation of the mind, not deduction but by an operation they called ***intuition,*** where one directly and immediately is aware of the truth of a proposition, or as they sometimes put it, these propositions are self-evident: they cannot be denied without contradiction; e.g., the propositions of logic, arithmetic and geometry are known to be true by intuition. Intuition and deduction are thus rational operations of the mind and provide, for the rationalist, the mechanisms for acquiring knowledge.

If the epistemic operations by means of which one acquires knowledge are mental operations, what about the material of knowledge—those ideas, concepts, or propositions about which the mind reasons?—where do they come from? Although this has never been maintained in quite the following form, one can say that the rationalist also believed these ideas came from the operations of the mind, that they were **innate.** The rationalists denied that our ideas are acquired or learned; they denied that they come from the senses; instead they thought that, in some sense, we are born with them.

Empiricism

Such a view is in direct opposition to empiricism. The term 'empiricism' comes from the Greek word for experience, which maintains the following:

> **All knowledge comes from and is reducible to experience (typically sense-experience).**

For the empiricist, typical representatives of which are John Locke (1632-1704) and David Hume (1711-1776), all our knowledge comes from and is reducible to what they variously called sense-impressions or sensations, the kind of information supplied by the senses—colors, sounds, shapes, smells etc. Hence they maintained that our ideas (concepts) are not innate but acquired—we learn them *subsequent to* sense-experience. They claimed, for example, that one could not have the idea of red unless one first had a sensation of red.

All of this was true of what they called *simple ideas,* these were the basic or atomic ideas into which everything else we know is reducible. In addition, there was what they called complex or *compound* ideas, which were combinations of simple ideas.

All compound ideas, they claimed, were analyzable into simple ideas and all simple ideas come from simple sensations. Hence, all our knowledge was reducible to sense-impressions. But how is it that simple ideas give rise to compound ideas? The empiricist could not bring into play any complex operation of the mind, otherwise this would be a major epistemological concession to their arch-enemy, the rationalists. The "glue" therefore had to be something very simple, involving at most a very low-level mental mechanism or operation. Although they disagreed somewhat among themselves over this basic combining operation, the concept of *association* is perhaps the best example of what they had in mind: simple ideas get associated together to form compound ideas; all compound ideas are thus associations of simple ideas.

As we have presented it, the issue between the rationalist and the empiricist involves a sharp contrast between their respective views. Certainly they disagreed about several things but probably no one has maintained either of these extreme views in this way. Rationalism and empiricism seem to be two ideal types representing the extreme end points along a continuum with respect to the question of the source of knowledge: is reason the source of knowledge or is experience?

Students typically want to respond by saying: both are sources of knowledge. But although this is a reasonable compromise, it is not easy to show how one can combine rationalism and empiricism. Such a question was the task that Immanuel Kant took on in the 18th century with mixed success.

Kantianism

Immanuel Kant (1724-1804) was the greatest philosopher of the Enlightenment. He attempted to synthesize the basic insights of rationalism and empiricism.

Knowledge starts from experience—on this point the empiricists were correct. But knowledge is not reducible to sense-impressions since it requires a great deal of interpretation by means of the rational operations of the mind (just as the rationalist had maintained). The senses provide the *material* of knowledge, its content, but in interpreting this material, the mind provides the *form* or structure of knowledge. This activity of the mind—of transforming the raw data of sensation into the complex structure we call science—occurred in two stages. First, the raw material is passed through the two cognitive filters of space and time, which are innate features of the knowing mind. Then, these sensations are brought under more complex cognitive categories and operations—categories such as causality, substance, etc.—which result in our ordinary world of objects interacting with each other as described by Newtonian physics.

Many thinkers of the cognitive revolution that occurred in the 1960s are inclined to think that Kant is describing how the mind works: it takes sensory input and produces cognitive output. But Kant is not making a factual claim that the mind actually works a certain way; he was claiming that *it is necessary that the knowing mind works this way.* It is because he was making such a strong claim that his epistemology is termed a transcendental epistemology: he was claiming that the world as we know it is, in a certain sense, constructed by the knowing subject, that these categories and structure are rooted in the knowing subject, and are *a priori*. Kant's particular way of putting this point was to claim that much of our scientific knowledge is synthetic a priori knowledge; hence, science gives us certain and necessary knowledge, but it does so only by restricting its scope and domain to the realm of phenomena—human experience. Hence, Kant concluded, scientific knowledge could not give us knowledge of ultimate reality—what he called noumena (things in themselves); it could only give us knowledge of human experience (phenomena). This raises the important question: what is the scope (extent) of scientific knowledge: can it give us a true picture of ultimate reality, or is it limited to describing and explaining phenomena? Kant's epistemology raises other important questions concerning the nature of scientific knowledge: (1) Is such knowledge synthetic a priori knowledge or something else? (2) More generally, Kant's theory of scientific knowledge has raised the question of the respective roles of mathematics and experimentation in the scientists' quest for knowledge: is scientific knowledge based upon reason (mathematics), and if so, to what extent?; or is it based upon experience (experimentation), and if so, to what extent? If both, what is the relation between the two? How should one conceptualize the respective roles of mathematical proof and experimentation in the scientific enterprise? Both of these questions will be presently discussed. First, however, we must briefly end our tour of the major epistemologies by briefly discussing one last school of epistemology—pragmatism.

Pragmatism

In recent times (the 19th and 20th centuries), another epistemology has appeared, stressing a source of knowledge somewhat different from reason and experience—successful practice. According to pragmatism, associated with the names of Charles Peirce (1839-1914), William James (1842-1910) and John Dewey (1859-1952), the only legitimate source of knowledge is: what works, what leads to practical consequences. The pragmatists adopt a particular conception of reason, namely, reason is a tool with which to cope with the environment; hence the major function of the mind is to assist the individual in problem-solving, in constructing possible means for obtaining various ends. (This is sometimes called an instrumental theory of rationality.) In order to obtain a certain goal, various possible means are tried out as alternatives—these are hypotheses—and the one that actually results in the attainment of the goal is the one that works and leads to the goal. Such a successful hypothesis embodies knowledge—how to reach a certain goal.

For the pragmatism, everything is to be evaluated in terms of its efficacy, its power in producing certain consequences. Something is valuable if it is useful, if it has utility. This applies to all ideas, concepts scientific theories, hypotheses, laws, etc., which are, as Dewey said, "instruments of actions". Dewey's advice is always to think of a theory as an instrument or tool by means of which one is able to reach a certain end. One knows something, therefore, if one *knows how* to attain a certain goal in a certain kind of situation.

In a sense, of course, pragmatism incorporates the insights of both the rationalist and the empiricist: (1) the pragmatist is a rationalist, but a certain kind of rationalist: eschewing the mathematical ideal of reason—pure, theoretical reason— associated with Euclid and Descartes, the pragmatist interprets reason differently—as applied, practical reason; (2) the pragmatist is also an empiricist, insisting that every idea be empirically tested by means of its effects in one's environment but the pragmatist disagrees with the classical empiricist's interpretation of the nature of 'experience.' (3) the pragmatist also believes Kant had put his finger on an important point on the mind's contribution to the experience. But they prefer to interpret Kant in more empirical, evolutionary terms, in which a priori structures are those structures that have proved useful in the course of history.

Clearly, pragmatism is tied to an approach that is practical in its orientation, and this has naturally been associated with applied science and technology more than with pure science. Clearly, science has this dimension—of stressing consequences that result in a certain state. Hence, we may take it that successful practice is one of the sources of knowledge along with reason and sensory experience. Whether these three are ultimately different or not, what their precise relationship is remains debatable topics. But for now we can merely say that the major sources of knowledge appear to be reason, sense-experience, and practical success.

The Scope of Knowledge

A perennial philosophical question about the nature of scientific knowledge concerns the question: what is the extent or scope of scientific knowledge? Can science know ultimate reality, that reality which lies behind our ordinary human experience and that is more basic than this experience? Or is scientific knowledge limited to the realm of Kant's phenomena?

First of all, it seems to be fairly clear that scientific knowledge is limited to the realm of the natural—the world that exists in space and time, consisting of particular objects. Hence, scientific knowledge is committed to *naturalism.* If there is a super-natural realm—a realm of reality outside of or beyond space and time, one with no causal power to affect our natural world—science cannot know it, and hence can ignore it *for scientific purposes.* Science does not have to take a position on whether such a realm exists or not; it can be silent about this question. It is up to those individuals who believe in such a realm to make a case that it exists and that it is possible to know it. Throughout the history of philosophy and science, individuals have certainly attempted to prove such a supernatural realm exists and that some kind of supernatural knowledge is possible. Most of these attempts have not been very successful on either count. Some of these individuals have also tried to show that this kind of supernatural knowledge is superior to the naturalistic knowledge of science. But, again, these accounts have been notoriously unsuccessful (in the author'' opinion). Fortunately, we do not have to discuss such issues here, since we have restricted ourselves to scientific knowledge.

Science, therefore, is limited to the natural realm and scientific knowledge is limited to naturalistic knowledge. But having said this, a question that is surely to arise is the following: suppose there is something called scientific knowledge and that it deals only with the realm of the natural. Such knowledge is encapsulated, let us suppose, in our collection of scientific theories that are tested by means of human experience. Scientific knowledge thus covers the realm of human experience. But is there a naturalistic realm above and beyond that of human experience, and if so can science know it? Related to this is the issue of *idealism versus realism*: does science give us a true picture of this reality or not?

According to one version of empiricism (**phenomenalism**), since the knowledge of reality is limited to phenomena (appearances), reality itself consists of such phenomena. Hence there is no reality beyond the realm of human experience and perception. (Phenomenalism is thus a close cousin of idealism.) The function of a scientific theory, therefore, is to describe phenomena in a convenient and economical way. According to **realism**, there is a reality independent of (outside of) human experience and our theories. According to **scientific realism**, it is the job of science to discover what this realm is, and this will be the function of our theories—to give us the best picture of reality. Historically, scientific realism has been coupled with the correspondence theory of truth and all of its attendant difficulties concerning how one can know this independent reality.

A Priori Knowledge

Traditionally, the type of knowledge to be found in mathematics and logic has been distinguished from the kind of knowledge to be found in chemistry and biology. Although the propositions of the natural sciences may be true, they have a weaker kind of truth than do the propositions of mathematics and logic. Lemons may be sour, snow may be white, gases may expand when heated., but these are only *contingently true;* the truth of 2+2 =4, and all triangles are three-sided are *necessarily true.*

> **Necessary truth: a proposition is necessarily true if (1) it is true, and (2) it *has* to be true.**

> **Contingent truth: a proposition is contingently true if (1) it is true and (2) it could be false.**

What it means to say that something has to be true is debatable, but according to one influential interpretation, that of Leibniz, something is necessarily true if it is true *in every possible world,* whereas something is contingently true if it is true of our actual world. In our world, therefore, lemons are sour and snow is white, but there are other possible worlds in which these things wouldn't be true; there are worlds, for example, in which Mendel's laws of heredity would not be true, but in every possible world, A=A.

The truths of logic and mathematics have been thought to be necessary (necessarily true) whereas the laws of biology seem to be contingently true. The truths of logic and mathematics also seem to possess another property the propositions of the natural sciences do not possess: they are known to be true in an a priori way whereas the propositions of the natural science are known a posteriori.

> **A priori: a proposition is known a priori if it known to be true independently of sensory experience**

> **A posteriori: a proposition is known a posteriori if it is known to be true as a result of sensory experience**

The distinction between a priori and a posteriori is the distinction, prior to versus posterior to, but this should not be taken in a temporal sense; rather it should be interpreted in terms of the method of verification: if a proposition is verified to be true only by means of sensory experience, then it is known a posteriori but if it is validated independently of experience, it is a priori. The basic question to be asked here is this: in order to know whether a proposition p is true, does one need to proceed to verify by sensory experience or not? If one does, it is a posteriori; if not, it is a priori.

A third crucial distinction is that between analytic propositions and judgments and synthetic propositions and judgments. According to

Kant, who was strongly influenced by Leibniz, *an analytic proposition is a proposition (judgment) in which the predicate term is contained in the subject term*, whereas *a synthetic proposition is one in which the predicate term is not contained in the subject term.* Thus, consider the proposition:

All unmarried males and unmarried.

This is an analytic proposition because the predicate term, "unmarried" is contained in the subject term "unmarried males." To know that it is true, all one has to do is to analyze the subject term, and one will discover the predicate term. Such a proposition is also a priori.

Now consider the proposition

All unmarried males are young

This is a synthetic proposition, because the predicate term "young" is not contained in the subject term "unmarried male;" it is synthetic because the predicate term is added to (goes beyond) the subject term, and in this case, such a proposition is false.

Although Kant made the analytic – synthetic distinction a crucial part of his epistemology, epistemologies tend to think his definition is not adequate and prefer the following semantic distinction:

Analytic: a proposition is analytic if it is true in virtue of the meaning of its terms.

Synthetic: a proposition is synthetic if it is not true in virtue of the meaning of its terms.

The Kantian revolution in epistemology consisted (partly) in raising the question of the existence of what he called synthetic a priori truths. Some propositions are *analytic a priori*: known to be true independently of experience because true by virtue of the meaning of its terms. This comprises many propositions of mathematics and logic. Some propositions are *synthetic a posteriori*: known to be true subsequent to experience and which are not true in virtue of the meaning of their terms. Here we find most of the propositions about the natural world: snow is white, dinosaurs became extinct because of a large meteor, bad money tends to drive our good money, etc. But what about the other two possibilities? Can there be analytic a posteriori and synthetic a priori propositions? According to most people, analytic a posteriori propositions seem to be impossible, for these would be propositions that were true by virtue of the meaning of their terms, but whose truth would have to be established by subsequent empirical investigations, and this has seemed to be contradictory: if it is true by virtue of the meaning of its terms (if the predicate is contained in the subject), then one would not have to verify it by subsequent experience; one would already know it is true by simply analyzing it.

Synthetic a priori propositions are a different matter, however (according to Kant), for these are propositions that are not true by virtue of the meaning of their terms—hence they were non-trivial propositions—but, ones that are known to be true *independently of subsequent sensory experience.* Needless to say, empiricists have found this notion impossible to believe and have routinely rejected it. But Kant (and rationalists) believe there are such propositions and that they make up the most important class of the propositions of science! Although Kant is not terribly clear about this, the kind of example it had in mind were propositions like: *every event has a cause*, *substances persist through time*, *space is Euclidean*, *the principle of inertia* and other laws of Newtonian mechanics, etc.

We may quarrel over Kant's examples of synthetic a priori propositions, but the question of their possibility is important in the context of scientific knowledge. For if scientific knowledge contains synthetic a priori propositions, this would be an invaluable part of an adequate epistemology of science and would provide a (partial) answer to the question of the nature of scientific knowledge.

The Presuppositions of Science

It is also important for another reason. Kantians have argued that certain kinds of propositions are absolutely necessary to science but are not verified by scientific investigation. As it is sometimes put, there are *presuppositions* of science, propositions that seem to be necessary for the very activity of doing science but whose verification does not seem to consist of performing empirical studies. Why? Because the very performing of such a study presupposes their truth; hence, such a study could not verify them on pain of circularity. One such example of such a presupposition of science would be the law of causality (or law of sufficient reason): *for everything that happens in nature, there is a sufficient explanation (or cause)*; another would be: the assumption of determinism, that *every event is subject to an exceptionalness law.*[1] Still another one is the uniformity of nature: *nature does not change in its mode of operation, so that if it has proceeded a certain way in the past, it will continue to proceed that way in the future.* Another one is: *there are natural kinds in nature*, i.e., natural phenomena occur together in clusters and patterns, these patterns invariably belong together as part of the category to which this thing belongs (e.g., gold, dog). (See Chapter 5 on the presuppositions of induction.)

Perhaps enough has been said to point out the importance the Kantian epistemology for the epistemology of scientific knowledge and the philosophy of science. Although most individuals believe he was mistaken in fundamental ways, nevertheless, many also believe he pointed out an important part of scientific knowledge.

Kinds of Knowledge

The above discussion naturally leads us to another closely related question: *what are the different kinds of knowledge,* and *what is their nature?* Traditionally, three types have been mentioned.

Knowing-How

Knowing-how is the type of knowledge stressed by the pragmatist. It is the ability (skill, craft) to obtain a certain end-state by employing suitable means. An individual knows how to ride a bicycle or to speak a language in the sense that he is able (he has the ability or skill) to perform this action in a certain kind of way in a variety of situations. Likewise an experimental scientist knows how to conduct an experiment with care and skill in ways very much like the way a skilled carpenter knows how to make a piece of furniture. Similarly, a mathematical scientist knows how to construct an elegant mathematical proof if she can manifest this ability in the appropriate situation. Knowing-how is thus a kind of instrumental or technological knowledge, exhibiting not only a knowledge of technique (techne) but also creativity and individuality.

It has been claimed by several individuals that such a craft (or artisan) tradition was a necessary ingredient in the rise of modern science since the practice of experimental science—see chapter 13—requires such an ability and tradition. To be able to perform an experiment requires a great deal of "know-how," an ability to use ones hands and instruments as tools of manipulation. Such a tradition was present in ancient Greece and was the basis of art and poetry, but it tended to be insufficiently appreciated as the basis of scientific knowledge.

Knowing-That

There were experiments performed in ancient Greece and there were experimental scientists in the ancient world. But the predominant tradition in Greek philosophy and science stressed the role of theory (contemplation), which the Greeks saw as a superior form of knowledge. The kind of knowledge contained in theories has been called **knowledge-that.** Knowing-that is knowing that something is the case, where what is the case is some truth, e.g., that the boiling point of water is 100° C, that the Battle of Hastings occurred in 1094, that the square root of 16 is 4, and so forth. Knowledge-that is concerned with the knowledge of propositions (sentence-like entities beginning with the word "that"). These propositions are factual-like in nature and the mode of awareness here is a kind of mental contemplation in which one sees what is the case. (Dewey called this the spectator theory of knowledge.), a paradigm-case of which is mathematical knowledge, in which one "sees" what is true and which propositions follow from other propositions.

Knowledge-that is one type of knowledge to be found in science, namely that type of knowledge found in theoretical knowledge. Theoretical knowledge consists of knowledge of propositions because theories consist of sets of propositions or laws, e.g., the laws of Mendelian genetics. But it also covers case of particular factual knowledge to be found, for example, in ordinary life and in history. Because this kind of knowledge has been so highly regarded in the history of philosophy (as compared to knowing-how), it has been the subject of a great deal of epistemological theorizing, in fact, it has become something of the paradigm-case of knowledge.

Acquaintance

Finally, a third type of knowledge is that of **acquaintance.** One knows something by acquaintance when one has experienced the object or state of affairs directly by means of the senses. Thus, one is acquainted with a particular person if one has met her and one knows Paris if one has first-hand experiential knowledge of the city. When one has a sensory experience of a certain kind, e.g., sees the litmus paper turn red, one is acquainted with this phenomenon.

Epistemologists have stressed the role of sensory experience in the acquisition of knowledge; empiricists, have stressed the crucial role of sensory acquaintance in knowledge. According to them, this is the only method of knowing. Sensory acquaintance constitutes evidence from our senses, empirical evidence, and is the basis of all of our knowledge of the natural world of phenomena.

Just as the two preceding types of knowledge are important in science, so is the third case. In fact, all three types of knowledge are to be found in science. Acquaintance gives one basic data; knowing-how is the ability to accomplish certain kinds of things required in experimentation; and knowing-that is the type of intellectual contemplation present in scientific theorizing. Not surprisingly, it is also the case that the epistemologies of empiricism, pragmatism and rationalism have stressed one of these kinds of knowledge: the empiricist stresses acquaintance, the pragmatist knowing-how, and the rationalist knowing-that.

The Nature of Propositional Knowledge

As we have indicated, knowledge-that has not received a great deal of attention in most discussions of the nature of scientific knowledge; whereas knowledge-that and acquaintance have been rather extensively discussed. Since we will discuss the nature of acquaintance in Chapter 7, we will briefly examine the nature of knowing-that here.

Throughout the history of epistemology and philosophy of science, one can find a famous account of knowing-that—propositional knowledge (knowledge by description). According to this account, there are three necessary and sufficient conditions for someone S to know that something p is the case:

(1) S believes p
(2) p is true
(3) S has adequate evidence for p.

All three of the conditions are necessary in order for an individual S to know that p is the case. First of all, S must believe p: it would be very odd to say: "S knows that p but S doesn't believe that p" (although some individuals have maintained this is possible). So, in order for an individual scientist S to know that something p is the case, S must be in the psychological state of belief that p is the case.

But belief is not a sufficient condition for knowledge. There are lots of things an individual may believe, but later come to discover was wrong. So belief is not equivalent to knowledge. We often ask things such as: "you believe it, but do you know it?", and we often say things like "I'm sorry. I thought I knew the answer, but I guess I didn't." So, in short, belief and knowledge are not the same. Hence there can be systems of beliefs but these are not necessarily systems of knowledge. Of course, the crucial question is: what is knowledge over and above belief?

Secondly, p must be true. One can't know that p unless p is the case. Although one can believe what is false (e.g., that the earth is flat), one can't know what is false. If water does not boil at 100°C, then one cannot know that it does; if one knows the planets in our solar system travel in ellipses, then it has to be true that they do. Hence, in order to know that p, p must be true.

But just because something is true, it doesn't follow that it is known. Presumably many things have been true for many, and yet no one has known them. In fact, there are reasons to believe that humans will never know everything that is true. So, being true is a necessary condition for knowledge but it is not a sufficient condition.

Finally, there must be adequate evidence in support of p. Suppose someone S claimed to know something p; suppose it turned out that S believed p, and p was true. For example S claims to know which horse will win the Kentucky Derby next year and we discover that S believes it and next year the horse he picked won. Did S know ahead of time which horse would win? If so, then all one needs for knowledge is true belief.

But something more is needed for knowledge than just truth and belief. We are all familiar with lucky guesses, cases in which we really don't know that something is the case but where we are right—accidentally right. For example, I read last Sunday's newspaper and believe today is Sunday. It turns out that today is Sunday but it is seven days later than the newspaper said. In such a case, I have a true belief but I don't know that today is Sunday. Such cases are disqualified from being cases of knowing, because the truth is not related to the belief in the right kind of way. In order to know p, S must be justified or warranted in his belief that p, and ordinarily we take this to mean there must be evidence e in support of p, and S must stand in the right relation to this evidence. But again, just because there is evidence e for something p, it does not follow that S knows that p. Since it is the hallmark of science to possess evidence for its claims, this condition has seemed to be necessary to just about everyone.

It should be pointed out that some individuals have objected to the above account. Some have argued that each of these conditions is not really necessary, and some have argued that although they are jointly necessary, they are not collectively sufficient since something else has been left out.

The most widely held view among epistemologists and philosophers of sciences is that they are necessary but not sufficient. In recent epistemological discussions, therefore, a great deal of attention has been spent on trying to revise these criteria so as to make them sufficient, or to supply a fourth condition of knowledge so that the four conditions are collectively sufficient for knowledge. Much of this discussion is not directly relevant to the nature of scientific knowledge, and so it will be passed over. The student is advised to consult the relevant sources for further information.

Belief

It has seemed obvious to most individuals that in order for an individual S to know that p, S must believe p. So it has seemed that knowledge requires belief and, consequently, that *scientific knowledge requires belief*, that there can't be scientific knowledge unless individuals have certain kinds of beliefs, e.g., Newton's theory is not an example of knowledge unless scientists believe Newton's theory.

Although this may appear to be correct as an account of what it is for an individual person to possess knowledge, at least one philosopher of science has argued this is an illusion when it comes to scientific knowledge. According to Karl Popper, belief is not necessary for scientific knowledge. Popper distinguishes subjective knowledge (knowledge in an individual person) from objective knowledge (public knowledge). Subjective knowledge may require belief, Popper admits, because subjective knowledge requires a knower. But this is not true of objective knowledge, which does not necessitate knowers. Think of the scientific knowledge contained in theories. Theories consist, we may suppose, of sets of propositions; these theories can be found in books, articles, etc., which in turn may be found in libraries or on computers. These theories, Popper maintains, contain objective knowledge. Newton's theory, for example, is an example of scientific knowledge since it describes, explains and predicts a great number of facts. It would do this even if there were no knowers to contemplate this theory! It is real objective knowledge existing in a world by itself, and it is this kind of knowledge we are after in science—impersonal and public knowledge—and this does not require a human mind to believe in them.

Although most individuals have not accepted Popper's suggestion that scientific knowledge consists of ideal, abstract entities existing in a world by itself, they do seem to think that Popper has put his finger on an important point about scientific knowledge. We do often speak of scientific knowledge in a kind of impersonal, objective way, one that does not seem to require individual beliefs. We say, for example, "the latest scientific knowledge shows. . . ." where we seems to be referring to something like Popper's objective knowledge. If so, then belief does not appear to be necessary for knowledge.

Truth

Of all the subjects of epistemology and philosophy of science, the subject of truth is the most controversial, and the one that produces the most heated arguments. Must a theory be true in order for it to be a case of knowledge? What does it mean for something to be 'true'? How can we know if something is true?

First of all, we must distinguish between two tasks of a theory of truth: (1) providing a *definition* of what it means for something (e.g., a statement or belief) to be true versus (2) providing a *criterion* for determining when something is true. A *definition* of truth indicates what it *means* for some statement to be true, e.g., it specifies the necessary and sufficient conditions for some statement to be true. A *criterion* of truth is a mark, sign, or indication of when a statement is true. The two need not be the same. For example, a barometer may be an excellent sign of when a storm is about to appear, but the barometer reading is not the definition of a storm. A criterion allows one to determine when an individual is to be classified as a certain kind of thing, but this is not the same as a definition of that thing. In the context of science, therefore, we need to distinguish the question of *what it is for a theory to be true* from the question of *how we can know that it is true*. The latter question has typically been thought to involve the notion of *evidence*, i.e., degree of confirmation, so that we know that a theory is true when we have an adequate amount of evidence in its favor. Hence we may have evidence in favor of a theory but this theory may actually be false. There are many examples in the history of science where a particular theory or hypothesis had evidence in its favor but was (apparently) false, e.g., throughout the ages individuals had accumulated a wealth of data that indicated space was Euclidean (three-dimensional). But this theory appears to be false. The Ptolemaic theory of the solar system, the phlogiston theory of heat, the steady state theory of the universe, the theory of spontaneous generation all had evidence in their favor, but all these theories appear to be false.

Some individuals may respond that we never have sufficient evidence to warrant our claim that a particular theory is true. From this they sometimes proceed to argue that one can never know what is true, and hence that speaking of truth in science is a mistake. But in response to this we can say that although we may never have sufficient evidence to know for certain, we can speak of "what seems to be true," of what the evidence indicates about what is true. In short, our access to truth is usually by means of our evidence. We must use it as our best guess or indicator of what is true. This is especially true of scientific theories that deal with the hidden structure of reality, with *unobservable entities* like quarks, black holes, and curved space-time. We can never directly experience these entities the way we can experience the color of litmus paper, the path of a particle in a bubble chamber, or the behavior of an animal. But if we can know these things without directly experiencing them, then we can know what theories are true. In short, we can not dispense with the concept of truth in science.

We do recognize the very distinction between true and false theories in science and in everyday life. It is false, most of us would say, that the earth is flat because we have such overwhelming evidence in favor of such a belief. Likewise the Ptolemaic theory of the universe is false (and so is the theory of the ether, the theory of phlogiston, etc.). In short, we do find it convenient to speak of false theories and beliefs; although we do, however, seem more reluctant to say "a theory is true," although we do say that the theory *seems* to be true or *is likely to be* true.

We also commonly say that a false theory give us a certain degree of knowledge. The Ptolemaic theory wasn't totally wrong, nor was catastrophism, Lamarckianism or the ether theory. These theories possessed *some* knowledge, which means they were partially true.

Sir Karl Popper and several others have suggested that there are *degrees of truth* and an *approximation to the truth:* Some theories are closer to the truth than other theories are (even though both are off the mark). Hence, Einstein's theory of relativity is closer to the truth than Newton's theory is, which, in turn, is closer to the truth than is Aristotle's theory. Hence, according to these individuals we should abandon thinking of truth in absolute terms, as black or white (either-or), and think of it as possessing degrees.

We can also talk about approaching or approximating what is true. According to Popper, the history of science can be described as getting closer to the truth: as our theories come and go, a subsequent theory make progress because it gets closer to the truth (even though ultimately it may be false). Hence, Harvey's theory of the circulation of the blood is closer to the truth than is Galen's theory of the circulation of the blood.

Throughout the history of epistemology, several theories of truth have been proposed (at least half a dozen). We will mention the three most influential theories: the pragmatic theory, the coherence theory, and the correspondence theory.

According to the **pragmatic theory of truth** (championed in various forms by William James and John Dewey),

> **a proposition (hypothesis, theory) is true if it works, i.e., if it is successful in allowing one to reach a certain end state.**

If a hypothesis allows one to solve a problem and to attain a certain end state, then this hypothesis is true. On all of the other theories of truth, says the pragmatist, there is no way to verify that something is true; hence, these theories have no value. For the pragmatist, therefore, success is not merely a criterion for deciding when a theory is true, it is the very meaning of truth.

Several individuals have argued that this account if wrong, since there can be false theories that work. For example, taking a certain plant (e.g., aspen leaves) may have a certain soothing effect, but the underlying theory explaining why it has this effect (it is because of a divine spirit in it) is false. Similarly, Newton's theory may allow us to fire a rocket to the moon successfully, but, strictly speaking, it is false. Hence, just because some theory works, this does not mean that it is true. Until recently, most individuals have rejected such a pragmatic theory of truth (although many accept it as a criterion of truth).

Secondly, there is the **coherence theory of truth:**

> **A theory (hypothesis, idea) is true if it coheres with the rest of one's theories (hypotheses, ideas).**

In order for a theory to be coherent, it must be logically consistent with the remainder of one's knowledge and beliefs; there must not be any logically inconsistencies between the items in an overall set of beliefs. Furthermore, there must be a kind of general theoretical harmony and coherence to one's set of beliefs: belief must mutually support each other and provide explanatory support to the rest of one's beliefs. If this is the case, then a particular belief is true.

But, again (critics have said), just because a hypothesis is coherent with the rest of one's beliefs, this need not make it true. The paranoid schizophrenic, for example, may have a very coherent system of beliefs about aliens from space who are out to steal his soul. But this belief is normally considered to be false. Hence, a coherent set of beliefs need not be true. Likewise, a scientific or mathematical system may be internally consistent and theoretically unified, and yet this does not, in itself, make it true. For example, a mathematician can construct an elegant and consistent mathematical system of space, say, a 24-dimension model of space. But such a model need not be true of what space in our world is like. Coherence, therefore, is not a definition of what truth is, although it is a good criterion of when a system is true or false.

What such coherent systems lack is *the right kind of connection to the real world.* Lacking this connection, the theory is a description of *a possible world* but it is not a description of our actual world. Abstract mathematical systems often have this property of being interesting possible worlds. But if they attempt to describe our actual world, such accounts may be wide of the mark.

The core sense of truth, therefore, requires the right kind of connection to the real world. This notion is perhaps best captured in the **correspondence theory of truth** (whose defenders include Aristotle, John Locke, Bertrand Russell and Ludwig Wittgenstein):

> **a proposition (hypothesis, theory) is true if it correspond to the way the world is.**

On this account, a theory is a representation (picture, map) of reality. If it represents the way reality is, then it is a true representation (picture, map); it agrees with (it is in conformity with) the way it is. If a theory does not match to the way the world is, then it is false (although it can be approximately correct if it partially corresponds to the world).

The correspondence theory of truth provides, in our opinion, the best account of the nature of truth, of what it is for something (e.g., a theory, model, law) to be true. Of course, such a theory must give us some account of the nature of this correspondence, and, up to now, nobody has offered a completely satisfactory account.

Some individuals have suggested that a theory corresponds to reality if it pictures reality in a certain way, if, for example, there is a one-one correspondence between items in the theory and parts of reality, between relati0ons in our theory and relations in reality, etc. Unfortunately, the metaphor of "picturing" was rejected at an early date in the 20th century with very little to take its place.

It must be remembered that the correspondence theory of truth is not been offered as a criterion of truth, of how someone can know that two things correspond to each other. If one wanted to know this, one way would be to compare the two things the way in which, for example, I could compare a photograph of an individual with the individual herself. Here, I must be aware of both of these terms; they both must be items of my experience, so that I relate the similarity of the two in my experience. But if I can experience the state of affairs a theory is about, then I would not seem to need a theory of it; I could just directly experience it. But most states of affairs in science are not directly experientiable; hence I cannot compare both in one act of conscious awareness. How, then, can I ever know if a scientific proposition corresponds to reality? One way I can is by means of the following: suppose I can deduce a testable state of affairs as a predicted consequence from my theory. I can then compare this prediction with the actual predicted outcome. If they agree, then my theory is confirmed and its probability is increased. Hence, on this account, I can come to know what is true via its degree of confirmation.

Both the pragmatic and the coherence theories of truth offer partial answers to the question: What are the criteria of truth?

Consider this analogy: suppose one discovers an alleged picture of Alexander the Great and one wants to know if this corresponds to the way Alexander the Great really looked. Since you do not have direct access to the person himself, what can you do? Of course, you could compare this picture with other replicas of Alexander, e.g. sculptures. But suppose you had none. Well, you might compare the picture with descriptions given by ancient sources concerning how Alexander looked. But here we seem to be using coherence—the coherence between two types of symbolic representations—as a criterion of truth.

On the other hand, suppose theories, models, and laws should be seen not as pictures of reality but rather as maps. A map of Athens is a true map of the city if it corresponds to the way the city is. How do we know if it is a good map? We look on the map and it locates the Acropolis southeast of the Agora: we then find our way to the Agora, head southeast, and then we find the Acropolis. So the map trulyrepresented where the Acropolis was, it corresponded to the way reality is. (Notice here we seem to be using the pragmatic test as a test of truth.) Theories (hypotheses, models, laws) thus seem to be like maps more than seem to be like pictures.

Evidence

Finally (and most importantly), there cannot be scientific knowledge unless there is evidence in favor of them. Of all the conditions necessary for scientific knowledge, this requirement seems to be the least controversial, since it is the hallmark of scientific knowledge to be backed by evidence, typically empirical evidence.

What is it for something to be evidence for something else? In the most general sense, something e is evidence of something h if the presence of e makes a difference to the probability of h being true, the presence of e increases the probability of h compared to the probability of h without e (positive evidence); or the presence of e decreases the probability of h (negative evidence); if the probability of the hypothesis is the same by itself as well as with the evidence, then the evidence is irrelevant to its truth.

The first point that needs to be made is that there is a difference between *direct evidence* and *indirect evidence*. Something is direct evidence for something else if it is a verification of the thing in question, e.g., I have direct evidence for the existence of a Jaguar XKE if I can verify its existence by, say, observing the car in my driveway. Most of the medium-sized objects of ordinary life have been verified to exist by means of direct evidence. But we have evidence for many things that is indirect, i.e., where we cannot see the object itself but can observe one or more of its effects. For example, there is only indirect evidence for the existence of black holes, electrons, the Ice Age, Alexander the Great, the Unconscious, etc. In order to have scientific knowledge, it is not necessary to have direct evidence; one can also have scientific knowledge if one indirect evidence for it.

But how much evidence does one need before it is adequate? This is a difficult question for which there is no accepted answer. Partly it is a matter of what the scientific community believes is required in order to establish a case of scientific knowledge; there seems to be adequate evidence that black holes exist, but whether there is adequate evidence for global warming is perhaps not so clear.

Some individuals have attempted to quantify the idea of adequate evidence in the following way. Letting 'e' stand for evidence, 'h' for hypothesis, and 'P' stand for probability. Then we can represent the statement *the probability of h* as P(h), and *the probability of h given (on the assumption of) e* as P(h/e). Assuming that such a representation can take on numerical values ranging from 0 (complete impossibility) to 1 (complete certainty), we can speak of the degree to which e gives h evidence, e.g., P(h/e) = .9 (very nearly certain) or .2 (very nearly uncertain). If .5 means h has a 50-50 chance of being true, then some individuals have suggested that a minimum requirement for e being adequate evidence for h is P(h/e) ≥ .5 (with the degree of adequacy increasing as the value of the quantity increases). Others have argued that a value of .5 is arbitrary and fails to match scientific practice; so there seems to be no consensus here as to how strong the evidence must be before it is adequate.

Science and Common Sense

One question that needs to be discussed concerns the issue of the relation between scientific knowledge and other forms of knowledge. According to many people, there are several ways of knowing: there is scientific knowledge, religious knowledge, artistic knowledge, literary knowledge, etc. How is science related to these forms of knowledge?

According to one view, which we will call **scientism,** the only really genuine kind of knowledge is scientific knowledge. Everything else may represent a belief system, a mode of discourse, but it fails to qualify as knowledge. Hence religion, literature, art, and common sense may be systems of representation, but they do not come up to the high standard of knowledge.

Scientism must face the problem of *demarcation*, of showing how one can demarcate science from these other systems of belief by providing a criterion, a principle by means of which we can separate them. This has been difficult to do because every proposed criterion has been shown to be inadequate largely on the grounds that there seem to be examples of science that do not satisfy the criterion proposed or there are criteria satisfied by non-scientific examples. For example, it has been proposed by Popper that one can demarcate science from non-science by the criterion of **falsification**: unlike these other belief systems, science can be distinguished by the fact that its theories and hypotheses are falsifiable—they make predictions such that if the predictions are false, the theories or hypotheses are false, and scientists abandon such theories. Unfortunately, it is widely recognized that falsificationism is not an adequate criterion of demarcation because several good examples of science, e.g., Newtonian mechanics, did not operate this way. There may be other, more adequate criteria of demarcation (e.g., verifiability), but up to now no one has produced a very successful candidate.

Another possible approach to this question is not to demarcate science from non-science in a sharp way—a difference in kind—but in a graded way—a difference in degree. On this view, science has produced a *greater degree of knowledge* than these other systems have (largely because scientific knowledge is more testable, objective and quantifiable than these other forms of knowledge are). But how do we grade these other forms of knowledge? Does literature give us more knowledge than art? Are both of these to be evaluated as lower than religion, which has given us a higher kind of knowledge? No one has come up a satisfactory account of how we grade cases of non-scientific knowledge. Furthermore, what are we to say of other, more controversial claims of knowledge, e.g., myth, astrology, palmistry, etc. Are they also to be classified as examples of knowledge? Does everything get classified as a kind of knowledge? If not, then we need to have some criterion of demarcation separating these from genuine knowledge.

Many individuals have looked to common sense knowledge as a form of knowledge superior to these other forms, in fact, rivaling or even surpassing scientific knowledge. How, then, is science different from common-sense knowledge? Two views have been prominent. First, some individuals (e.g., Bertrand Russell, Sir Arthur Eddington), who have championed a form of scientism, have argued that they are radically different, a difference in kind, such that science is superior to common sense. Second, others (e.g., William James, John Dewey) have argued that science is an extension of common sense: scientific knowledge is not radically different from common sense knowledge; instead, it is based upon common sense, but goes beyond it in terms of degree of precision, exactitude and amount of evidential backing. Nevertheless, science is rooted in common sense and in the ordinary procedures of knowing prevalent in common sense, which regularly employ the scientific method in attempting to solve problems and answer questions. This view, which we can call *critical common sense* is the view we will adopt in this book. (See also Chapter 6.)

The Aims of Science

What is the goal of scientific knowledge? What is the ultimate purpose of knowing the world? Traditionally, science has been said to have several goals: description, explanation, prediction and control.

First of all, science aims to give a **description** of the world. It describes phenomena, objects, events, etc., but it does so in a different language. As we will see in more detail later (Chapters 8 and 9), when one gives a description of something, one is engaged in an act of classification. On the one hand, one can give a description of one's experiences or of phenomena in the language of common sense, which is largely Aristotle's approach. Here one uses the qualitative language of smells, tastes and looks we employ in ordinary life. On the other hand, a scientific description uses a language different from common /sense, a technical language that is more precise and exact—one that is typically mathematical—and one that ignores many features of one's experience. Ordinarily, but not always, a scientific description of the world will not mention the ordinary smells and tastes of common sense but will be couched in a language far different. Consider, for example, how an astronomer describes a sunset, how a physicist describes the path of a leaf in flight, how a chemist describes a vintage wine, or how a psychologist describes love. Rarely will these descriptions match those of the market place!

Secondly, science aims not only to describe phenomena but to **explain** them. Science not only describes what happens, it explains why it happens. In the 19th century, one group of philosophers and scientists claimed that science is nothing more than description. According to them, science doesn't explain, it merely describes—it describes in a more convenient and economical way. But such an approach does violence to how most scientists view their subject matter, and runs counter to examples in the sciences. For example, Darwin first discovered and described the basic phenomena of evolution by natural selection. But he had no adequate explanation of why it occurred. Later Mendel proposed certain laws of heredity, which did explain certain features of inter-generational phenomena. Still later Mendel's laws were, in turn, explained by the molecular genetic of Watson and Crick, whose model of the gene gave a good account of Mendel's laws.

The above is a hierarchical model—some have called it the chocolate layer-cake model—where one level is explained by another and deeper level and this in turn is explained by still another deeper level. Typically this kind of level-explanation involves what can be called *causal mechanisms*: these causal mechanisms involve internal "gears and pulleys" explaining the transition from input data to output data by means of intermediate processes. A case in point is the explanation of the gas laws (Boyle's Law, Charles Law) by reference to the kinetic theory of gases, a theory involving the behavior of molecules in a container.

Thirdly, science aims to **predict** events and phenomena before they happen, e.g., a forthcoming ellipse, the return of Halley's comet, global warming, an epidemic, who will win the next election, etc. According to some people, however, prediction is not in and of itself the goal of science, but rather a test of when one has an adequate theory. On this view, prediction is the criterion of a good theory, but the goal of a theory is description and explanation. Others would deny this.

A relevant distinction here is that between pure science, applied science and technology. **Pure science** aims at a general theoretical understanding of the world, the type of knowledge contained in our best scientific theories. Its interest is *understanding* why things happen the way they do and how the basic furniture of the world explains this. The basic sciences here are physics, chemistry, biology, psychology, etc. **Applied science** has the goal of providing knowledge aimed at improving the human condition by indicating what means to employ to intervene in nature and to change it. Consider, for example, the goals of engineering, medicine, psychotherapy, social work. The aim of these sciences is practical—to improve the well-being of people by providing them with water and electricity, health, happiness, and social improvement. The kind of knowledge found here is instrumental knowledge, the knowledge contained in means-ends connections, the knowledge of what leads to what. Finally, **technology** is concerned with providing knowledge concerning how actually to create the means that will result in certain desirable outcomes. This technological knowledge involves the construction of material artifacts, e.g., the construction of dams, medical instruments, new fertilizers and machinery; in addition, technology is concerned with providing knowledge in the form of *techniques* necessary to produce these means, e.g., psychotherapeutic techniques, cooking techniques, educational techniques, etc.

On this account, therefore, pure science aims to describe and explain, applied science aims at prediction and control, and technology aims at implementation. In order to know how to create the physical means necessary to produce something (e.g., good health), one has to know what leads to what (e.g., a certain kind of nutrition leads to health). This applied knowledge in turn depends upon theoretical knowledge (e.g., anatomy, physiology, biochemistry).

The aims of prediction and control are thus closely connected to those of description and explanation. To control an event—to produce it or prevent it—presupposes being able to predict when an event of a certain kind will occur. This in turn depends upon being able to describe and explain the course of nature in terms of general laws and principles.

If science has the goals of description, explanation, prediction and control, it appears that pure science has the goals of description and explanation, examples of knowing-that. Applied science predicts and controls and seems to be concerned basically with knowing-how: knowing how to obtain certain kinds of results by employing certain kinds of means. But technology is also concerned with knowing-how—knowing how to create the means necessary to obtain a certain result.

For Further Reading

There are several good introductory textbooks on epistemology. See, for example, Hospers (1988) for an elementary introduction and then read, e.g., O'Connor & Carr (1982) or Moser, Mulder, & Trout (1998). There are numerous anthologies in epistemology, e.g., Moser & vander Nat (1995). An excellent introduction to the philosophy of science is Hung (1997), with a more advanced but clear discussion being Salmon, et al (1992). Unfortunately, there are no anthologies in the philosophy of science at an elementary level; but several at the intermediate level, e.g., Boyd, Gasper & Trout (1991).

Notes

[1] Both of these assumptions have been questioned in contemporary quantum physics.

PART TWO

Logic

3

Introduction to Logic

Introduction

An essential part of science involves reasoning. For example, after performing an experiment concerning the effects of saccharin on the incidence of cancer in a certain strain of rats, a scientist will draw a conclusion about what the study showed, e.g., "the results of our study indicate that. . ." and he will fill in the blank with a statement such as: "saccharin increases the frequency of cancer in rats (if taken in sufficiently high dosage)." The statement is the conclusion of the study; it is what the experiment showed.

Similarly, when reviewing the current status of a theory in a select field of specialization, a statement such as the following will be made: "On the basis of all the evidence so far available, we may conclude that the big-bang model of the origin of the universe is very well confirmed". Here again a scientist is drawing a conclusion from some evidence. A colleague may raise the question, "Why did you use a sample of only two individuals in your study?" "Because," it might be replied, "this was only a pilot study and one doesn't need such a large sample in such cases." Here, an action is being questioned and the response consists of giving reasons for what was done.

In these examples, scientists are **reasoning**: they are giving reasons for their beliefs and actions, and from these reasons they are drawing conclusions. A fundamental question that arises here is, When are these reasons good ones and when are they not?

The particular subject matter that investigates such questions is **logic** (more precisely **formal logic**). Logic has a long and distinguished history, going back at least as far as the Greek thinker **Aristotle**. It has continued up to the present day as an important field of study. The basic question of logic can be formulated as, *When are the reasons given for a conclusion good ones?* In order to explain what this question means, we will need to clarify several important concepts.

Arguments

It is customary in logic to distinguish several types of linguistic entities: sentences, statements and propositions. A **sentence** is a meaningful set of words (either written or spoken) that expresses an assertion, question, command, request, and so forth. Thus, the following are sentences: *Shut the door, it's freezing in here! Why in the world did you go skiing in the middle of May?* The concept of a sentence is very broad and covers a variety of things: the expression of feelings, emotions, and attitudes; the giving of an order or command; the asking of a question; the statement of a fact; and so on. Logic, as traditionally conceived, deals with only one of these kinds of things, namely, sentences expressing an assertion. (Recently, however, several branches of logic have begun to investigate commands, imperatives, etc., but these are not relevant for our concerns.)

To single out this one particular kind of sentence, we can call it a **statement** and say that *a statement is a sentence that makes an assertion*, that is to say, *it is a sentence that is either true or false.* Thus, the sentence "the temperature outside is 15°C" is an assertion. It is a sentence that asserts that something is the case, that it is true that the temperature outside is 15°C. Of course, the sentence might be false. The person uttering it might be misinformed, he may have misread the thermometer, etc. But that doesn't matter, since a statement is just something that is either true or false. Likewise, it is not important that we *know* whether it is true or false. If someone were to say "the temperature on the back side of the moon is 0°C," we may not presently know whether this statement is true or false, just as we may not know whether the statement "this rock is harder than a diamond" is true or false. Of course, we may be able to find this out easily (by making the appropriate test) but sometimes we will not be able to determine the truth or falsity of a statement. Hence, we may not know whether it is actually true or false. The important thing about a statement, however, is that it is either true or false, not that we know whether it is true or false. In this way, statements are different from questions and commands, which are not statements! A request to "bring me another beer!" is not true, but neither is it false. Likewise the question "why do birds sing?" is itself neither true nor false.

Propositions

Although the subject matter of logic is more akin to statements than to sentences, we have not yet clearly delineated the precise subject matter of logic. The reason for this can be seen if we consider the following pair of statements: "it is raining" and "es regnet." Both of these are statements, since they are either true or false. But they seem to be saying the same thing only in different languages (just as "Charlie Brown hit Lucy" and "Lucy was hit by Charlie Brown" say the same thing). The same thing they both are asserting is what logicians call a **proposition**. A proposition is what a statement asserts; it is the logical content of the statement. A proposition is a very abstract concept, one far removed from actual sentences consisting of ink on paper, but such an ideal entity is the subject matter of logic just as numbers, which are abstract ideal entities, constitute the subject matter of arithmetic.

We can say, therefore, that logic is concerned with evaluating propositions, with determining when a proposition is justified, when there are good reasons for a proposition, when a proposition warrants a conclusion, and so forth. As will become clear later, logic--at least that dominant branch of it known as formal logic--is concerned with the *formal* relations between propositions and with determining when one set of propositions logically entails another proposition, with the relation of implication. For this reason it is often useful to abbreviate a proposition such as "elephants love peanuts" by a symbol such as *p* or *q*. Formal logic is concerned with the structural relations between symbolically abbreviated propositions such as:

if p then q; and not q; therefore, not p.

To explain what this last expression means, we must examine the nature of an argument.

As everyone knows, people often have arguments. They disagree about something; they yell and scream at each other, they stalk off in an angry mood, etc. Although this may be part of the ordinary meaning of an *argument*, it is not what logicians mean by that term.

As we saw above, logic is interested in determining certain kinds of *relations* between propositions. It is concerned with determining when something is a good reason for something else, when something is justified by something else, etc. Clearly we are dealing here with relations between two things:

. . . a good reason for- - -
. . . justified by- - -.

One of the propositions is called a **conclusion**--that which is supposed to follow from the other proposition(s)--and the other class of propositions is called the **premises** (reasons, or grounds)--that which warrants or proves the conclusion. In an argument, it is being claimed that the premises and conclusion are related in a certain way:

premises: { *statement* / *statement*

⇓

conclusion: *statement*

where the arrow indicates a particular kind of relation that is supposed to hold between them, namely, the relation *is a good reason for*, *warrants, justifies, entails*, etc.

We have, therefore, three kinds of things to consider: (1) the premises, (2) the conclusion, and (3) the relation between them. (The crucial question for logic concerns the relation between the premises and the conclusion.) These three things together constitute what is called an **argument**, which we may define as:

> *a set of propositions, one of which is the conclusion, at least one of which is the premise(s), such that the conclusion is claimed to follow from the premises.*

If we abbreviate the conclusion by 'q' and the premises by 'p,' then an argument is something that claims:

q follows from p
q is a consequence of p
q is justified by p

or in general

q because p.

We can, however, view the relation in the opposite way--as being:

p entails q
p necessitates q

p justifies q
p is evidence for q
p is good reason for q

or in short,

p, therefore q.

Thus, "q, because of p" and "p, therefore q" are really mirror images of each other and assert the same thing, namely, there is a particular kind of relationship between the premises (p) and the conclusion (q).

Inference Indicator Words

In deciding what the author of a passage of a written text or an oral presentation is claiming, one must identify the set of premises and the conclusion. This is not always easy to do and sometimes even the author himself may be unclear about this. To facilitate the process of identifying the structure of an argument, it is useful to use words such as *because* and *therefore.* These words are often called ***inference indicators***, because they indicate what the relation is between the premises and the conclusion.[1]

There are two kinds of inference indicators, those indicating the conclusion *(conclusion indicator words)* and those indicating the premises *(premise indicator words).*

Conclusion Indicator Words. Each of the following words usually show that the statement that *follows* is a conclusion:

therefore. . .
hence. . .
thus. . .
so. . .
implies that. . .
entails that. . .
which shows that. . .
proves that. . .
indicates that. . .
consequently. . .
allows us to conclude that. . .
we may deduce that. . .
points to the conclusion that. . .
suggests very strongly that. . .

leads me to believe that. . .
from which it follows that. . .
bears out my point that. . .

Thus, in the argument

> A child's reaction to its environment can only be instinctive, animal-like; it is, therefore, from the point of view of evolution, regressive

the conclusion follows the inference indicator word *therefore*. The conclusion is: It [a child's reaction] is, from the point of view of evolution, regressive. The statements preceding the word *therefore* are the reasons (premises). If we let 'p' abbreviate 'a child's reaction to its environment can only be instinctive, animal-like,' and 'q' abbreviate 'it [a child's reaction to its environment] is from the point of view of evolution, regressive,' the argument is: p, therefore q. We can use an arrow to represent the structure of the argument:

$$\begin{array}{c} p \\ \Downarrow \\ q \end{array}$$

Premise (Reason) Indicator Words. Each of the following words or phrases usually shows that the statement that *follows* is a reason:

for. . .
since. . .
because. . .
for the reason that. . .
assuming as we may that. . .
may be inferred from the fact that. . .
may be deduced from. . .
as shown by. . .
as indicated by. . .
as is substantiated by. . .

In "p, therefore q," q is the conclusion and p is the reason. Thus, consider the argument: spiders are not insects, because insects have only six legs. The word <u>because</u> indicates that what follows is the reason. But what is the conclusion? Obviously the first statement: spiders are not insects. The key here is the following. In

q, because p

q is the conclusion and p is the premise. It can be represented as

$$p \Downarrow q$$

Consider the following: whales are not fish because fish are cold-blooded and whales are warm-blooded. Here the *because* indicates that what follows is the reason for the conclusion. In this case, two reasons are given. The conclusion is: whales are not fish.

Exercise 3.1

In each of the following passages, circle all inference indicators, underline premises (reasons) and bracket the conclusion.

1. Logic is a good subject to study, for it constitutes the intellectual backbone of Western science.

2. Punishment, when speedy and specific, may suppress undesirable behavior, but it cannot teach or encourage desirable alternatives. Therefore, it is crucial to use positive techniques to model and reinforce appropriate behavior that the person can use in place of the unacceptable response that has to be suppressed.

—Walter Mischel & Harriet Mischel, Essentials of Psychology

3. Because logic is one of the principal means which assures intellectual discipline and integrity, it can, if properly applied, only promote the achievement of desirable social ends.

---Ilmar Tammello, Outlines of Modern Legal Logic

4. Since the drive for profits underlies the very existence of business organizations, it follows that a most important function of an accounting system is to provide information about the profitability of a business.

—Walter Meigs & Robert Meigs, Accounting

5. The earth is spherical in shape. For the night sky looks different in the northern and southern parts of the earth, and this would not be so if the earth were not spherical in shape.

---Aristotle, Physics

6. The true distinction between these forms. . .is, that, in a democracy, the people meet and exercise the government in person; in a republic, they assemble and administer it by their representatives and agents. A democracy, consequently, must be confined to a small spot. A republic may be extended over a large region.

---James Madison, The Federalist

7. How do education and training affect lifetime income? Are they worth their cost? The evidence answers, decidedly yes. Men who never finish eight grades of school earn scarcely $3,800 annually; college graduates do three times as well. Unemployment among school dropouts exceeds that of graduates by a growing margin.

---Paul Samuelson, Economics

Analyzing More Complex Arguments

Rarely are arguments as simple as these. There are several ways in which real arguments are more complex.

First, an author may cite more than one reason for a conclusion. She may, for example, argue:

> Einstein's theory of general relativity is currently the best theory going because considerable evidence supports it and because it is simpler than all the others.

Here two premises are being offered as evidence and we have two inference indicators (two examples of 'because'). It can be analyzed as:

> [Einstein's theory of general relativity is currently the best theory going] because considerable evidence supports it and because it is simpler than all the others.

Its structure is: p because q and r. Again, we can construct what is called a **tree diagram** of this in the following way (where '+' represents 'and'):

```
q + r
------
  ⇓
  p
```

It would be nice if authors would always use inference indicator words. Unfortunately they do not: sometimes they omit inference indicator words altogether. For example, an author may argue:

> Logic is a good subject to study; it constitutes the intellectual backbone of Western science.

What is the conclusion here and what are the premises? The conclusion is: *Logic is a good subject to study*. The premise is: *Logic constitutes the intellectual backbone of Western science.* How can one determine this? One way is the following: try inserting the word *therefore* between the two statements and see if the argument makes sense; then insert the word *because* and apply the same test.

Another difficulty in analyzing arguments occurs when premises and conclusions are implied rather than explicitly stated. Often, arguments contain missing premises, assumed but not stated. Such premises are called *suppressed premises*; arguments containing them are called **enthymemes**. Ordinarily (but not always), the context makes clear what these unstated premises are. A good test for determining what premises are implicitly being assumed is to ask: "do the premises adequately support the conclusion, or are other premises needed to link the premise to the conclusion?" For example, in the argument

Skinner is a determinist
But I believe in moral responsibility
Therefore, I can't agree with Skinner's views

a missing premise is: "Determinists do not believe in moral responsibility." Another one is: "I believe in moral responsibility." Similarly, in the argument, "Computers can't be used in the humanities because the humanities are not quantitative," the missing premise is: "computers can't be used with nonquantitative subjects." In the above arguments, the missing premise was easily spotted. But in real-life examples of argumentation, there may be several missing premises.

Some passages contain more than one argument. This will typically be shown by the presence of more than one conclusion indicator word. Consider the following passage (from R. G. Collingwood, *The Idea of Nature)*

> Matter is activity, and therefore a body is where it acts; and because every particle of matter acts all over the universe, every body is everywhere.

One argument consists of the inference:

> matter is activity; therefore, a body is where it acts.

The second argument is:

> because every particle of matter acts all over the universe, [the conclusion is that] every body is everywhere.

The student should not be misled by the lack of a conclusion indicator word in the second argument. In place of such a conclusion indicator word you have two other clues: first, the presence of a reason indicator word and secondly a comma separating two propositions. It is common in ordinary and scientific discourse to find the argument:

> because of p, q

or even just

p, q.

In many cases, however, the grammatical punctuation will give you the clue as to what the argument is supposed to be: this is especially true of commas, semicolons, and colons.

Exercise 3.2.

Each of the following passages contains more complex arguments. Circle all inference indicator words, underline premises and bracket conclusions. If an indicator word has been left out, supply one.

1. Since the secondary light [from the moon] does not inherently belong to the moon, and is not received from any star or from the sun, and since in the whole universe there is no other body left but the earth, what must we conclude? What is to be proposed? Surely we must assert that the lunar body (or any other dark and sunless orb) is illuminated by the earth.

---Galileo Galilei, The Starry Messenger

2. The year 1859 is perhaps the most important one in the history of biology to date. In that year Charles Darwin published his theory of evolution by natural selection, which has deeply affected not only biology, but other branches of human thought as well.

---L. C. Dunn and T. H. Dobshansky, Heredity, Race and Society

3. Women of the working class, especially wage workers, should not have more than two children at most. The average working man can support no more and the average working woman can take care of no more in decent fashion.

---Margarette Sanger, Family Limitations

4. Since private property helps people define themselves, since it frees people from mundane cares of daily subsistence, and since it is finite, no individual should accumulate so much property that others are prevented from accumulating the necessities of life.

---Leon Baradat, Political Ideologies, Their Origin and Impact.

5. Archaeology is truly an infant discipline. Its parent, anthropology, is scarcely a century old; as the child, it is even younger. Stratigraphic excavation was not done systematically until the early years of the twentieth century; carbon-14 dating is but 18 years old, and the corpus of anthropological theory upon which sound archaeology must rest has only become truly useful during the past 30 years.

--James Deetz, Invitation to Archaeology

6. Sedatives do not merely induce sleep, and indeed people who suffer from a transient insomnia are ill advised to take several drinks and then sleeping pills, particularly if they are over fifty. In combination the drug effects are enhanced, and the doses a person may consider moderate can even prove fatal.

---Gay Luce and Julius Segal, Sleep

The Complexity of Scientific Reasoning

As we have just seen, arguments in science (and in everyday life) are very complex, often involving several arguments strung together in a series. Inference indicator words are also often lacking. Such features of reasoning are not the only features that make it difficult to understand a scientific argument. Consider, for example, the following example of an argument advanced by a prominent scientist:

> Perhaps you may think that the whole question of the creation of the Universe should be avoided in some way. But this is not so. To avoid the issue of creation it would be necessary for all the material of the Universe to be infinitely old, and this it cannot be for a very practical reason. For if this were so, there could be no hydrogen left in the Universe. . . Hydrogen is being steadily converted into helium throughout the Universe and this conversion is a one-way process--that is to say, hydrogen cannot be produced in any appreciable quantity through the breakdown of the other elements. How comes it then that the Universe consists almost entirely of hydrogen? If matter were infinitely old this would be quite impossible. So we see that the Universe being what it is, the creation issue simply cannot be dodged.
>
> --Fred Hoyle, "Continuous Creation and the Expanding Universe" (1971)

Upon encountering this argument, the student may be at a loss to understand what the conclusion is and what the premises are. Indeed, its structure appears to be quite complex. But if indeed such passages are frequently encountered in the scientific literature, how can the student proceed to analyze such an argument? Although we cannot go into any more detail concerning the process of argument analysis, we will analyze the above example in order to illustrate how argument analysis can be done.

Let us begin by circling inference indicators, bracketing conclusions, underlining premises and supplying missing indicator words. We will also introduce the convenient device of numbering all the statements. Often an author will make a point by asking a question, when he really wants to make an assertion. In such cases, we must rewrite the sentence as a statement.

(1) [The whole question of the creation of the Universe cannot be avoided.] because (2) To avoid the issue of creation it would be necessary for all the material of the Universe to be infinitely old, and (3) the Universe is not infinitely old for a very practical reason. For

> (4) if the Universe were infinitely old, there could be no hydrogen left in the Universe. . . (5) Hydrogen is being steadily converted into helium throughout the Universe and (6) this conversion is a one-way process--that is to say, hydrogen cannot be produced in any appreciable quantity through the breakdown of the other elements. (7) The universe consists almost entirely of hydrogen. (8) If matter were infinitely old this would be quite impossible. So we see that (9) the Universe being what it is, the creation issue simply cannot be dodged.

We must rewrite the first sentence. We will also rewrite the third and fourth sentences somewhat (for the sake of clarity).

To begin with, (1) is a conclusion and (2) and (3) are reasons for it. So we have:

(2) + (3)

⇓
(1)

But we are told that (3), in turn is a conclusion, supported by (4). Before proceeding to (5) it should be pointed out that the author at this point is really relying upon what is called a tacit (suppressed) premise to complete his argument. That is to say, he is covertly assuming there is hydrogen left in the universe. In fact, later in the argument this premise takes explicit form in (7). Hence we have:

(4) + (7)

⇓
(3)

(5) and (6) in turn are reasons for (4). So:

(5) + (6)

⇓
(4)

(8) is really not very new since it is claiming virtually the same as (4). Likewise (9) is the same as (1). In summary, therefore, the argument is:

(5) + (6)

⇓
(4) + (7)

⇓
(2) + (3)

⇓
(1)

This argument has a complex form. But most real scientific arguments are equally complex and require considerable analysis to determine their underlying structure. Sometimes there will be disagreement about what diagrammatic representation best captures the underlying structure. In this case, the only thing that can be done is to offer the best argument one can in favor of a particular interpretation.

Exercise 3.3

In the exercises in 3.2, number premises and conclusions. Construct a tree diagram for each argument. Where necessary rewrite sentences as statements.

Deduction and Induction

Logic, we have said, is concerned with the *relation* between one set of propositions--the reasons or premises--and another proposition--the conclusion. Logic is concerned with determining whether this relation is a good one or not. Do the premises really result in the conclusion?, Do they provide adequate support for it?

The link between the premises and the conclusion can vary in its *strength.* Some arguments have zero degree of justification and are called **fallacies**. Other arguments have the maximum degree of warrant possible and are said to be deductively valid. One branch of formal logic--**deductive logic**--is concerned with determining when an argument has this maximum or absolute amount of justification. Other arguments fall in the middle range between these two extremes: their premises give *some* support to the conclusion but less than an absolute amount and more than zero amount. These arguments are often said to be more or less *probable* and as such they belong to another branch of logic--inductive logic. **Inductive logic** is concerned with determining when an argument gives *some* support to the conclusion. Combining these two forms of logic, deductive logic and inductive logic, one can say that

> *logic is concerned with studying the strength of the (evidential) relation between the premises and the conclusion of an argument.*

Deductive logic (unlike inductive logic) is an all-or-none affair. An argument (in deductive logic) is either completely valid or it is not valid at all. An argument in inductive logic, however, is not an all-or-none affair but a more-or-less affair. If we represent the two degrees of strength a deductive argument can have as 0 and 1—zero validity versus complete and absolute validity—then we can represent the domain of inductive logic as being that which falls in between 0 and 1: it is concerned with evaluating the adequacy of an argument where the adequacy is somewhat greater than 0 but less than 1; for example, an inductive argument might have the value of .1, .4, .6, .8, etc. Obviously the greater this value, the better the argument so that an inductive argument with a value of .9 or .99 or .999. would be about as strong as one could get; it should be remembered, however, that an inductive argument can never have a value of 1.0 (although it can approach it).

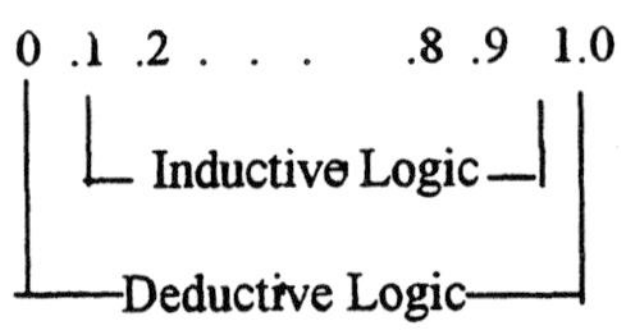

Figure 3.1. Degrees of Support

In a **deductive argument**, the premises warrant the conclusion to such an extent that the premises are said to *entail* the conclusion, to make the conclusion *necessary*. When this occurs, the argument is said to be **valid** (a technical term of formal logic that should not be confused with its ordinary meaning). In an **inductive argument**, by contrast, the premises provide evidence for the conclusion but an inductive argument can *never* entail the conclusion or make it necessary. An inductive argument is always less than valid. (Technically speaking, we could say that inductive arguments are always invalid, but this would be somewhat misleading.)

Because deductive logic is concerned with the question:

> *Do the premises entail the conclusion?*

whereas inductive logic is concerned with the question:

Do the premises provide adequate evidence for the conclusion?

we can distinguish deductive and inductive logic in the following way.

An argument is ***deductively valid*** *if and only if it is impossible for the premises to be true and the conclusion to be false.*

An argument is ***inductively strong*** *if and only if it is improbable that the premises are true and the conclusion false, and it is not deductively valid.*

We can represent this in the following way:

Premises { *statement* / *statement* }	*Premises* { *statement* / *statement* }
Prove ⇓	*Warrant* ⇓
Conclusion: statement	Conclusion: statement
Deductive Logic	*Inductive Logic*

The crucial questions concerning inductive strength are these: when is it improbable that the premises are true and the conclusion false?, and how improbable is it? Put in a more positive way, one of the fundamental questions of inductive logic concerns the notion of the *probability* that the conclusion is true, given that the premises are true or, in short, *How strong is the evidence supporting a conclusion?*

Just as there are indicator words in deductive logic, so there are indicator words in inductive arguments. These indicator words share the feature that they make a guarded claim, namely, that the evidence is less than absolutely certain but the conclusion is warranted to a certain degree.

Conclusion Indicator Words: the following words typically indicate that the statement that follows is a conclusion:

probably
appears that
seems that
the evidence shows

studies have shown
it is plausible that
usually

Reason Indicator Words: the following words typically indicate that the statement that follows is a reason:

this is evidenced by
the evidence is
this is supported by
this is warranted by

In addition to the presence of inductive indicator words, there are several other ways in which to identify an argument as an inductive one. Generally, in all of them there is evidence given in the premises that is **empirical**; that is, the evidence is obtained by a scientific study such as an interview, a survey, a questionnaire, an observation, an experiment, etc.

1. Particular to General.

One characteristic feature of an inductive argument is the presence of a premise about some particular evidence and a conclusion concerning a generalization. For example:

> Galileo, Newton and Einstein relied heavily upon mathematics in advancing their theories. Thus it is part of the essence of scientific theorizing to employ mathematics.

2. Sampling

Another characteristic feature of an inductive argument is the presence of a premise mentioning some sample This is evident in the following argument:

> [The Assistant Secretary for Health of the Department of Health and Human Services] released the results of the National Household Survey on Drug Abuse, which examined a national sample of Americans over 12, and the High School Senior Survey. The surveys showed that one out of three Americans over age 12 has tried some kind of illicit substance.

3. From Past to Present

When the evidence cited mentions particular cases of some event happening in the past, and then proceeds to draw an inference about something that will happen in the present or in the future, an inductive argument is being advanced.

> The Farmer's Almanac, which has an excellent record on such matters, predicts that we are in for a hard winter this year. So there is going to be a tough winter ahead.

4. Analogy

When an argument involves mentioning some similarity or analogy between two or more things, an inductive inference is being drawn. For example:

> These mushrooms have a very similar appearance to those growing in the garden and the ones in the garden are edible. Thus these mushrooms are probably edible.

5. Causal Generalization

When something is cited as the (likely) cause of something else, this typically will be a clue that the argument is an inductive one. The following argument by Charles Darwin is an example of a causal argument:

> When we compare the individuals of the same variety or sub-variety of our older cultivated plants and animals, one of the first points which strikes us is, that they generally differ more from one other than do the individuals of any one species or variety in a state of nature. And if we reflect on the vast diversity of the plants and animals which have been cultivated, and which have varied during all ages under the most different climates and treatment, we are driven to conclude that this great variability is due to our domestic productions having been raised under conditions of life not so uniform as, and somewhat difference from, those to which the parent species had been exposed under nature.
>
> ---Charles Darwin, The Origin of Species

Exercise 3.4

In the following passages, circle all inference indicators, underline premises and bracket conclusions, indicate which passages are deductive and which ones inductive and why. Construct a tree diagram of the underlying argument.

1. If the fetus is a human life, then there are only a few circumstances where abortion is justified, and if the fetus is just a mass of cells, the all cases of abortion are justified. Either the fetus is a human life or just a mass of cells. Hence, either there are only a few circumstances where abortion is justified or all cases of abortion are justified.

2. A gardener who cultivates his own garden with his own hands, unites in his own person the three different characters, of landlord, farmer, and labourer. His produce, therefore, should pay him the rent of the first, the profit of the second, and the wages of the third.

—Adam Smith, The Wealth of Nations

3 A new drug, acebutol, already available in Great Britain and now under study here, looks excitingly good for dangerous heart rhythms. At Stanford University not long ago, doctors tried it for a man with a potentially lethal arrhythmia that resisted all the standard antiarrhythmia drugs such as quinidine and procainimide, and had required repeated electrical shocking of his heart. In the eight months he has been on acebutol, he has had no recurrence of arrhythmia.

4. Statistics reveal that 80 percent of those who receive flu shots do not get the flu. Jack received a flu shot one month ago. Therefore, he should be immune, even though the flu is going around now.

5. The governor's commission strongly recommends that use of marijuana continue to be illegal since marijuana leads to the use of heroin. This is evidenced by the fact that 94 percent of 1,000 heroin addicts interviewed reported that they had smoked marijuana before becoming addicted to heroin.

6. Barbara's skin has a peculiar yellowish hue, and she complains of loss of energy. Barbara must have hepatitis.

7. Farmer Jones plowed up two plots of ground, applied a coat of lime to one and then planted both with corn. The limed plot yielded ten bushels per acre more than the other. Probably the extra yield was due to the application of lime.

8. Two contrary psychic tendencies exist universally in the father-son relationship, i.e., submission and rebellion, and both derive from the Oedipus complex. Since most patriarchal religions also veer between submission to a paternal figure, and rebellion (both submission and rebellion being sexualised), and every god, like a compulsive super-ego, promises protection on condition of submission, there are many similarities in the manifest picture of compulsive ceremonials and religious rituals, due to the similarity of the underlying conflicts.

—F. Fenichel, The Psychoanalytic Theory of Neuroses

Validity

In formal deductive logic, the fundamental question concerns the validity of an argument. Validity is thus a concept of paramount importance. We will give several definitions of validity, all of which can be shown to be equivalent.

An argument is ***valid*** if and only if:

[1] the premises *entail* the conclusion
[2] it is *impossible* for the premises to be true and at the same time for the conclusion to be false
[3] if the premises were to be true, the conclusion *would* also have to be true.

To say the premises *entail* the conclusion is to say that the truth of the premises *necessitate* or *require* the conclusion to be true. Another way of putting this same point is to say (as definition [2] does), that it is impossible for the premises to be true and for the conclusion to be false, and this in turn can be explained (as in definition [3]) as meaning that if the premises *were* to be true, then the conclusion *would* also have to be true.

This third definition contains words of a special kind, not merely those (like "have to") which indicate a sense of necessity, but also words like "were to be true" and "would also. . . ." These words do not say the same thing as "is true" and "are true". The latter words tell you what happens to be true right here and now or (as it is sometimes put) what happens to be true but only contingently true of our world (in the sense that it could be otherwise). Thus, it happens to be true that mature tomatoes are red but mature tomatoes don't have to be red; we can imagine another world--a world of science fiction for example--where there are mature tomatoes that are of a different color, say, blue. So, it happens to be a fact of our world that mature tomatoes are red but this is not necessarily true.

When we do imagine the case of tomatoes being blue, we are considering the truth of the proposition *tomatoes are blue* in a different way, a way logicians term **'contrary to fact'** or **'counter-factually'** (since tomatoes being blue is counter to actual fact in our world). So we can say: suppose (counter-factually) that tomatoes are blue, or "if (contrary to fact) tomatoes were blue". . . . The use of the term "were" indicates we are in this counter-factual mode, just as the words "would," "should," also indicate that. Our third definition thus claims that if (counter-factually) the premises were true, then the conclusion would also have to be true.

Students are often puzzled about why logic is concerned with arguments that are not actually true, that is, with counter-factually true premises. They often say: "Why not just be concerned about what is true right now in our world and forget about all of this nonsense about other possible worlds or things that might be true." The answer to this is quite simple: in ordinary life and especially in science, we often reason about things that are not actually true but could be true. We deal, in other words, with hypothetical situations. For example, suppose I say: "If the Denver Broncos had (contrary to fact) won the Superbowl, then they would have got a different draft pick and they would have drafted so-and-so." Here I am reasoning hypothetically (contrary to fact). As the example indicates, football coaches often reason in the hypothetical way. In logic, what we are concerned about is the implicatory relation between premises and conclusion, what we want to know is whether the conclusion is entailed by the premise(s) or not.

Similarly, one often reasons counter-factually in science. We say things like: "suppose something could go faster than the speed of light," "suppose one can escape the singularity of a black hole," or "suppose this hypothesis about cancer *were* true." One of the essential features of science is reasoning *hypothetico-deductively*. But it is the hallmark of this method that one does not know whether the hypothesis is actually true or not; in fact, the whole point of using this method is to decide *if* the hypothesis is true or not. One can do this only if one first *supposes* the hypothesis were true and then formally deduces logical consequences from it. In all of these cases, we are hypothetically supposing certain things and then proceeding to inquire what would follow as a consequence. Questions about what would follow is a question about the validity of the argument.

Validity and Truth

What then is the relation between validity and truth? Validity and truth are not the same concepts. To see this, consider the various possibilities concerning the truth of the premises and the truth of the conclusion in an argument. There are just four possibilities: true premises and true conclusion, true premises and false conclusion, false premises and true conclusion, and false premises and false conclusion:

Premises	True	False	False	True
Conclusion	True	True	False	False
	(a)	(b)	(c)	(d)

In which of these cases--(a), (b), (c) or (d)--can an argument be valid? It's clear, first of all, that case (a) is the case we ordinarily think of as being a good logical argument. This intuition, it will turn out, is correct. What about (c)? Can one have a valid argument in which there are false premises and a false conclusion? Although this may seem counter-intuitive, it is certainly possible as the following example, using what is called a **Categorical Syllogism**, illustrates.

All philosophers are wealthy men
All wealthy men are Republicans

Thus, all philosophers are Republicans

Here the premises are actually false and the conclusion is actually false: as philosophers will attest, they are not wealthy and, as a matter of fact, there are many wealthy men that are not Republicans. Finally as a matter of fact most philosophers are Democrats. So all three propositions are false. And yet the argument is valid. You can test for its validity in several ways but one way is the following: suppose the premises were (counter-factually) true. Then would it follow that the conclusion would also have to be true? The answer is yes. So, it is possible to have a valid argument in which there are false premises and a false conclusion.

What about case (b) in which we have false premises and a true conclusion? Again, this is possible as the following example illustrates.

All spiders are deadly creatures
All deadly creatures have eight legs

Thus, all spiders have eight legs.

Here, the premises are both actually false but the conclusion is true. Furthermore, it is a valid argument, which you can test by applying the above test for validity. So, we can have a valid argument with false premises and true conclusion.

Finally, what about the last case (d)? Can we have a valid argument with true premises and a false conclusion? No. It is not possible precisely because it violates our definition of validity--definition [2]. Hence it is not possible for one to have a valid argument in which the premises are true and the conclusion false.

Soundness

At this point, the astute student might raise this question: if you can have valid arguments with either true or false premises, what's the point in having true premises at all? Why not just construct valid arguments that have false premises? Why, in short, should we prefer case (a) to case (b) or (c)? Although there are several answers that could be given to this question, the short of it is that *only with valid arguments involving true premises will you have a true conclusion.* (Logicians speak of this property as that of *truth-preserving;* namely, the truth of the premises is transferred to the truth of the conclusion.) Well, our astute student might respond, so what? Why prefer true conclusions to false ones if all you are after is validity? The answer is that we want more than mere validity in an argument--we want what is called **soundness**. But to show why we want more and why validity is not sufficient for the evaluation of an argument, let us consider a practical goal of logic.

Logic is concerned with good reasoning and logical persuasion. Suppose, for example, there is a debate occurring between two individuals and suppose they are arguing about the issue of feminism. Suppose one individual (Gretchen) is trying to convince the other individual (Rolf) that women are equal in intellectual ability to men. This, in other words, is the conclusion she is trying to establish and it is the one which she wants Rolf to accept as true (since this will amount to convincing him about this point). That is often the point of an intellectual debate--to convince an opponent (by logical means) to accept the truth of the conclusion. Now, how can she do this? Of course, she could employ various tricks or use threats or bribery but we said that we were concerned with the case in which she employed logical means to do this; and in logic this means that she must give him a good argument. At the very least a good argument has to be a valid one. So the ideal case we are considering is one in which we have a valid argument with a *true conclusion.* (This is because Gretchen believes the conclusion true and wishes Rolf to believe it is true also.) However, this is consistent with both case (a)and (b). Why prefer (a) to (b)?

The answer is simple. Gretchen wants Rolf to accept the truth of the conclusion. She will accomplish this, we have agreed, by giving him a valid argument. Now suppose he agrees to accept it as valid. Will he also have to accept the conclusion as true? No, because Rolf can respond that although the argument is valid, he need not accept the conclusion as being true, for he could claim that the argument is valid but that the conclusion is false. He can do this only if he has case (c) in mind. Hence Rolf can avoid admitting the truth of the conclusion if he insists the premises are false (for in that case you have a valid argument with false premises). So, in order for Gretchen to avoid this possibility, she must convince Rolf that the premises are also true. If she does this, that is, if she convinces him the premises are true and the argument is valid, then Rolf has no other option than to accept the truth of the conclusion (which was the desired goal in the first place). In short, once Rolf admits the truth of the premises and the validity of the argument, he is necessarily bound (by the truth-preserving property of arguments) to admit the truth of the conclusion. So, if Gretchen wants to convince Rolf of the conclusion, she must do two things: she must give a valid argument and she must give true premises. Such an argument is said to be sound.

An argument is ***sound*** *if and only if*
(l) it is valid, and
(2) its premises are true.

Now suppose one were Rolf, who refuses to accept the truth of the conclusion. One can do so by employing two strategies: either show the argument is formally **invalid**—by showing it is possible for the premises to be true and for the conclusion to be false—or by showing the premises are not true. Either of these strategies, if successful, will show the argument to be unsound, and if the argument is not sound, then one does not have to accept the truth of the conclusion.

Exercise 3.5

In the following deductive arguments, identify the premises and the conclusion as being actually true or false (to the best of your knowledge). Then determine whether the argument is valid or invalid. Finally, determine whether each argument is sound or unsound, and indicate why.

1. Since *Moby Dick* was written by Shakespeare, and *Moby Dick* is a science fiction novel, it follows that Shakespeare wrote a science fiction novel.
2. If George Washington was beheaded, then George Washington died. George Washington died. Thus, George Washington was beheaded.
3. Chicago is a city in Canada and Canada is a part of the United States. Therefore, Chicago is a city in the United States.
4. All triangles are squares. All squares are rectangles. Thus, all rectangles are triangles.
5. All dogs are carnivores. All carnviores are meat eaters. Therefore, all dogs are meat eaters.

Determining Unsoundness

How do you show an argument is invalid? You can do this basically by becoming familiar with the rules of formal logic--those rules which allow you to determine when an argument is valid and when a **formal**

fallacy is being committed, etc. (You will also need to be able to identify **informal fallacies**.) So, a knowledge of logic will be helpful in determining when an argument is fallacious. Secondly, one can show that the premises are not true. How does one do this?

To determine whether a premise is actually true or not, one must (typically) cite empirical evidence that is relevant for determining (to a reasonable degree) whether the premises are true or false. Although the issue is very complicated, we can say that something is **evidence** for something else just in case its existence would increase the likelihood of the latter's existing (or if it is a statement, of it being true). Empirical evidence would be evidence that is obtained by the aid of the senses--naturalistic observation, experiment, etc.

Where does one go to find such evidence? Obviously, common experience is a major fund of such evidence: we all know it is true, for example, that water is wet, fire burns, etc. Another source is science. If it is a question, say, of whether a certain chemical pollutes the air or not, this is a question chemistry and environmental science can probably answer. On the other hand, if it is a question about whether men are aggressive and competitive, one may not need to cite scientific evidence for this; instead, one may rely upon common experience for an answer.

We can diagram this situation in the following way.

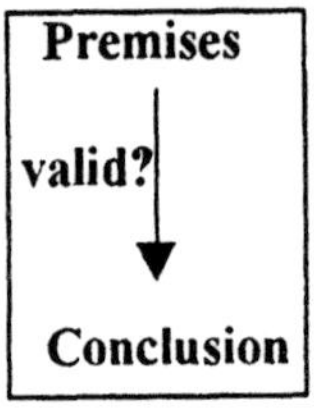

DEDUCTIVE LOGIC

The arrow in the box concerns the relation of entailment (validity). We want to express the relation that evidence has to our premise such that the premises are true or confirmed. If we express this evidence as a set of statements about the evidence, then we can investigate the logical relationship between evidence statements and premise statements (such a relationship belonging to inductive logic instead of deductive logic). Continuing the convention of representing relations between statements as being contained in a box, we might then imagine inductive logic as operating on a horizontal dimension (analogous to the vertical dimension of deductive logic represented above).

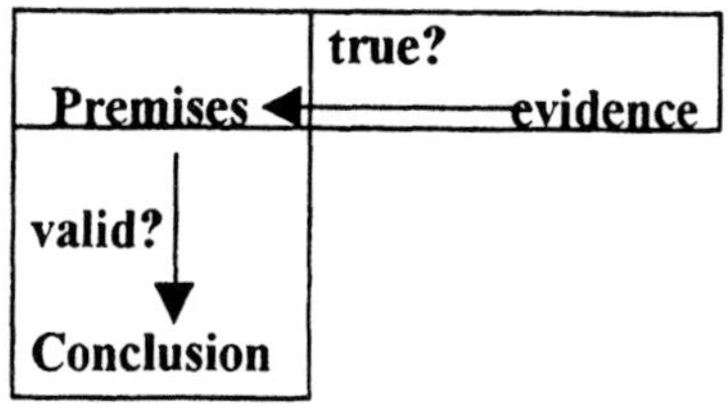

Inductive Logic

Deductive Logic

Hence the relation between inductive logic and deductive logic is very complex: when investigating the concept of soundness, inductive logic is clearly relevant since its is that branch of logic concerned with the question of how one can know whether the premises are true or likely true. In effect, therefore, both branches of logic are necessary in order to evaluate the degree of evidential support premises give to the conclusion.

Another question that arises here is, what is the relation between actual empirical evidence and statements about the evidence? Such a relation would be between an experience and a statement and hence would not be contained in the box since it would be a relation between the world and language. How one should conceptualize such a relation is a deep epistemological and metaphysical question. Although logicians and philosophers of science have discussed the question and proposed answers (a currently popular answer is that the relation is that of causality), no consensus has emerged.

For Further Reading

There are scores of good textbooks on logic. A widely used one is Copi & Cohen (19886), which covers all branches of logic.

Notes

[1] A note about terminology: one statement may imply another statement but one statement does not infer another statement (people make inferences).

4

Deductive Logic

Compound Propositions

*U*p to now we have been dealing with simple or **atomic propositions**, those propositions that do not contain other propositions as components. Propositions that contain other propositions as components are called **compound propositions** (or molecular propositions). If simple atomic propositions are propositions like *John went skiing*, *John broke his leg*, *John failed his logic class*, then a compound (molecular) proposition would be something like: *John went skiing and John broke his leg **and** John failed his logic class,* or ***if** John went skiing **and if** John broke his leg, **then** John failed his logic class.* In such compound propositions, we have atomic propositions joined together to make compound propositions by means of words like "if. . .then. . .," "and," and so forth. These are called **connectives** and there are several kinds:

Negation: not (abbreviated ~)
Disjunction: or (abbreviated v)
Conjunction: and (abbreviated &)
Implication: if--then (abbreviated ⊃)
Biconditional: if and only if (abbreviated ≡)

These produce several kinds of compound propositions: negative propositions, disjunctive propositions, conjunctive propositions, hypothetical propositions, and biconditional propositions.

Negation

The negation of an atomic proposition (e.g., not-p), which we will represent as ~p is a simple matter. The negation of 'snow is white' (p) is ~p, which can be read as: it is not the case that snow is white. How do we decide when the negation of an atomic proposition is true? Obviously ~p is true when p is false; if p is false, we can infer ~p.

A device very useful in representing these matters is the *truth table.* Suppose we have an atomic proposition p. Since a single proposition can be either true or false, logicians say that p can have two **truth values**, namely true (T) or false (F). We can thus represent this situation in the following way:

p
T
F

If we have a compound proposition containing two propositions (p, q) we have four possible combinations of these individual truth values:

p	*q*
T	T
T	F
F	T
F	F

With three propositions (p, q, r) we have eight possible combinations, and so forth.

The truth table for negation is:

p	~p
T	F
F	T

This defines the case when ~p is true or false, namely ~p is true if p is false and ~p is false if p is true. We will have more to say about negation later.

Disjunctive Arguments

A **disjunctive proposition** is a proposition containing at least two simpler propositions (called the *disjuncts*) joined by an operator called disjunction--*or. Either I'll buy a Porsche or I'll buy a Corvette* is an example of such a disjunction, which can be symbolized as: p or q. We will also symbolize 'or' as 'v' so that the above can be represented as: p v q.

Just as in the case of negation, we can determine if a disjunction is true by reference to the truth table for disjunction. A disjunction is true (it is valid to infer a disjunction from its disjuncts) when at least one of the disjuncts is true: a disjunctive proposition is false when both disjuncts are false. Accordingly, the truth table for disjunction is:

p	q	(pvq)
T	T	T
T	F	T
F	T	T
F	F	F

This says that a disjunction (p or q) is true under three conditions: when p is true and q is true, when p is true and q is false, and when q is true and p is false. When p is false and q is false, then the disjunction is false. The truth table thus defines what it means for a disjunction to be true.

A **disjunctive argument** we will define to be *an argument consisting of three statements (two premises and one conclusion) such that one premise is a disjunction, one premise affirms or denies one of the disjuncts, and the conclusion affirms or denies the other disjunct.* A disjunctive argument can either be valid or invalid.

A *valid* disjunctive argument (called a **disjunctive syllogism**) is an argument in which one premise denies one disjunct and the conclusion affirms the other disjunct. In symbolic form:

p or q
not-p
----
q

For example:

> either Newton's theory of gravity is true or Einstein's theory of gravity is true; but Newton's theory of gravity is not true; therefore, Einstein's theory of gravity is true.

A disjunctive argument is *invalid* (in particular, it commits a formal fallacy) when a premise affirms one of the disjuncts and the conclusion denies the other disjunct. In symbols:

p or q
p
------
not-q

This is a formal fallacy. The reason this is a fallacy can be seen if we give a different example of a disjunctive argument. Suppose I say:

> Either Newton was a brilliant scientist or Einstein was a brilliant scientist
> Newton was a brilliant scientist
> Therefore, Einstein was not a brilliant scientist.

This is obviously an invalid argument since it is possible for both premises to be true and for the conclusion to be false. According to our second definition of validity, therefore, this argument is not valid. The reason it is not valid, of course, is that it is possible for both men to be brilliant scientists. Consequently, logicians distinguish two kinds or senses of disjunction: an inclusive sense and an exclusive sense.

Exclusive versus Inclusive 'Or'

In the *exclusive* sense of 'or,' one proposition is true but the other proposition cannot be true--its truth is excluded by the truth of the first; in the *inclusive* sense of 'or,' at least one of the propositions must be true but the other proposition might also be true--its truth is not excluded by the truth of the first.

When one uses 'or' in an exclusive sense, one is claiming that one (*and only one*) disjunct is true, i.e., that one but not both disjuncts are true. For example: either she is in Ft. Collins or she is in Denver. Since one cannot be in two places at one time, this would be an exclusive sense of 'or.'

For reasons that will not concern us here (but basically because it includes more kinds of valid argument, logicians have agreed to use the weaker, inclusive sense of 'or' in all cases of disjunctive reasoning. This is simply a convention that has been adopted. Hence, if one were to reason

> Either Newton's theory of gravity is true or Einstein's theory of gravity is true
> Newton's theory of gravity is true
> Therefore, Einstein's theory of gravity is false

one would be committing a fallacy. We could, however, turn this above invalid argument into a valid one by incorporating the implicit assumption mentioned above into the argument itself:

> Either Newton's theory of gravity is true or Einstein's theory of gravity is true *but not both*
> Newton's theory of gravity is true
> Therefore, Einstein's theory of gravity is false

This argument is now valid. The point to remember is this: if one means the exclusive sense of 'or' build it explicitly into the premises; otherwise, the inclusive sense of 'or' will be meant.

Conjunctive Arguments

A **conjunctive proposition** is a proposition containing at least two propositions (called conjuncts) joined by the connector word *and* (abbreviated by '&'). In addition, conjunction can abbreviate all of the following words: 'but,' 'however,' 'moreover,' 'although,' 'yet,' 'still,' 'even though' (and the comma and semicolon).

A conjunctive proposition is true under the condition that both conjuncts are true; otherwise, it is false. Similarly, we can say that a **conjunctive argument**, which is an argument containing the two conjuncts as premises and the conjunction (p and q) as conclusion, is valid when both conjuncts (p, q) are true; otherwise, it is false. Using the truth table method, this reads:

p	q	(p&q)
T	T	T
T	F	F
F	T	F
F	F	F

There are two valid types of inferences involving conjunction--conjunction and simplification.

p q ——— (p and q) *Conjunction*	p and q ——— p *Simplification*

Since these are obvious, we will not spend any time discussing them.

Conditional Propositions

A **conditional proposition** (also called *hypothetical propositions*), such as, *if we have another inch of rain in August, then it'll set a new record,* is a compound proposition consisting of two propositions (*we have another inch of rain, it'll set a new record*) joined by an if-then connective. The proposition that precedes the "then" is called the **antecedent** and the proposition that follows the "then" is called the **consequent**.

We will use the horseshoe '⊃' to represent 'if-then,' so that 'if p then q' is symbolized as: 'p ⊃ q'. 'If-then,' it should be noted, can be used to represent a great variety of sentence connectives, including, 'provided that,' 'on the condition that,' 'in case that,' etc.

The Truth Table for Conditionals

The question of when a conditional proposition is true is a very complicated question and one which students find puzzling. Clearly, a conditional proposition is *true* when both the antecedent and the consequent are true. When is a conditional false? It is false when the antecedent is true and the consequent is false. But what about the case in which the antecedent is false and the consequent is true, and the case in which the antecedent is false and the consequent is also false? For reasons we will not go into,[1] logicians have argued that in these two latter cases, the conditional proposition is also true. Our truth table thus looks like:

p	q	$(p \supset q)$
T	T	T
T	F	F
F	T	T
F	F	T

A **conditional argument** is an argument consisting of three propositions (two premises and a conclusion) such that one premise is a conditional proposition, one premise either affirms or denies the antecedent or consequent, and the conclusion either affirms or denies the antecedent or consequent not mentioned in the second premise. If we represent our conditional by $p \supset q$, then the different possible second premises would be: ~p, ~q, p, q and the different possible conclusions would be: ~q, q, ~p, p. We thus have the following possible conditional arguments:

[a]	[b]	[c]	[d]
$p \supset q$	$p \supset q$	$p \supset q$	$p \supset q$
p	p	q	q
q	~q	p	~p

[e]	[f]	[g]	[h]
$p \supset q$	$p \supset q$	$p \supset q$	$p \supset q$
~p	~p	~q	~q
q	~q	p	~p

Some of these possible conditional arguments make little sense (cases [b], [d], [f] and [g]). We will focus, therefore, on the remaining four: [a], [c] [e] and [h]. These four argument forms will prove to be of special importance throughout the remainder of the book. So the student should learn them well. Two of them are valid conditional argument forms, and two of them are invalid formal fallacies. Because these four are so important they have names attached to them. The valid argument forms are:

$p \supset q$
p

---

q
Modus Ponens

This is the most obvious case. Consider, for example:

If the Broncos win the Superbowl, then they obviously have the best quarterback
The Broncos (did) win the Superbowl
Therefore, they have the best quarterback

Consider, however, the following example:

If the Broncos win the Superbowl, then they obviously had the best quarterback
They have the best quarterback
Therefore, they (did) win the Superbowl

Is this argument valid? No. How do you decide? Try the test of validity: if the premises were true, would the conclusion also have to be true? No. Is it possible for the premises to be true and the conclusion to be false? Yes. How? Well, the Broncos might lose the Superbowl because their defense was terrible. So, it is possible for the premises to be true and for the conclusion to be false, just as this is possible in the following argument;

If a train runs over you, then you'll be dead
You are dead

Therefore, a train has run over you

The premises can be true and the conclusion false in many ways; for example, you may have been hit by lightning or have eaten a pound of arsenic in the dormitory's soup.

The above argument is invalid and commits a fallacy called *the fallacy of affirming the consequent.* This has the following form:

$$\begin{array}{l} p \supset q \\ \quad q \\ \hline p \end{array}$$

Fallacy of Affirming the Consequent

The name of the fallacy describes precisely what the source of the problem is: affirming the consequent. *Whenever* a premise affirms a consequent, the argument will be invalid.

Another fallacy results when one denies the antecedent. For example: if this is a Chrysler, then it costs a lot of money; it is not a Chrysler; therefore, it does not cost a lot of money. Again, it is possible for the premises to be true and the conclusion, for example, this car is not a Chrysler but a Mercedes. So, the argument is invalid because it denies the antecedent:

$$\begin{array}{l} p \supset q \\ {\sim}p \\ \hline \quad {\sim}q \end{array}$$

Fallacy of Denying the Antecedent

The final conditional argument form we will deal with is a valid one: *Modus Tollens.* It consists of denying the consequent and then denying the antecedent:

$$\begin{array}{l} p \supset q \\ \quad {\sim}q \\ \hline {\sim}p \end{array}$$

Modus Tollens

For example:

If this is Legal's Seafood Restaurant, then the fish is fresh
The fish is not fresh
Therefore, it is not Legal's Seafood Restaurant.

This is a valid conditional argument, and its validity can be determined by seeing if it is possible for the premises to be true and for the conclusion to be false. Clearly it is not.

One can see this in another way. Suppose someone said: the premises are true and the conclusion false. Now, from the truth of the conditional premise (*if this is Legal's Seafood Restaurant, then the fish is fresh*) and the alleged truth of the conclusion (*it is Legal's Seafood Restaurant*--which is the same thing as claiming the conclusion is false), one can derive, via *Modus Ponens*, the new conclusion (*the fish is fresh*). But this conflicts with the agreed upon assumption that the original premise was true, i.e., that the fish is not fresh. Hence, we have a contradiction. Since assuming the truth of our conclusion resulted in a contradiction, this indicates our conclusion must really be false. (This is a form of reasoning called *reductio ad absurdum:* if by assuming the falsity of the conclusion, one derives an absurdity, e.g., a contradiction, then the conclusion cannot be false and hence must be true.) But this is precisely what is being claimed in Modus Tollens.

In summary, we have the following four cases the student should commit to memory:

$p \supset q$	$p \supset q$	$p \supset q$	$p \supset q$
p	$\sim q$	q	$\sim p$
——	——	——	——
q	$\sim p$	p	$\sim q$
Modus Ponens	**Modus Tollens**	**Fallacy of Affirming the Consequent**	**Fallacy of Denying the Antecedent**

Combining Connectives

One final thing must be said about the above connectives: they can be combined in increasingly complex ways. For example, if I say: "If the recipe calls for fresh tarragon, then either I'll have to go to the Market again or our dinner will not be as good as I promised" (and letting 'r,' 'm,' and 'g' stand for the three propositions), one can symbolize this as: $r \supset (m \vee \sim g)$. If I were to say: "it's false that if the recipe calls for fresh tarragon, then either I'll have to go to the Market again or our dinner will not be as good as I promised," this is symbolized as: $\sim[r \supset (m \vee \sim g)]$.

Exercise 4.1

In the following passages, symbolize each sentence by a conditional: "the market rises" (p), "we sell our railway stock" (q), "we invest in oil stock" (r), "we put our money in the bank" (s), "we invest in insurance stock" (t), and "we buy commodities" (u). In the above examples, assume "the market rises" (p), "sell our railway stock' (q), "invest in oil stock" (r) are true, and "put our money in the bank" (s), "invest in insurance stock (t), and "buy commodities" (u) are false. Determine the truth values of these compound sentences by the truth tables (be sure to indicate whether they are true or false).

1. If the market rises, then we will sell our railway stock and invest in oil stock.

2. If the market does not rise, then either we will sell our railway stock or put our money in the bank.

3. If the market does not rise, then we will sell all our railway stock, but we will not put our money in the bank.

4. It is not the case that if the market rises, then we will invest in insurance stock and we will not buy oil stock.

5. If the market rises, then if we do not put the money in the bank, then we will invest in oil stock.

6. Either the market rises or it's false that if we invest in oil stock, then we'll buy commodities.

Necessary and Sufficient Conditions

In interpreting the meaning of conditional propositions, it is important to understand the differences between "if p then q," "p only if q," and "p if and only if q." A useful way of approaching these distinctions is by means of the distinction between necessary and sufficient conditions. This distinction is one we will encounter later in our discussion of Mill's methods of experimental inquiry, and it is one that will prove to be of inestimable value in understanding and evaluating scientific arguments.

There are various ways of expressing the difference between something's being a sufficient condition and something's being a necessary condition.

p is a sufficient condition for q =

p suffices to produce q
p guarantees the truth of q
p causes (produces) q
p can't be true unless q is true
you can't have p without q

q is a necessary condition for p =

q must obtain in order for p to obtain
you have to have q in order to have p
if q is not true, then p is not true
p can't be true unless q is true
you can't have p without q

In general, we can say that "p is a sufficient condition for q" can be translated as "if p then q" and "q is a necessary condition for p" can be translated as "if not q, then not p." For example, in the proposition *if you eat a pound of arsenic, then you'll surely die*, eating a pound of arsenic is claimed to be a sufficient condition for dying. However, if you were to claim that eating a pound of arsenic was a necessary condition for death, what would this statement claim? Not that eating a pound of arsenic will (by itself) produce death--since this would be the claim that eating arsenic is a sufficient condition for death. To say that something is necessary for something else means that if you do *not* have this first condition, then you will *not* have the second condition. It means that if you do not eat a pound of arsenic, then you will not die. Clearly, this statement is false and hence eating arsenic is not a necessary condition for death. Why? Because there are other ways of dying. So:

p is a sufficient condition for q = if p, then q
q is a necessary condition for p = if ~q, then ~p

To give an example of the difference between something's being necessary versus it's being sufficient, consider the following two examples. Suppose I announce to my class that there are three ways a student may obtain a grade of an "A": either if one scores 98% or higher on the final examination, or if one writes a term paper and receives a letter grade of an A, or if one gives an excellent class lecture. Here I am claiming either one of them by itself is a sufficient condition for an A:

98% on the final examination (f) is a sufficient condition for an A (a)
Writing a term paper and receiving a letter grade of an A (p) is a sufficient condition for an A (a).
Giving an excellent class lecture (d) is a sufficient condition for an A (a)

$f \supset a$
$p \supset a$
$d \supset a$

Suppose I am an avid gardener, and someone asks me how I raise such fine corn. I respond: in order to produce good corn, you must have seed, sunlight, water, soil, and fertilizer. What I am claiming first is that seed is a necessary condition for the growth of corn (since if you don't have it, you can't have corn). But it is not sufficient since, by itself, it clearly will not produce corn. Similarly, I am claiming that sunlight is a necessary condition for the growth of corn just as water, soil, and fertilizer are each necessary conditions for the growth of corn:

Seed (a) is a necessary condition for the growth of corn (c)
Sunlight (s) is a necessary condition for the growth of corn (c)
Fertilizer (f) is a necessary condition for the growth of corn (c)
Soil (d) is a necessary condition for the growth of corn (c)
Water (w) is a necessary condition for the growth of corn (c)

How should we diagram this? If we have a complete list of necessary conditions, then the conjunction of them would be sufficient.

$$\left\{\begin{matrix} a \\ s \\ f \\ d \\ w \end{matrix}\right\} \Rightarrow q$$

Exercise 4.2

Which item in column A is a sufficient condition for the corresponding item in B? Which item in column A is a necessary condition for the corresponding item in B?

A	B
1. Birth	Death
2. Zygote	Pregnancy
3. Being married	Pregnancy
4. Sexual Intercourse	Pregnancy
5. Being at least	

35 Years old	President of U.S.
6. Presence of Virus	Influenza
7. Fever	Influenza
8. Feeling sick	Influenza
9. Having a Parent	Being a Child
10. Having an Uncle	Being a Nephew
11. Being a Male	Being an Uncle
12. Being a Male	Being a Nephew
13. Education	B.A.
14. College Diploma	B.A.
15. Oxygen	Fire
16. Match being lit	Fire

Logical Notation

In translating sentences into hypothetical propositions expressing necessary and sufficient conditions, it is crucial to distinguish between propositions of the form *if p, then q*, *p, if q*, and *p only if q*. *p, then q* claims p is a sufficient condition for q. *p, if q* is the same proposition as: *if q, then p,* which claims that q is a sufficient condition for p. *p only if q* claims that q is a necessary condition for p, e.g., *you will get an A in class only if you study hard* claims studying hard is a necessary condition for an A in class.

Finally, in the proposition *p if and only if q*, we have a biconditional (see below) in which we have a conjunction of two conditionals: *p, if q* and *p only if q*. This, in turn, is equivalent to saying that q is a sufficient condition for p and q is a necessary condition for p or, in short, that q is a necessary and sufficient condition for p. It is difficult to find examples of such propositions in daily life and in science, but perhaps the best examples of them come from *definitions* in which one thing (e.g., 'water') is defined in terms of other properties (e.g., 'H_20'). In this case, a state of water would be a necessary and sufficient condition for being H_20.

In the example of *water is* H_20, it was being claimed that water is a necessary and sufficient condition for H_20 or was it being claimed that H_20 is a necessary and sufficient condition for being water? Actually, both are correct since you can prove logically that if p is a sufficient condition for q, then q is a necessary condition for p, and if p is a necessary condition for q, then q is a sufficient condition for p. This is because *p only if q* (read *if ~q then ~p),* is equivalent to *if p then q* (by the logical principle of transposition--see below) and the latter claims p is a sufficient condition for q. Thus, in a biconditional, where something p is claimed to be a necessary and sufficient condition for q, it is being claimed that p is a necessary and sufficient condition for q and that q is a necessary and sufficient condition for p.

One final comment about the word *unless*. Suppose the football coach says: *there'll be no Superbowl championship unless we get a better running game*. What is being claimed here? The coach is claiming that a better running game is a necessary condition for a Superbowl championship. He surely is not claiming that a better running game is a sufficient condition for a Superbowl championship since many other things are also necessary. So, one standard way of translating *no p unless q* is *q is a necessary condition for p* (since *no p unless q* means *unless q, then not-p*, which means *if ~q then ~p*. (Think of reading "unless" as "without" as in: *there'll be no Superbowl championship without a better running game.*)

Suppose the coach says: *we'll win the Superbowl unless Elway gets hurt* (p unless q). What is being claimed here? Following the above notation, *p unless q* should be read *unless q, then p*. This, in turn, should be translated as: *if not-q, then p*, which means not-q (not Elway getting hurt) is a sufficient condition for p (winning the Superbowl). So the phrase 'unless p' should be translated as 'p is a necessary condition' or 'p is a sufficient condition' depending on whether the consequent has a negation in it or not: if it has a negation, p is a necessary condition; if it does not have a negation, p is a sufficient condition. The point to remember is the logical form: *if not-p, then not-q* means p is a necessary condition for q (and q is a sufficient condition for p); *if not-p, then q* means not-p is a sufficient condition for q (which, in turn, means q is a necessary condition for not-p, that is, *if not-q, then p*. In short:

no p unless q $= \sim q \supset \sim p$

p unless q $= \sim q \supset p$

Exercise 4.3

A. In each of the following statement, determine whether a sufficient condition, a necessary condition, both, or neither is being claimed.

1. If the big cities voted Democratic, then the Democratic candidate will win.
2. Bill's success depended upon Ross's helping him.
3. Bill couldn't have succeeded without Ross's help.

4. You gotta be a football hero to make it with beautiful girls.
5. There'll be no dance unless we get a band.
6. If you go home, you'll get something to drink.
7. You'll get something to drink only if you go home.
8. To get to New York from Detroit, all you have to do is to go through Canada.
9. A good library is a *sine qua non* for greatness in a University.
10. All who use big words have abstract general ideas.
11. Only doctors and hospital administrators are eligible.
12. Anyone who helps an unfortunate person helps Jesus.
13. The patient will die unless we do a heart transplant.

B.

1. Express the conditions below in terms of (l) the "if. . .then___" form of the conditional, and (2) the ". . .only if__" form of the conditional.

(A) is a sufficient condition for (B) } entails If. . . then___ = . . .only if___ condition

(C) is a necessary condition for (D) } entails If. . . then___ = . ..only if__

2. (a) "(X) is a necessary and sufficient condition for (Y)" implies:

Check each correct answer
( ) if X, then Y
( ) X only if Y
( ) if Y, then X
( ) Y only if X

(b) "Not r unless s" (e.g., the pump will not work unless it has been primed") implies:

Check each correct answer
( ) if r, then s
( ) s is a necessary condition for r
( ) If not s, then r
( ) r only if s

(c) "n is a necessary condition for m" entails:

Check each correct answer

( ) if m, then n
( ) n only if m
( ) If n, then m
( ) m only if n
( ) not m unless n

C.

1. A lighted match is thrown near a pile of leaves by a hunter and a forest fire develops. What were the necessary and the sufficient conditions?

2. Many psychologists believe cognitive conflict (disequilibrium) is necessary in order for learning to occur. Others believe it is sufficient but not necessary. Explain the difference between these two claims. How could an experiment decide which of these two claims was correct? If an experiment shows it is not necessary (sufficient), can a scientist still claim it is sufficient (necessary)? How could she do this?

D. In the following passages, symbolize what is being claimed in the form of a conditional sentence.

1. Bartender: there'll be no booze [p] unless you've got an I.D. [q]
2. There'll be no vacation at Christmas [p] unless you get the time off [q].
3. There'll be class on Friday [p] unless there's snow [q]
4. I'll go to Mesa Verde over Spring Break [p] unless I have to write a term paper [q].

Hypothetical Syllogism

Another argument form often discussed in conjunction with conditional arguments is the *hypothetical syllogism*. Here we have an argument consisting of three conditionals, e.g., *if microbes are contagious, then drug X should kill the microbes*; *if drug X kills the microbes, then the experiment will be successful*; therefore, *if microbes are contagious, then the experiment will be successful*. In symbols this reads:

If p, then q
If q, then r

If p, then r

One easy way to see that this is valid is to think of the conditional "if-then" as an arrow →. Then we have:

$p \rightarrow q$

q → r
----
p → r

For, after all, if p → q and q → r, then p → q → r and we can simply bypass q and go directly from p to r: p → r. (This is known as a *transitive* inference.)

Although the hypothetical syllogism is valid when it is expressed in the above symbolic way, other arguments, which seem to be cases of hypothetical syllogism, are not valid. For example:

If p then q
If p then r

If q then r

Biconditional Propositions

A biconditional proposition is a proposition of the form: *p if and only if q*, which can be represented symbolically as p ≡ q. It is a conjunction of two conditionals: (if p then q) and (if q then p). For example: *Harry will get an A in Logic if and only if he scores 98% on his final* is equivalent to: *If Harry will get an A in Logic then Harry scores 98% on his final* and *if Harry scores 98% on his final then Harry will get an A in Logic*. A biconditional is true when both of these conjuncts are true, that is to say, when both conditionals are true. It is false if either the one conditional is false or the other conditional is false. This is represented in the following truth table:

p	q	(p ⊃q)	(q ⊃ p)	(p ≡ q)
T	T	T	T	T
F	T	T	F	F
T	F	F	T	F
F	F	T	T	T

This in turn can be abbreviated as:

p	q	(p≡q)
T	T	T

T	F	F
F	T	F
F	F	T

In short, a biconditonal is true just in case both propositions are true or both propositions are false.

Exercise 4.4

Are the following arguments deductively valid? If so, which valid argument form do they exemplify? If not, what fallacy is being committed?

1. If the theory of relativity is true, objects should appear more massive as their velocities increase. As a matter of fact, objects do appear more massive as their velocities increase. This shows the theory of relativity is true.

2. If the patient has tuberculosis, the test will give a positive result. The test gave a positive result. This shows the patient has tuberculosis.

3. If the drug is taken as prescribed, the patient will recover. If the patient is smart, the drug will be taken as prescribed. Hence, if the patient is smart, the patient will recover.

4. If Pluto, according to Halliday's calculations, had a diameter of more than 4200 miles, then an occultation would have occurred at McDonald Observatory, and the records clearly indicate it did not. Thus Pluto must be that size or smaller; it cannot be larger.

   ---Thomas Nicholson, "The Enigma of Pluto"

5. Either Jane will take ballet or Jane will take ice skating lessons. Jane will take ballet. Therefore, she will not take ice skating lessons.

6. Either Israel will remain neutral or Syria will go into Lebanon. Syria will not go into Lebanon. Thus, Israel will remain neutral.

7. If that bar is magnetized, it will attract iron filings. Since it does not attract iron filings, it isn't magnetized.

8. If Darwin's Theory of Evolution is correct, then we should find fossils that reveal sequential changes throughout the ages. We do find fossils, which reveal sequential changes throughout the ages. Therefore, Darwin's Theory of Evolution is correct.

9. This substance is not sodium because it burns with a green flame, which it would not do if it were sodium.

10. The car stopped in the intersection. The car was involved in an accident. Hence, the car stopped in the intersection and was involved in an accident.

More on Negation

Suppose we have a disjunctive proposition (p or q) where the parenthesis indicates that what occurs within it is a single unit. When we negate this, that is, when we assert (p or q) is false, what are we asserting? We are asserting that it is false that either p or q is true, and this in turn is equivalent to saying neither p nor q is true, that is to say, not-p is true and not-q is true. Thus

$$\sim(p \vee q) = (\sim p \,\&\, \sim q)$$

For example, *it is not true that (either Today is Thursday or it is Friday)* is equivalent to *it is false that either Today is Thursday or Friday* which, in turn, means *neither is today Thursday nor Friday* or *today is not Thursday and it is not Friday.*

Suppose we negate a conjunction, say, *she plays tackle for the CSU Rams and she is homecoming queen.* This becomes: *it is false that (she plays tackle for the CSU Rams and she is homecoming queen).* But this in turn means that it false *both* that *she plays tackle for the CSU Rams and that she is homecoming queen.* But this in turn means that *either she does not play tackle for the CSU Rams or it is false that she is homecoming queen.* In symbols:

$$\sim(p \,\&\, q) \equiv (\sim p \vee \sim q)$$

As a memory device, remember that negating a conjunction results in a disjunction and negating a disjunction results in a conjunction.

What is the negation of a conditional, e.g., *if Sally runs for president, then the chances of world peace are increased?* Its negation is: *it is false that (if Sally runs for president, then the chances of world peace are increased).* This in turn means that the conditional is false. The truth table tells us that a conditional is false just in case the antecedent is true and the consequent is false. Thus, it is false that (if Sally runs for president, then the chances of world peace are increased) is equivalent to *Sally runs for president and the chances of world peace are not increased.* In symbols:

$$\sim(p \supset q) \equiv (p \,\&\, \sim q)$$

Finally, since a biconditional is a conjunction of two conditionals, the negation of a biconditional is equivalent to claiming that either one conditional is false or the other conditional is false:

$$\sim(p \equiv q) \equiv [\sim(p \supset q) \vee \sim(q \supset p)]$$

Exercise 4.5

Put the following statements into symbolic form (put parentheses around atomic propositions). Then, assuming 'Cobalt is a solid' (c), 'Racial disturbances continue in South Africa (p),' 'The government will move in to stop the conflict (q),' 'Deklerk will remain neutral (r),' are true and 'Cobalt is a liquid' (l), 'Mandela will step into the conflict (s),' 'the U.S. will pledge support to the Zulu (t),' and 'Russia will pledge support to the Afrikaners (u),' are false, determine the truth value of each of the compound statements. Indicate whether the compound proposition is true or false and why.

1. It is false that the U.S. did not pledge support to the Zulu.
2. It is not the case that racial disturbances will continue in South Africa and Mandela will step into the conflict
3. It is false both that the U.S. pledged support to the Zulu and Russia pledged support to the Afrikaners.
4. It is not true that Deklerk will remain neutral and it is the case both that racial disturbances will continue in South Africa and Mandela will step into the conflict.

Exercise 4.6

What is the negation of the following propositions?

1. If this is Thursday, then we must be in Belgium.
2. We are either in Belgium or we're in France.
3. Sally and Susan are both good candidates for president.
4. If Darwin's Theory of Evolution is correct, then we should find fossils that reveal sequential changes throughout the ages.
5. Louise will win the Coors Classic if and only if she can beat Karen once more.

Logical Relations

It is very useful to know how one proposition is logically related to another proposition, e.g., when the form of one proposition is logically equivalent to the form of a different proposition.

Propositions can be related to each other in various kinds of logical ways. In this section, we give several kinds of logical relations between propositions. This will take the form of several short definitions:

> ***Logical Dependence*** **(logical relation).** *p is logically related to q [q logically depends on p] = the truth (or falsity) of p (q) influences or determines the truth (or falsity) of q (p).*

On the other hand, two propositions p and q are ***logically independent*** if the truth (falsity) of the one does not influence the truth (falsity) of the other. The relation of implication is one kind of logical dependence--a strong relation--in which the truth (falsity) of the premises influences the truth (falsity) of the conclusion. Other examples abound where the truth/falsity of one or more propositions influences (to some degree) the truth or falsity of another proposition. For example, the positive evidence gathered in an experiment is (to a certain degree) logically related to the truth of a hypothesis under test.

> ***Logical Equivalence.*** *p is logically equivalent to q = p implies q and q implies p (i.e., if p then q and if q then p).*

This is precisely the relation we were expressing above when we claimed the following:

$$\sim(p \vee q) \equiv (\sim p \ \& \sim q).$$

~(p v q) is logically equivalent to *(~p & ~q)* because *~(p v q)* implies *(~p & ~q)* and *(~p & ~q)* implies *~(p v q)*. Since *implies* is a concept expressed by a conditional proposition, we are clearly expressing a biconditional. Thus, when two propositions are logically equivalent to each other, their logical equivalence takes the form of a biconditional

$$[\sim(p \vee q)] \equiv (p \ \& \sim q)\ ]$$

We have simply been following the convention that '=' can also be represented by a biconditional (≡) and vice versa, we have been assuming that '=' and '≡' were logically equivalent.

Similarly, the negation of p and q (~[p & q]) is logically equivalent to (~p v ~q) and the negation of a disjunction (~[p v q]) is equivalent to (~p & ~q).

~(p & q) ≡ (~p v ~q)
~(p v q) ≡ (~p & ~q)

***Logical Incompatibility**. p is logically incompatible with q = not both p and q, i.e., ~(p & q).*

This means that if p is true, then q is false or if q is true, then p is false.

***Logical Contradictories**. p and q are logical contradictories = not both p and q are true and not both p and q are false, that is, ~(p & q) and ~(~p & ~q). Thus either p or q is true.*

***Logical Contraries**. p and q are logical contraries = not both p and q are true but both might be false, i.e., ~(p & q).*

For example, *this ghost is white* and *this ghost is not-white* are logical contradictories since both can't be true and one of them must be true. This is different from the case of *this ghost is white* and *this ghost is black* since both can't be true but both might be false in case, say, the ghost was red. Thus it is important to distinguish when two propositions are contradictories and when they are contraries.

Transposition

Another logical equivalence that should be mentioned is the principle of transposition. Suppose you have the conditional (p ⊃ q). What you get by the principle of transposing p and q is (~q ⊃ ~p). To perform the operation of transposing, merely replace all occurrences of p by q and all occurrences of q by p and reverse the positive and negative signs. Thus (p ⊃ q) becomes first (q ⊃ p), and then, since p and q were originally affirmative, make the subsequent q and p negative: (~q ⊃ ~p).

This same principles applies when you have something like (p ⊃~q). Reverse the p's and q's to get (~q ⊃ ~p) and change the original signs: since q was originally ~q, this now becomes ~~q which is equivalent to q, and since p was just p, change it to its negative form ~p. This gives you (q⊃ ~p.)

Hence we have the following *contrapositives*:

(p ⊃ q) ≡ (~q⊃ ~p)
(p ⊃ ~q) ≡ (q ⊃ ~p)
(~p ⊃ q) ≡ (~q ⊃ p)
(~p ⊃ ~q) ≡ (q ⊃ p)

Some other useful logical equivalences:

If p is sufficient for q, then q is necessary for p: (p ⊃ q) ≡ (~q⊃ ~p)

If p is sufficient for ~q, then ~q is necessary for p, i.e., if ~~q then ~p = if q then ~p: (p ⊃ ~q) ≡ (q ⊃ ~p).

Truth Tables and Arguments[2]

As we have already indicated, truth tables can be used to define the conditions under which a compound proposition is true. This applies not only to propositions like (p or q); it also applies to more complex propositions involving several connectives. Consider for example the proposition: ~(p v q) ≡ (~p & ~q). We can use the truth table method to *prove* this proposition to be true. How? By means of the fact that a truth table for this very complex proposition gives the conditions under which it is true. If it turns out that this proposition is always true under any assignments of truth values, the proposition is a **tautology**.

How do we proceed? In the following steps.

1. List all possible truth-value assignments for the atomic propositions.

2. Determine the truth-value of the smallest compound proposition (using the truth tables for the connectives).

3. Next, move to the next larger component, repeating step 2 until you reach the main (largest) truth connective.

4. If the truth value of the largest connective is true under all entries, it is a tautology and we have proved it.

In the proposition ~(p v q) ≡ (~p & ~q), the first step results in:

p q	~(p v q) ≡ (~p & ~q)
T T	T T F F
F T	T F F T
T F	F T T F

F F	F F T T

Next, determine the truth value of the smallest compound proposition—the disjunction and the conjunction:

p q	~(p v q) ≡ (~p & ~q)
T T	T T T F F F
F T	T T F F F T
T F	F T T T F F
F F	F F F T T T

Repeating step 2 with the next larger compound proposition, ~(p v q), results in:

p q	~(p v q) ≡ (~p & ~q)
T T	F T T T F F F
F T	F T T F F F T
T F	F F T T T F F
F F	T F F F T T T

Next, repeat step 2 with the largest compound proposition, the biconditional. This results in:

p q	~(p v q) ≡ (~p & ~q)
T T	F T T T T F F F
F T	F T T F T F F T
T F	F F T T T T F F
F F	T F F F T T T T

Because the biconditional has nothing but values of 'true' under it, it is a tautology and we have proven the biconditional to be true.

Exercise 4.7

Prove the following propositions to be true by means of the truth table method discussed above.

1. ~(p & q) ≡ (~p v ~q)
2. (p ⊃ q) ≡ (p v ~q)
3. ~(p ⊃ q) ≡ (p & ~ q)
4. (p ≡ q) ≡ [(p ⊃ q) & (q ⊃ p)]

The truth table method can also be used to test for the validity of an argument. Consider, for example, the valid argument form Modus Ponens:

$$\begin{array}{c} (p \supset q) \\ p \\ \hline q \end{array}$$

The truth table can show this to be a valid argument form in the following way. First, conjoin all the premises into a conditional proposition: [(p ⊃ q) & p]. Then form a conditional, with the preceding as the antecedent and the conclusion as the consequent--this is called the corresponding conditional.:

{[(p ⊃ q) & p] ⊃ q}

Then use the above method to show the conditional is a tautology:

```
{[(p ⊃ q) & p] ⊃ q}
-------------------
  T T T T T T T
  T F F F T T F
  F T T F F T T
  F T F F F T F
```

If, however, the entries under the largest compound proposition have any entities with 'F' under them, the argument is not valid. For example, the argument

(p ⊃ q)
Therefore
(~p ⊃ ~q)

is an invalid argument. This can be shown in the following way. First, form the corresponding conditional:

[(p ⊃ q) ⊃ (~p ⊃ ~q)]

Substituting truth values gives you

$[(p \supset q) \supset (\sim p \supset \sim q)]$

T	T	T	T	F	T	F
T	F	F	T	F	T	T
F	T	T	F	T	F	F
F	T	F	T	T	T	T

Because the corresponding conditional has one entry (row three) which has an "F" entry, this argument is not valid. Why? Remember the definition of 'validity.'

Exercise 4.8

Use truth tables to prove the validity or invalidity of the following arguments:

1. p, therefore, (p v q)
2. [~p v (q & p) and ~q], therefore, [(p ⊃ q) & (q ⊃ p)]
3. [(p ⊃ q) and (q ⊃ r)], therefore, (p ⊃ r)
4. (p ⊃ q) and ~q, therefore, ~p
5. ~q and (p ⊃ q), therefore, ~p.

Notes

[1] Since p ⊃ q is false when (p & ~q) is true, ~(p & ~ q) would be the denial of its falsity, which would be equivalent to asserting the truth of p⊃ q. But ~(p &~q) is equivalent to (~p v q). So p ⊃ q is true when p (the antecedent) is false--cases 3 and 4--or when q (the consequent) is true--cases 1 and 3. Combining these, p ⊃ q is true in cases 1, 3 and 4.

[2] Shaded material represents optional material, usually material of a more advanced nature.

5

Inductive Logic

Introduction

An inductive inference is never formally valid since the premises of an inductive argument never logically entail the conclusion; it is always possible for the premises to be true and for the conclusion to be false. This is because inductive inferences always go beyond the information contained in the premises and hence are **ampliative.** Inductive arguments, therefore, can never attain the necessity characteristic of deductive arguments.

This does not mean, however, that inductive arguments cannot be evaluated as to their logical adequacy. In fact, one branch of logic--inductive logic--has as one of its goals the formulation of principles and criteria for the evaluation of the adequacy of inductive arguments.

Inductive logic is thus similar to deductive logic. But inductive logic has not and perhaps can not attain the certainty and exactitude of deductive logic. Partly this is due to historical reasons and partly because of its subject matter. Unlike deductive logic, which has reached a level of maturity and rigor in the twentieth century unforeseen by earlier logicians, inductive logic is a relatively young field in which there is disagreement concerning many of the foundational principles. We will present an account of inductive logic that is not only plausible (from a logical point of view) but, more importantly, one followed by many scientists in their actual reasoning.

Validity versus Inductive Strength

A necessary first step is defining what a good inductive argument is. This in turn involves defining what an inductive argument is.

An **inductive argument** is an argument in which the premises do not entail the conclusion but render the conclusion likely (probable) to some degree.

It should be pointed out that the usual way of defining inductive arguments is inadequate. Many textbooks and dictionaries define induction as "arguing from particular to general," whereas deduction is "arguing from general to particular." As a definition of induction, this account is inadequate. Admittedly, there are many examples of induction and deduction that satisfy the above account, but these are not correct definitions of the two types of argument. Some inductive inferences are from particular to particular, e.g., *Sally is a philosophy major; therefore, she must be pretty smart.* (This is an example of eduction.) On the other hand, many deductive inferences are from general to general. Hence, even though many examples of inductive inferences may be inferences from particular to general, this is not an adequate definition of inductive inference. We need, therefore, a better way to define the nature of induction versus deduction. A better way to differentiate the two is in terms of *validity versus warrant.*

Although the crucial concept in deductive logic is validity, this concept is not appropriate for inductive logic. However, there does seem to be a concept analogous to validity, that of *inductive strength.*[1] We can say, therefore,

> *An argument is* ***inductively strong*** *if and only if*
> *[1] the premises render the conclusion likely;*
> *[2] it is improbable that the premises are true and the conclusion false;*
> *[3] if the premises were true, then the conclusion would be likely.*

Of course, how improbable or probable an inductive argument is depends upon one's evaluation of the strength of the premises; hence, inductive strength, unlike validity, has *degrees.* The degree of inductive strength depends on *how probable* the conclusion is, given the premises.

Assessing Inductive Strength

Although we will have a great deal more to say about how one goes about assessing the inductive strength of an argument, at this point we will merely cite several generally accepted rules for determining inductive strength.

1. **Strength of Conclusion**. The first thing to consider is how strong the conclusion is, that is, how strong the conclusion indicator is. Some inductive conclusion indicator words are quite strong ("seems virtually certain"), whereas others are weak ("seems"). Clearly, the stronger the conclusion being drawn, the greater the chance of it being wrong and, hence, the weaker the argument.

2. **Scope of the Conclusion.** A second criterion for assessing the strength of an inductive argument is the scope of the conclusion: in drawing the conclusion, how far out on a limb, so to speak, does one go? From a study involving 20 year-old College Sophomores, an individual may draw a conclusion about all 20 year-old College Sophomores, all College Sophomores, all 20 year-olds, all college students, all individuals in the U.S., all humans beings currently alive, all humans who have ever lived, all biological organisms, etc. Clearly, the narrower the scope of one's conclusion, the less chance one has of being wrong and, hence, the stronger the inductive strength of an argument.

3. **Amount of Evidence**. Next, the greater the evidence cited in the premises, the stronger the conclusion. If after citing one individual who has a particular property, one then infers that a larger group has that property, the argument is weaker than if several individuals are mentioned in the premises. In short, the greater the quantity of evidence cited, the stronger the argument.

4. **Variety of Evidence**. An argument which has a great deal of evidence supporting it is an inductively strong one. But arguments can also be assessed as to the variety of evidence mentioned. For example, suppose I claim that Freud's theory of unconscious motivation is an excellent scientific theory and in support of this conclusion I cite evidence drawn from the treatment of neurotic behavior. No matter how many neurotic patients have recovered as a result of Freud's theory, this evidence is less adequate than the following additional types of evidence: (1) evidence from the interpretation of dreams, (2) evidence from slips of the tongue, (3) evidence from jokes and humor, and (4) evidence from neurophysiology. Here the evidence does not consist of just one kind but is

heterogeneous in nature. In short, the greater the variety of the evidence, the stronger the inductive conclusion.

Exercise 5.1

In the following inductive arguments, circle inference indicators, identify conclusion and premises, and indicate how strong each argument is in terms of the above four criteria and how strong the overall argument is.

1. A new drug, acebutol, already available in Great Britain and now under study here, looks excitingly good for dangerous heart rhythms. At Stanford University not long ago, doctors tried it for a man with a potentially lethal arrhythmia that resisted all the standard antiarrhythmia drugs such as quinidine and procainimide, and had required repeated electrical shocking of his heart. In the eight months he has been on acebutol, he has had no recurrence of arrhythmia.

2. That Hamilton ever held any considerable sum in securities seems highly improbable, for he was at no time a rich man and at his death left a small estate.

---Charles Beard, *An Economic Interpretation of the Constitution of the United States*

3. The results of this experiment seem to run counter to the theory that putting an emotionally upset child in with a group of well-adjusted children will improve the emotional development of the child. Thus, for example, Redl introduced a boy with considerable masochistic tendencies into a group of normal boys who were all friends and members of the same club. In a short time the disturbed boy had stirred up in the others "more sadistic-pleasure temptations than they could cope with," and their violent aggressions disrupted the social relations of the entire group.

4. Two contrary psychic tendencies exist universally in the father-son relationship, i.e., submission and rebellion, and both derive from the Oedipus complex. Since most patriarchal religions also veer between submission to a paternal figure, and rebellion (both submission and rebellion being sexualised), and every god, like a compulsive super-ego, promises protection on condition of submission, there are many similarities in the manifest picture of compulsive ceremonials and religious rituals, due to the similarity of the underlying conflicts.

---F. Fenichel, *The Psychoanalytic Theory of Neuroses*

5. The governor's commission strongly recommends that use of marijuana continue to be illegal since marijuana leads to the use of heroin. This is evidenced by the fact that 94 percent of 1,000 heroin addicts interviewed reported that they had smoked marijuana before becoming addicted to heroin.

6. A remarkable feature of malaria is the precise timing of its recurrent attacks, which are always at some multiple of 24 hours. Since the cause of malaria is a small protozoan (single-celled) parasite that grows in the red

blood corpuscle, the attacks show that this rather simple organism has a highly accurate biological clock.

--Frank Hawking, "The Clock of the Malaria Parasite"

7. Barbara's skin has a peculiar yellowish hue, and she complains of loss of energy. Barbara must have hepatitis.

Types of Inductive Inference

There are several types of inductive inference. Some of these are of considerable importance and will receive extended treatment. Other types are less important and will only be mentioned in passing. Some inferences are called 'inductive' but are really not inductive at all. For the sake of completeness, therefore, we will provide a listing of the various types of inferences that have been called "inductive inference."

Eduction

We have already mentioned one type of inductive inference--eduction. Here there is a direct inference from a premise (or set of premises) mentioning a particular (or class of particulars) to a conclusion mentioning another particular. For example:

Mars is a solar planet
The earth is a solar planet
The earth is inhabited
Thus, Mars is inhabited

Because such inferences are infrequent in science, they will not be discussed in the remainder of this book (although they do occur when there are inferences involving analogy). They are mentioned only for the sake of completeness and as an example of inductive reasoning in which the inference is not from *particular to general.*

Intuitive Induction

A second type of induction sometimes encountered is intuitive induction. According to Aristotle, after experiencing a number of particular instances of a class, e.g., a number of human beings such as Socrates, Plato, etc., we can come to know their essential nature by an act of intuition: an immediate cognitive grasp of the underlying essence or form of something. This seems to be an example of induction proceeding from particular to general, but this is called intuitive induction since it involves a particular rational faculty of our minds: intuition.

According to most logicians, intuitive induction is really not a case of inductive inference at all. Unlike ordinary induction, intuitive induction is alleged to give us *necessary* knowledge (see Chapter 2); but it is not really an example of inductive argument since it does not consist of a set of premises related in a certain way to a conclusion. Instead, it is a cognitive grasp of a universal property present in many particular things. Hence, although intuitive induction may be an important concept in the history of scientific method—since this type of induction seems to have been employed by Descartes, Galileo—and Newton, it is currently not in much favor.

Mathematical Induction

Another kind of induction often encountered in courses in logic and mathematics is mathematical induction. In mathematical induction one reasons in the following way:

A particular number—e.g., the number 1—has a property p
If any natural number n has the property p, then n + 1 has the property p
Thus, all natural numbers have the property p.

The crucial step is step 2, sometimes called the inductive step, since this shows that property p holds for the successor of any number.

It is now generally accepted that mathematical induction is not really a case of induction at all but rather a case of deduction. Although there is an inductive step in mathematical induction, it is not similar to the type of induction found in the empirical sciences. In fact, according to many, it is not really inductive at all since the conclusion of such an argument is not merely probable but necessarily true--as is the case with mathematical proofs in general. But if the conclusion is necessarily true, an argument containing it cannot be inductive.

Perfect Induction

Another type of induction, derived from the work of Aristotle, is perfect (or complete) induction. Here, a universal proposition (*All F is*

G) is established by an exhaustive enumeration of all the instances of that class. For example:

Venus' orbit around the sun is an ellipse
Neptune's orbit around the sun is an ellipse
Saturn's orbit around the sun is an ellipse
Jupiter's orbit around the sun is an ellipse
Mars' orbit around the sun is an ellipse.
Mercury's orbit around the sun is an ellipse
Earth's orbit around the sun is an ellipse
Uranus' orbit around the sun is an ellipse
Pluto's orbit around the sun is an ellipse

Therefore, all the orbits of the planets around the sun in our solar system is an ellipse

In this kind of inductive argument there are two features that are important. First, the conclusion is universal in nature (All F is G), constituting what we will call a **universal generalization** since it indicates that every individual in the subject class (F)—the class of planets traveling around the sun--is also in the predicate class (G)—elliptical orbits. By contrast, a statement claiming that some Fs are G (or more precisely n% of the Fs are Gs) will be called a **statistical generalization**, since it indicates that a percentage of the Fs are Gs.

In perfect induction, the conclusion contains a universal generalization, but it covers a finite number of individuals, in our example, nine planets. Hence, the premise constituting the evidence contains information about *every single member* of the universal generalization; there are no unexamined instances.

The tacit or suppressed premise implicitly contained in the above inference is that Neptune, Saturn, Venus, Mercury, Uranus, Mars, Earth, Jupiter, and Pluto are *all* the planets in our solar system. If we include this proposition explicitly among the premises, then the argument becomes a deductive one. Because of this, perfect induction is not really an inductive argument at all. Consequently, we will have no more to say about it.

Statistical Syllogism

Some inductive inferences involve premises about an entire population and a conclusion about a sample from that population. For example:

> 75% of all smokers (in the population) have or will have lung cancer
> S is a (random) sample from the population
> Thus, 75% of the smokers in the sample have or will have lung cancer

Here we have what is called a statistical syllogism. It has the form:

> n% of all Fs in the population are G
> Sample S is a (random) sample from the population
> Thus, n% of sample S that are F will be G

A statistical syllogism resembles a syllogism in deductive logic. But because it contains a statistical generalization as the key premise, it is a statistical syllogism. A limiting case of a statistical syllogism is one in which the sample S has only one individual.

> 75% of all smokers (in the population) have lung cancer
> Sir Walter Raleigh is a smoker from the population
> Thus, Sir Walter Raleigh has (or will get) lung cancer

In this example, it must be remembered that it is only likely to the certain degree that the conclusion Walter has (or will get) lung cancer is true. According to most individuals, the probability that Sir Walter Raleigh has or will get lung cancer is .75. To be more cautious, such a inference would include a certain *margin of error*, namely, the probability that Sir Walter Raleigh has or will get lung cancer is .75 plus or minus a certain percent. This comes out even more clearly in our earlier example, since one would not expect that exactly 75% of the sample who smoked has or would get lung cancer; we would expect that it would be close to 75%, but not exactly. So, we would include a certain margin of error, say, plus or minus 3%. What the margin of error is depends upon several factors covered under the topic of statistical inference (see Chapter 14).[2]

Induction by Simple Enumeration

In induction by simple enumeration, there is an inference from a number of observed instances in a class--the sample--to a conclusion about the entire class--the population--where the sample is less in number than the population. For example, we might argue:

All observed samples of gold have a melting point of x degrees C.

Therefore, all gold has a melting point of x degrees C.

It is because such arguments refer to the number of cases in the sample, i.e., counting them, that they are called *enumerative*. Furthermore, all the information contained in the premises is positive (supporting) evidence.

It is characteristic of enumerative induction that it proceeds from knowledge about a particular number of individuals in the sample to a conclusion about all cases. The conclusion can take the form of a *universal generalization*: For example:

217 specimens of dinosaur remains of Tyrannosaurus Rex have had a bone structure of type S

Therefore, all specimens of dinosaur remains of Tyrannosaurus Rex have a bone structure of type S.

On the other hand, the conclusion might be a *statistical generalization*. Suppose, for example, that instead of observing 217 dinosaur specimens and noting that all had a particular kind of bone structure, one discovered that only 90% of them did and from this information, one concluded that 90% of all remains of T. Rex have bone structure S. Here we would have what is often called a statistical generalization.

n% of the examined Fs are G
Thus, n% of all Fs are G

The basic form of a statistical generalization is to proceed from premises mentioning characteristics of a sample to a conclusion which is a statistical generalization. Unfortunately, both the conclusion and the entire argument form are called statistical generalization. Ordinarily, however, this process-product ambiguity should cause no problems.

Induction by simple enumeration and statistical generalizations are inferential patterns proceeding from "some observed instances" to "all instances." The observed instances constitute a sample and the entire class the population. Whether induction by simple enumeration and statistical generalizations are good inductive inferences or not depends on the nature of the sample, in particular, whether it was representative or not. (The same applies to statistical syllogisms.)

There are various techniques of sampling which determine whether or not the samples are *representative* of the population. If one has a sample that is a biased indicator of the population, one's evidence will not be adequate as a basis for drawing an inference to the population. The most desirable technique (in general) is that of a random sample in which every individual or subgroup in the population has an equal chance of being selected. However, random sampling is difficult to obtain and one must sometimes resort to non-random sampling. (Random samples can occur with or without the sample drawn being replaced in the population.)

Secondly, the *size* of the sample is important. A sample size that is too small will not be an adequate basis upon which to draw an inference about the population. Ordinarily, it is desirable to have as large a sample as possible, but this may not be possible from a practical point of view. What should be done? That will depend upon the situation in question, but as a general rule of thumb and for statistical reasons, many individuals suggest a sample size of about 30 (or more) is advisable. This is largely for the purposes of statistical analysis and inference.

If we include these two conditions among our premises, then induction by simple enumeration would take the form:

All observed Fs in sample S are G
Sample S was a representative sample of size m
Thus, all Fs are G

A statistical generalization would have the form:

n% of all observed Fs in sample S are G
Sample S is a representative sample of size m
Thus, n% of all Fs are Gs

Ordinarily, one does not expect that exactly n% of all Fs will be G because there are various kinds of error that may occur (measurement error, sampling error, etc.). Hence, one will include an error term in one's conclusion—a term indicating the margin of error (±k):

n% of all observed Fs in sample S are G
Sample S is a representative sample of size m
Thus, n% ±k of all Fs are Gs

Statistical generalizations—inferences from a statistical sample to a statistical population—lie at the heart of statistical inferences. (Because statistical inferences are so important to so many of the sciences, we will devote a later chapter to it.)

A statistical inference has a complicated logical structure, which can be (roughly) summarized as follows:

n% of all observed Fs in sample S are G
S is a random sample of size m
there is a margin of error of ±k
Thus, n% ±k of all Fs in the population are Gs

This form is readily apparent when one infers (estimates) characteristics of the population from these same characteristics in the sample (statistical estimation—see Chapter 9). An even more complicated form of statistical generalization involves statistical hypothesis testing in which one must also add to the premises information about the statistical hypotheses being tested and the significance level of one's test. This form of statistical generalization will be discussed in Chapter 14.

Eliminative Induction (Mill's Methods of Experimental Inquiry)

The philosophers John Stuart Mill and Sir Francis Bacon advanced another type of inductive argument: eliminative induction. Unlike enumerative induction, eliminative induction proceeds by *eliminating* a hypothesis on the basis of the evidence contained in some sample (in this respect, it is similar to statistical hypothesis testing—see Chapter 14). When all hypotheses except one have been eliminated, the remaining one is the correct hypothesis. John Stuart Mill gave five different types of methods applicable in this type of reasoning: Agreement, Difference, Joint Method of Agreement and Difference, Residues and Concomitant Variation. (Sir Francis Bacon gave several

"Tables.") The structure common to these various forms is the following:

Either Hypothesis H_1 or H_2 or H_3 is correct
The available evidence indicates H_1 and H_2 are probably false

---

Thus, H_3 is correct

Here evidence consists of a number of examined instances, which possess or lack certain features. Unlike induction by simple enumeration, the premises can contain both positive and negative evidence. From this evidence one then infers that one or more of the hypotheses is probably false and are consequently eliminated. Chapter 10 discusses these and related issues.

Analogical Generalization

One form of enumerative induction is sufficiently important to warrant separate treatment: analogical generalization. For example, a scientist will often perform experiments on rats to determine the effect of a particular drug on humans. Unfortunately, the effect of this drug is often harmful to the rats. The scientist will often then argue by analogy that such a drug is also likely to be harmful to humans. The basis for this inference is the assumed similarity between rats and humans. On the basis of the alleged similarity, the scientist may feel justified in concluding what the likely effects of the drug on humans will be. Here the scientist is reasoning in the following way:

Rat #1 has properties P, Q and R
Rat #2 has properties P, Q and R

Rat #n has properties P, Q and R
Humans have properties P and Q
Therefore, humans probably will also have property R

Analogy, analogical generalizations, and analogical inferences will be the subject of Chapter 11.

Confirmation: The Hypothetico-Deductive Method

There is one other form of induction that has been widely discussed throughout the centuries—what we will call confirmation. In many ways this type of inductive inference is the most important one and the most controversial of all. It is essential to one interpretation of the hypothetico-deductive method, a topic we will spend considerable time discussing.

All versions of confirmation employ, in one way or another, the **hypothetico-deductive method** of testing scientific hypotheses. According to this method, scientists begin with a hypothesis (H) about the likely explanation for some phenomenon. From H one then deduces a statement about some observational fact (O), which is a logical consequence of H—a prediction made by H. H is evaluated according to whether O turns out to be correct or not. The hypothesis (H) in question can take on a great variety of forms containing theoretical concepts such as *gravity*, *germs*, *electrons*. These hypothetical entities are things that can not be directly seen (by the naked eye). Nevertheless, such things explain a great deal about the features of our world. Germs, for example, can explain certain illnesses, and genes explain hereditary traits. If science were to do without these hypothetical entities, it would be impoverished.

In Mill's interpretation of eliminative induction, the hypotheses ordinarily involve causes detectable by the senses. Because of this, these methods are thought to be inadequate to handle certain kinds of inductive inference in science, namely, those involving the confirmation of hypotheses. Hence the hypothetico-deductive method appears to be more powerful than Mill's methods of eliminative induction and more flexible. (Mill disagreed about this.) According to most scientists and philosophers of science, the hypothetico-deductive method represents an essential aspect of scientific method. We will distinguish several kinds of inferences belonging to this class.

Corroboration. The initial step in the hypothetico-deductive method has the following form: If H then O. The second step consists in the actual observation of the predicted fact (O) or the failure to observe it (~O). In the latter case, our hypothesis is refuted via deductive logic:

$$\begin{array}{l} H \supset O \\ \quad\sim O \\ \hline \sim H \end{array}$$

Some philosophers, notably Sir Karl Popper, stop here and argue that the hypothetico-deductive method needs only deductive logic since one is attempting to falsify a hypothesis. This is known as falsificationism.

Suppose, however, our predicted observation (O) is correct. In that case, we have

$$\begin{array}{l} H \supset O \\ \quad O \\ \hline ?? \end{array}$$

What should we conclude in this case? According to Popper, the only thing one can say is that the hypothesis H has been corroborated. What this means (and the only thing that it can mean, according to Popper) is that the hypothesis H has so far not been falsified or refuted. One cannot say that the hypothesis is true, or that it is likely to be true, or that it has some degree of positive evidential support. One can only say that H has withstood the attempt to falsify it. Hence, Popper's interpretation of the Hypothetico-deductive method is committed to corroboration but not confirmation.

A Logic of Confirmation. Other philosophers, notably Rudolph Carnap and Carl Hempel, disagree. They claim that, in addition to deductive logic, the hypothetico-deductive method needs an inductive logic—a logic of confirmation—precisely in those cases in which our hypothesis H has not been refuted or falsified. When a hypothesis (H) makes a prediction (O) and O turns out to be correct, this evidence should count favorably towards the truth of H. On Popper's account, however, one can say nothing very positive about the hypothesis. By contrast, the suggestion of Carnap and Hempel is that in these cases we can say something positive, namely, H is confirmed (to some degree). Thus, we would have:

$H \supset O$
O

H is confirmed (to some degree)

Instead of using the word *confirm*, other words are sometimes used: warranted, probable, likely, presumed to be correct, plausible, etc., where all of these words indicate that the hypothesis is not true but that it has some evidence in its favor, and the likelihood of H being true has been increased.

It is the connection between the observation O and the hypothesis H that leads these thinkers to speak of a logic of confirmation. The question here is: *how well- confirmed is the hypothesis on the basis of the evidence*? As we will see in Chapter 6, many logicians have claimed it is possible to construct a logic of confirmation in which one can assign numerical values to the hypothesis on the basis of the evidence. If this program were to succeed, we would be able to say, for example: The degree of confirmation of hypothesis H (on the basis of evidence E) = .9(*where numerical values can range form 0 to .99999*). Several philosophers of science have attempted to construct such a quantitative logic of confirmation.

Bayesian Inference. Finally, no cursory discussion of the hypothetico-deductive method would be complete without at least mentioning the fact that there is an influential school of thought that interprets the hypothetico-deductive method in a still different fashion (according to some Bayesians, the method is actually rejected). This Bayesian interpretation of scientific method has become increasingly influential in recent years and attempts to combine certain principles of probability theory with its interpretation of the hypothetico-deductive method. Because it relies upon certain principles of the probability calculus, and because it attempts to attach numerical values to the likelihood of the hypothesis, it has more in common with those thinkers advocating a logic of confirmation than with those advocating corroboration.

Briefly put, a Bayesian believes that one should consider not only the observational evidence (E) predicted by the hypothesis (H); one also must take into consideration how likely H and E were on their own (as it were) and compare this to how likely the evidence is on the basis of the hypothesis. Their formula for calculating the *posterior probability* of a hypothesis H—its probability after its observational test E—is related to the *prior probability* of H—the probability of the hypothesis before the observational test E—as well as the probability of the evidence given the hypothesis (p(E/H))and the probability of the evidence E (p(E)) alone.

> The probability of the hypothesis after a confirmed prediction from the hypothesis = the prior probability of the Hypothesis x the probability of the Evidence, given the Hypothesis, divided by the probability of the evidence E

or in symbols

$$P(H/E) = \frac{P(H) \times P(E/H)}{P(E)}$$

We will have a great deal more to say about this important school of thought in later chapters. The only point to note at present is that Bayesian inductive logic is an attempt to construct a logic of confirmation based upon probability theory and certain features of the hypothetico-deductive method.[3]

Inference to the Best Explanation (Abductive Inference)

As our last example of inductive inference, we will briefly mention a very controversial but widely discussed form of inference. According to some logicians and philosophers of science, there is a form of inference different from the hypothetico-deductive method and yet one which is widely used in science. This method is sometimes called abductive inference (by the 19th century philosopher Charles Peirce), whereas others call it inference to the best explanation. (Others have argued that it is just another version of the hypothetico-deductive method and represents nothing unique.)

According to the advocates of this method, an abductive inference is an inference to the plausibility of a hypothesis based upon certain kinds of evidence. For these individuals, this form of inference is different from an inductive inference which proceeds from evidence to the hypothesis, namely:

Evidence E

If Evidence E, then Hypothesis H
Thus, Hypothesis H

It is also different from the hypothetico-deductive method

If Hypothesis H then Evidence E
Evidence E
Thus, Hypothesis H

In an abductive inference, we have:

Some phenomenon E is observed
If Hypothesis H were true, E would be explained
Thus, Hypothesis H is likely

The key difference between the hypothetico-deductive method and abductive inference is which comes first—the facts to be explained or the hypothesis doing the explaining. In the hypothetico-deductive method, the hypothesis comes first and the observational evidence comes second; in an abductive inference, the observational evidence comes first and the hypothesis is second.

Abductive inference has also been called *inference to the best explanation*, for in an abductive inference we are inferring what the best (or most plausible) explanation is given certain puzzling facts to be explained. Suppose, for example, that John, someone we have known for several years, begins to act in a new and rather strange way: he begins taking classes in subjects he had never before shown any interest in and that a certain young woman—Jennifer—is a student in all of these classes; John takes a job as a cook in Jennifer's sorority; John spends hours on the telephone talking to Jennifer about academic subjects; etc. Given all these facts, what is the best (or most plausible) explanation? The best explanation seems to be that John is in love with Jennifer (or is romantically attracted towards her). Of course, there are other explanations of this behavior--it may be a coincidence, it may be they are just friends, etc.--but intuitively the most plausible explanation seems to be one in terms of romantic attachment: it explains the patterning of behavior better than competing explanations.

The inference here has the following form:

Some surprising phenomenon E (John's unexpected behavior) is observed

If Hypothesis H (John is attracted to Jennifer) were true, E would be explained
Thus, Hypothesis H is likely

Such patterns of reasoning are widely prevalent in science, e.g., a patient reports fever, aches, chills, etc., and from these facts we infer that the most likely explanation is that the patient has influenza.

Exercise 5.2

In the following passages, first identify the conclusion and then indicate what type of inductive inference is present:

1. When the elaborate methods of preparation of some of the plants used to break down the monotony of life are studied, it becomes quite evident that primitive man must have possessed something other than chance to reveal to him the properties of food and drug plants. He must have been a keen observer of accidents to discover fermentation, the effects and localization of alkaloids and toxic resins, and the arts of roasting and burning a product to gain from it the desired narcotization or pleasing aromas (coffee).

—Lewis Mumford, *Technics and Human Development*

2. If a man's destiny is caused by the star under which he is born, then all men born under that star should have the same fortune. But masters and slaves and kinds and beggars [whose fortunes differ greatly] are born under the same star at the same time. Thus astrology—which claims that a man's destiny is caused by the star under which he is born—is surely false.

—Pliny the Elder, *Natural History*

3. [The Assistant Secretary for Health of the Department of Health and Human Services] released the results of the National Household Survey on Drug Abuse, which examined a national sample of Americans over age 12, and the High School Senior Survey.

The surveys showed that one out of three Americans over age 12 has tried some kind of illicit substance.

4. Tar (extracted from cigarette smoke), when smeared on the skin of laboratory mice causes skin cancer. Therefore, cigarette smoking causes lung cancer in humans.

5. Medical researchers are telling the truth when they report on the basis of sustained and intensive experimentation that they can identify no therapeutic powers in Laetrile that can destroy cancer. Certain patients are telling the truth when they report having recovered from cancer after having taken Laetrile. How is this possible? There are many kinds of cancer which are reversible under certain conditions, and the human body possesses a system which, if properly mobilized, is capable of overcoming various forms of cancer. Laetrile is successful because it is a placebo. Some who use it think it is effective and this belief puts their curative forces to work.

6. The annual rainfall in Minneapolis has been over 15 inches every year for the past 30 years. Therefore, the annual rainfall next year will probably be over 15 inches.

7. Only about 3% of cigarette smokers actually die from lung cancer, so my smoking won't result in death from lung cancer.

8. Barbara's skin has a peculiar yellowish hue, and she complains of loss of energy. Barbara must have hepatitis.

9. A person can have all the energy in the world, and sleep very little--yet once in twenty-four hours he must sleep. The most elemental explanation for this inevitable rest is given by mothers to tired children: "You must sleep to restore energy." Doctors pronounce the same message for patients in slightly more sophisticated language. For as long as scientists have worried about the purposes of sleep, the same theory has recurred. Sleep must be restorative, and vast expenditures of mental and muscular energy finally, so to speak, poison us into slumber so that the body may rebuild its supplies. There is one flaw in this plausible idea that bothers everyone. Why does an Olympic runner sleep no more than a lazy sales clerk, and why doesn't he fall asleep after his four-mile sprint, instead of waiting for nightfall?

---Gay Luce and Julius Segal, *Sleep*

The Problem of Induction

Deductive inferences are truth preserving: the truth of the premise is transferred to the truth of the conclusion without exception. This is because, in a deductive inference, there is no information contained in the conclusion not contained in the premises; in fact, the conclusion is already implicitly contained in the premises.

In an inductive inference, however, this is not true; in an inductive inference, there is more content or information in the conclusion than there is in the premises. Inductive inferences are thus ampliative. Hence, the truth of the premises is not transferred to the truth of the conclusion, and it is possible for the premises to be true and for the conclusion to be false. In the argument

Planet #1 travels in an ellipse
Planet #2 travels in an ellipse
Planet #3 travels in an ellipse
Therefore, all planets travel in ellipses

it is possible for the premises to be true and for the conclusion to be false (the next examined planet may not travel in an ellipse)

Of course, our evidence of three planets would not be very good evidence for our conclusion; our conclusion would be more likely if likely if we had observed a large number of planets, say, several hundred. But even if we had, our conclusion about all planets might be false even though our premises are true.

Hume's Problem

The 18th century British philosopher, David Hume, asked a question about inductive inference no one before him had asked: How can one justify inductive inferences? Given that we have observed several instances of elliptical planets, how can one rationally justify the conclusion that all planets have elliptical orbits? Given that every day in the past, the sun has risen, how can one rationally justify the conclusion that the sun will rise tomorrow? If such an inference is an inductive one, then it is possible for the premises to be true and the conclusion false. So, how do we know it isn't false? Of course, we *believe* the sun will rise tomorrow, but is this belief rationally justified on the basis of the evidence, and if so how is it rationally justified? By what principles is such an inference warranted?

This question has become known as Hume's question (or Hume's problem) and it has never been adequately answered (at least to everyone's satisfaction). This is the classical problem of induction.

Solutions to Hume's Problem of Induction

Hume's argument has seemed pointless to most people: "of course," they say, "we are justified in our belief that bread will nourish us, that fire will burn, and that the sun will rise tomorrow." But as to why we are justified, most individuals have been silent. For the most part, scientists, too, have not found Hume's question sufficiently bothersome to answer; they too assume that induction is justified, but have not provided anything like an answer to it. It is been the philosophers who have worried most about Hume's problem, and who have devoted the most time towards answering Hume.

The Uniformity of Nature. The most likely candidate for an answer to Hume's problem is the following: if something has been true every time in the past, it will be true the next time because nature is uniform in its mode of operation. This is known as the *Principle of the Uniformity of Nature* and this principle has taken on a variety of forms: there are invariable regularities governing all natural events such that there are no exceptions to them (determinism); the same causes always produces the same effect; nature does not change how it operates, etc.

If nature were uniform in its operation, things would not change; hence, what is true of the past would be true of the future. In this case, we could know that the sun would rise tomorrow, because the sun has always risen in the past.

But how can one establish the truth of the Principle of the Uniformity of Nature? We could show it is true either deductively or inductively. To show it deductively would amount to showing that it was the conclusion of a valid deductive argument, that such a principle was (somehow) necessary. But this does not seem to be very likely.

Could we show it is true on inductive grounds? Our inductive evidence for the Principle of the Uniformity of Nature would be past evidence that it is true, i.e., evidence that in the past, nature has always been uniform in its operation. Because nature has been uniform in the past, we conclude that it will continue to be uniform in the future. But how could we know that in the future Nature would be uniform?

John Stuart Mill offered an inductive justification of the Principle of the Uniformity of Nature. According to him, there are levels of inductive generality. On the bottom level, we inductively confirm that, say, all lemons are yellow by reference to particular examples. On the next higher level, and on the basis of these lower-level inductive conclusions, we draw a higher-order inductive conclusion, about all citrus fruit. We then draw inductive generalizations about all fruit. Then from inductive conclusions about all fruit, all fish, all mammals, etc., we draw a higher-order inductive generalization about a certain class of natural objects. Hence, we inductively confirm a variety of scientific laws. We continue this inductive process until we reach the conclusion: nature is uniform. This conclusion has been established by a process of abstraction—of generalizing concerning the common properties of objects—and induction.

Most philosophers have argued that such a justification of the Principle of the Uniformity of Nature is, unfortunately, still circular, since the most one can conclude is that such inductive inferences have been warranted in the past; but there is no reason to believe that this will continue to operate in unexamined cases.

Suppose, instead, we were to argue: an inductive inference is not certain, of course, because then it would be a deductive inference, not an inductive one. But an inductive inference can be *probable:* we can say, for example that since Nature has been uniform in the past, it is *probable* that it will be uniform in the Future. But before we can examine this reply, we must be clear about the nature of *probability* (see Chapter 6), and what interpretation of that concept is being offered by the defender of the Uniformity of Nature. Are they saying, for example, that one feels *subjectively very certain* that Nature will be Uniform in the future? No doubt, Hume would agree to this interpretation. But he would then ask if such a feeling was rationally warranted, and if so, how? If the subjectivist interpretation of probability is not adequate, then what other interpretation can be offered?

For example, suppose in the past that one thing E has often occurred when another thing C occurred (this is sometimes called the relative frequency interpretation of probability: the probability of E given C is just the relative frequency with which E occurs given that C occurs). Then we might say: in the past, it was probable that E given that C. But what is our justification for believing that *tomorrow* it will be true that E will be probable given that C? If C occurs tomorrow, we are inclined to say: it is probable that E. But what is our justification? It can only be, Hume said, that we are assuming the Uniformity of Nature, that what is probable in the past will continue to be probable in the future. But once again this would arguing in a circle.

A Priori (Deductive) Justification. Some individuals have tried to give a deductive justification of induction. Some (Williams, 1947) have argued that, using certain principles of mathematical statistics, e.g., the Law of Large Numbers (see Chapter 6), the Central Limit Theorem (see Chapter 14), etc., one can show deductively that certain kinds of inductive inferences are warranted using the statistical syllogisms. The structure of the argument is the following:

1. Samples match their population (Bernoulli' Theorem)
2. This is a sample from a population.
3. x% of the sample Q is P.
4. Probably, x% of the population of Qs are P.

From these *direct inferences* from population to sample, one can make *inverse inferences* from these samples to the population.

Unfortunately, these arguments are fairly technical, and a discussion of them would take us too far afield. Most philosophers seem to be of the opinion that although such direct inferences are unobjectionable, the corresponding inverses inferences are problematic since they are based upon certain assumption about the sample (e.g., that it is random) and that one can not know if this is true. In short, these deductive justifications of induction have not been very well received. Hence we will not pursue them any further.

Deductive Justification: Postulates. Some individuals have tried to justify inductive inference in a different way—by the introduction of postulates. A postulate is a premise or assumption one makes in order to show that something else is true. In order to justify induction, therefore, some individuals have suggested that certain postulates must be accepted if induction is to be justified. These individuals differ as to their choice of postulates, but one example would be that properties come in clusters and are not randomly distributed, that every event has a cause, or that mere spatial location does not make any difference to the operation of a scientific law.

In suggesting the need for such postulates, individuals have been influenced by Kant (see Chapter 2), who suggested that certain principles are necessary in order for science to exist and that these principles were synthetic a priori principles, i.e., principles that were not true by virtue of the meaning of their terms, but still necessary. The defenders of the postulational approach generally do not favor Kant's interpretation of what a postulate is. But how one can justify such postulates of induction remains unclear.

Kant seems to have thought that one could justify a postulate in the following way:

1. We have a premise p (e.g., the existence of science).
2. But a necessary condition for p is q (e.g., the Uniformity of Nature).
3. Therefore, q

(This is sometimes called a transcendental argument.)

There are several problems with such an argument, including the fact that we seem to be assuming, in premise 1, certain aspects of science, e.g., that it can make sound inductive inferences about the future. But this is precisely what the debate is all about; hence, the argument seems to be begging the question.

Inductive Justification. Another group of individuals have looked to induction itself to justify induction. As we have seen, the attempt to justify an inductive inference inductively seems to fail, since such an argument appears to be circular in nature. Being aware of this, these individuals have focused their energies on showing that, although there is a kind of circularity here, it is not a vicious, formal circularity, but a benign kind of circularity.

For example, Max Black (1954/1970) has argued that one can give a "self-supporting" inductive argument. Inductive arguments often have the following form:

> Most instances of As examined in a wide variety of conditions, have been B.
> Therefore, probably the next A to be encountered will be B

This inductive inference is an instance of (or is governed by) the rule R:

> (R) To argue from "most instances of As examined in a wide variety of conditions have been B" to (probably) "the next A to be encountered will be B."

Now, what about the rule R itself? How successful has it been? We can give evidence for the inductive success of R by means of a *second-order inductive argument*:

> In most instances of the use of R in arguments with true premises examined in a wide variety of conditions, R has been successful.
> Hence (probably) in the next instance to be encountered of the use of R in an argument with a true premise, R will be successful.

This second-order arguments uses the rule R in showing that R will be successful, i.e., will generate a true conclusion, but it also supports the rule R. However, the argument is not circular in a formal sense, since R does not appear as one of the premises.

Several critics have reponded, however, that although there is no formal circularity, there is another kind of circularity, and hence, such an inductive justification of induction is objectionable. (Black has argued that these objections are wrong.)

Rejection of Induction. One solution to Hume's problem championed by Sir Karl Popper is to concede that there is no way of justifying inductive inference—either deductively or inductively—and hence to conclude that induction plays (and/or should play) no part in science at all. According to Popper's account of the nature of scientific method, the only type of logic needed in science is deductive logic. The aim of scientific method is to test a scientific hypothesis or theory

theory by means of deducing from it a predicted test consequence and then to determether the prediction is correct or not. Since the only mode of deductive inference permitted here is Modus Tollens, it must be concluded that the aim of science is to falsify a scientific hypothesis by showing the predicted consequence to be false. If the outcome is positive, Popper claims, we can not say the hypothesis is probably true, since this would involve an inductive inference, and Hume has shown that inductive inferences are unwarranted. All that we can say, according to Popper, is that so far the hypothesis has passed this test (or a certain number of tests). This is a statement about the theory's past track record, which Popper calls the *corroboration* of the theory. But we cannot draw any conclusions about the veracity of the hypothesis or about how it will fare in the future.

The Pragmatic Justification of Induction. Several individuals have argued that Hume's dilemma—either deductive justification or inductive justification—is a false dichotomy, and that there are other kinds of justification, namely, a pragmatic justification of induction. For example, Hans Reichenbach (1949) argued that one cannot give an a priori proof that inductive inferences are warranted, nor can one show this a posteriori. Inductive inferences are always fallible, and might turn out to be wrong. But what is the alternative? (Some people would argue that this was also Hume's solution.) One must act, e.g., either one eats the bread, which has nourished one in the past, or one doesn't' eat the bread. But which choice should one make? You should use the results of the past as a basis or guide for your action. It is true that one can never know if these past results are an adequate basis for one's future action. But in acting, you are assuming inductive inferences are reasonable, and that there are certain recurrent regularities in nature. Either it is or it is not. If it is not, if the past is irrelevant to one's future actions, then no plan will be successful, and no action will be successful; on the other hand, if it is, then an inductive inference will be successful (in which case ones action will be warranted). So, one can give a pragmatic justification of inductive inferences, namely, they will be successful if any thing will be successful.

Induction is a Pseudo-Problem. The final, main solution offered is to reject the entire problem as a pseudo-problem, the result of fuzzy thinking and conceptual confusion. The question is: Is it rational to act on the basis of past experience or not? Is it rational not to stick one's hand in the fire because in the past you have been burned? Of course it is, these individuals (e.g., Strawson, 1953) answer, and the reason is that to be rational is just to act in accordance with one's past experience. Suppose one has often been burnt when one put one's hand in the fire. If one is to act rationally, then one should refrain from doing this in the future. If one were to continue to thrust one's hand into the fire, we would judge the individual to be irrational or insane. Reasoning inductively is precisely what being rational means. Hence to ask if it is rational to engage in inductive reasoning is a silly question, since it is to ask if it is rational to be rational.

Although we do not have the space to devote to a full examination of these proposed solutions and their various objections, perhaps enough has been said to give the student a feel fro what the varous options are.

There is little consensus concerning which, if any, of these approaches represents an adequate solution to Hume's problem. Philosophers remain divided on the question; hence no definitive answer can be given. The most popular solutions to the problem of induction seem to be the pragmatic approach and the "dissolution" approach—at least to judge by current discussion.

The problem of induction continues to remain unanswered or unsolved; it continues to be "the scandal of philosophy" that no one has been able to answer Hume successfully. This is not an idle, abstract philosophical problem, for inductive inferences lie at the heart of the scientific method. Any inductive inference that proceeds from a premise about a certain number of observed instances to a conclusion about an unobserved instance—and virtually all inductive inferences have such a form—is confronted by Hume's challenge to provide a justification for making such an inference. Any scientific law worth its salt has been empirically tested and presumably supported by a finite number of (varied) instances. But how can one justify such an inference? How can one justify a conclusion that Newton's laws apply to every region of space and time when (at most) we have tested it only on the earth and for only three hundred years. How do we know that such laws are (approximately) true of the other side of the moon or the further regions of space? Our evidence appears meager in this regard, and yet scientists routinely accept the universality of these laws. Hence, it is incumbent on any rational thinker to provide some kind of account that justifies or warrants such an inference. Needless to say, it remains very unclear how one would do this in the light of Hume's challenge and yet it is something that is necessary to do.

Goodman's New Riddle of Induction

One widespread kind of inductive inference is from a statement mentioning evidence to a conclusion about the truth of a scientific law. Related to this are those arguments that proceed from a series of empirical tests to a conclusion about the probable truth of a scientific theory; e.g., given all the current evidence, it is very probable that Darwin's theory of evolution is correct. Hume's problem of induction raises question about how such inferences are justified.

Another kind of inductive inference concerns a prediction—predicting what properties the next instance of a certain class of objects will have, e. g., that the next piece of copper will conduct electricity, that the next instance of a heated gas will expand, etc.

What predictions we make depends upon how the properties of objects are described. This has been shown rather remarkably by Goodman (1983) in his *new riddle of induction*, or what some people call *the second problem of induction* (with Hume's problem being the first problem of induction).

Consider the generalization

(A) All emeralds are green.

We have a vast amount of evidence in favor of such a generalization, and for the prediction that the next emerald encountered will be green. But now consider the property *grue*, which is defined as: Something is grue at time t (e.g., the year 3000) $=_{df}$ (1) x is green prior to t and t is 3000, or (2) x is blue at t and t is after 3000. Now consider the generalization

(B) all emeralds are grue.

Generalization B is supported by all the accumulated evidence we have obtained in support of A, since it is not yet 3000. But what happens when one makes a prediction about the color of the next emerald examined after 3000? B leads to the prediction that, since it will be grue, it will be blue. A leads to the prediction that, since it is green, it will be green . *We have two contrary predictions supported by the very same body of evidence.* But to suppose that emeralds will change their color from green to blue at the year 3000 seems wrongheaded. "Why should time make any difference to the color of an object?", they are likely to ask. What has been green in the past will continue to be green in the future—an inductive inference. But why is that inductive inference justifiable and not the inductive inference that what has been grue in the past will continue to be grue in the future, with the corresponding claim about the color changing? Remember: all the past inductive evidence favoring A also favors B. But if so, then why is one more reasonable than the other? As Goodman puts it, why is one predicted projectible but the other one is not?

Goodman's own answer to this question is that we project those predicates that are *entrenched* in our scientific practice. Green is much better entrenched than grue because it is based upon a greater number of past successful projections whereas grue has not been. Other philosophers of science reject Goodman's own solution and have proposed a variety of alternatives—now numbering about 20—including simplicity, no temporal reference, natural kinds, etc.

Although the problem of the new riddle of induction may appear to be a purely philosophical problem with little relevance to actual science, this is not altogether correct. As we have seen, scientists make inductive inferences all the time and Goodman's problem is nother example of that. There are several examples of this that are relevant (1) given a certain amount of past evidence e, one can predict the next emerald after t will be green (H_1) or grue (H_2). Either one is equally well supported by the evidence e. So, which prediction is to be preferred on rational grounds? (2) e supports the generalization (law) that all emeralds are green to the same degree that it supports the generalization (law) that all emeralds are grue. Which law statement should we accept? (3) In Chapter 9, we discussed the problem of curve-fitting: any set of data can be fitted by an indefinite number of curves (mathematical equations). Suppose three different curves can be fitted to the same set of data points. Which curve should be chosen as a basis of description of the three data point? Which should be chosen as a basis of extrapolation beyond the data points and of interpolation between data points? Ordinarily our preference is for the straight line. But how do we justify such a choice, since the data underdetermine it? Clearly, this question is just another version of the grue-green problem. (4) In Chapter 8, we discussed the nature of qualitative classification: a set of data needs to be described, classified, and grouped according to certain categories, concepts and predicates. But the grue-green problem shows there are various ways of qualitatively classifying and describing the same data. There is not single way of doing this. Are all of these ways equally good and justifiable, or are certain descriptive predicates better than others? If so, what make some label better than another one?

These dilemmas are not, therefore, just philosophical worries; they are also worries for the scientist, who must make a decision about them. But in making these decisions, the perennial question is: which decision is preferable on rational grounds?, and an answer to this question would seem to be an answer to Goodman's problem.

For Further Reading

Most standard logic textbooks (e.g., Copi & Cohen, 1986) have a section devoted to inductive logic (although these tend to be sketchy). Several textbooks are exclusively devoted to inductive logic. See, for example: Baird (1996), Coyne (1984), Gustafson (1994), Skyrms (1986). Skyrms is widely used.

Most of the older textbooks on logic have discussion of *eduction, intuitive induction*, and *perfect induction*. See Johnson (1924/1964), who introduced the concept of eduction, Joseph (1916) who coined the term intuitive induction.

For mathematical induction, consult a modern textbook on algebra (e.g., Lial & Miller, 1973, pp. 345-50). Then read a philosophical discussion such as Cohen & Nagel (1934, pp. 147-8) followed by a more advanced mathematical discussion, e.g., Pickert & Goerke (1986). Poincare (1905/1952) takes a different view.

Later chapters of this book discuss statistical generalization, analogical generalization, eliminative induction, the hypothetical-deductive method, and Bayesian inference. On Popper's theory of corroboration see his (1935/1965, 1965). Carnap's 'logic of confirmation' is contained in his (1962), which is difficult reading for the beginner. For Hempel's view see his (1966, ch 2-4). For Bayesian inference, see Howson & Urbach (1993). On abductive inference see Peirce (1957) and for inference to the best explanation see Harman (1965), who invented the term, and Lipton (1991).

A widely discussed criticism of the hypothetico-deductive method is Glymour (1980), which is technical in places.

For Hume's statement of the problem of induction, see his *Enquiry* (sec. 4) (Hume, 1748/191902, which can also be found in Luchenbach (1972). Bertrand Russell's statement of the problem (Russell, 1912) is widely accepted as brilliant and witty; it can be found in Foster & Martin (1966). John Stuart Mill's defends the Principle of the Uniformity of Nature in his Logic (Mill, 1872), which is reprinted in Foster & Martin (1966).

The most plausible defense of a deductive justification of induction is by Williams (1947)—fairly technical—and the postulational approach is by Russell (1948)—which is accessible to the beginner. Perhaps the best representative of the inductive justification is Black (1954/1970) .Popper's rejection of induction is contained in his 1935/1965, ch. 11). Reichenbach's pragmatic justification of induction is contained in his 1949. These accounts are somewhat technical but pretty accessible. Strawson argument that the problem of induction is a pseudo-problem is quite accessible to the beginning student. Selections from all of these thinkers can be found in Foster & Martin (1966). For discussion of these various approaches to the problem of induction, I would recommend Hung (1997), followed by Salmon (1992) or Skyrms (1986). In recent years, interest in the problem of induction seems to have inexplicably waned.

Notes

[1] There has been no discussion of the question of whether inductive arguments contain a notion analogous to *soundness*. The difficulty is the following: an inductive argument might be said to be *inductively sound* if and only if (1) it is inductively strong and (2) its premises are true. But since inductive strength varies in degrees of inductive strength, so would inductive soundness. Secondly, to determine whether the premises of an inductive argument are true or not would seem to involve inductive logic itself and this appears to involve an infinite regress.

[2] The key factors here would be the *size* of the sample, whether it was a *random* sample or not, whether the sampling distribution was *normal* or not, etc.

[3] It should be pointed out that there is considerable controversy between the defenders of the hypothetico-deductive method and the defenders of a Bayesian approach. Although several philosophers see them as incompatible, this seems to be an unduly restrictive view of the matter.

6

Probability

Introduction

Throughout the previous discussion of scientific inference, we have frequently made use of the term *probability*. We said, for example, that an inductive inference is one in which the premises make the conclusion *likely* (to a certain degree), that observational facts render a hypothesis *probable* or *confirmed* (to a certain degree), that one hypothesis is more *probable* than another hypothesis, and so forth. This cluster of words--probable, likely, confirmed, corroborated--all seem ultimately to involve the concept of *probability*. The concept of probability is thus central to any understanding of the logic of scientific method. In this chapter, we will discuss the nature of probability under two headings: first, the nature of the probability calculus--the set of rules governing probabilistic conclusions; and secondly, the interpretations of 'probability,' together with the question of how one determines the initial probability.

Several preliminary matters must first be discussed. First, 'probability' is a term applied to several things, to a hypothesis (theory), to a belief, to a proposition, to the existence of kinds of things, to an event (or series of events), to something causing something else, etc. We say, for example, that the probability of drawing an ace of spades from a deck of cards is 1/52 (an event or occurrence), the probability of the proposition 'that it will rain tomorrow' is .5, the probability that there are black holes in the universe is very small, the probability of a particular hypothesis or theory being true is not very likely. If we rather crudely divide the domain of discourse into two categories, linguistic-type entities (propositions, beliefs, theories) and things in the natural world (events, processes, things), then probability statements can be said to be applied to both of these sub-domains--propositions versus things. As we will see, the two dominant theories of probability focus on either one of these two cases: the relative frequency theory taking events in the natural world as primary, whereas the logical relation view considers propositions and relations between propositions to be primary. For the sake of convenience, however, we will usually speak of the probability of a proposition.

Secondly, some notation must be introduced. It is now common to represent the probability of a proposition, say, A, in the following way: **P(A)** (to be read: the probability of A). Similarly, **P(A or B)** is the probability of either A or B, **P(A and B)** is the probability of both A and B, and so forth. Since most individuals believe one can attach numerical values to probability statements, the statement 'P(A) = r' (where r ranges from 0 to 1—see below) can be read as 'the probability of A is r.'

Suppose we wish to symbolize the probability of something A on the assumption that something else B. We may wish to ask: if B is true, what is the likelihood that A is true? For example, given that it snowed last night, what is the probability that the streets are now wet?; or supposing that a patient has been given a particular drug, what is the likelihood that she will recover? These statements are represented as: **P(A/B)**: the probability of A *given* B or the probability of A *if* B is true (this is also called **conditional probability** of A).

The Probability Calculus

The probability calculus is a set of rules that are analogous to the rules of inference in logic. In logic, we want to know what conclusion follows from a set of premises. Likewise, in probability theory, suppose we are given a statement concerning the probability of something. Then the rules of the probability calculus can be invoked to decide what further probability statements follow from these premises. These rules of inference are deductive rules of inference, which have been cast into mathematical form.

General Rules[1]

First, the probability of a statement can take on values ranging from 0 to 1, but nothing can have a probability less than 0 or more than 1:

Rule 1. $0 \leq P(A) \leq 1$

A probability of 1 is, by definition, the greatest degree of probability anything can have, amounting to absolute certainty, and a probability of 0 is the least probability anything can have (amounting to absolute uncertainty). It is, therefore, meaningless to suggest that the probability of a proposition could be less than 0 or more than 1

Rule 2. If a proposition A is a tautology, then P(A) = 1.

This is because a tautology is absolutely certain (incapable of being false); hence, its probability must be 1. It follows, in particular, that a tautology such as 'A or not-A' must have a probability of 1:

P(A v ~A) = 1

For example, what is the probability of either drawing an ace of spades or not drawing an ace of spades? The answer must be: absolute certainty, i.e., 1.0.

From the above we can also prove (by virtue of the special disjunction rule--see below) that

P(A) + P(~A) = 1

It follows that P(two on die) + P (not-two on die) = 1, which in turn means P(two) + P(one, three, four, five, or six) = 1. This in turn is equivalent to: P(two) + P(one) + P(three) + P(four) + P(five) + P(six) = 1. From this, it also follows that:

Rule 3. (Negation): P(~A) = 1 - P(A)

For example, the probability of drawing a non-ace from a deck of cards is equivalent to 1 minus the probability of drawing an ace. Since the probability of drawing an ace is 4/52 (4 aces divided by the total number of cards in a deck), the probability of drawing something other than an ace is: 1 - 4/52 = 48/52.

If the probability of a tautology is 1, then the probability of a contradiction must be 0:

Rule 4. If a statement A is a contradiction, then P(A) = 0

In particular,

P(A & ~A) = 0

The probability of drawing an ace and a non-ace on a single draw, for example, must be 0 since it is impossible.

The next rule is self- evident:

Rule 5. If two statements A and B are logically equivalent, then P(A) = P(B)

Disjunction Rules

Though the above rules are self-evident, the next rules may not appear to be quite so obvious. The first two rules address the question of *adding* the probabilities of two propositions, whereas the next set involves rules for *multiplying* the probabilities of two propositions.

Consider first the following situation. The probability of drawing an ace is 4/52, which is the same probability as drawing a king. Now suppose someone draws a card. One can ask not only, What is the probability that it was an ace?, but one can also ask, What is the probability that it was *either* an ace or a king? Clearly the answer will involve *adding* the two probabilities together: P(Ace or King) = P(Ace) + P(King), which equals 4/52 + 4/52 = 8/52. This is possible, however, only on the assumption that the two events described in the propositions [P(A), P(B)] are mutually exclusive.

Two events are ***mutually exclusive*** *$=_{df}$ both events cannot occur, i.e., if one event occurs, then the other event cannot occur.*

This gives us:

Rule 6. (Special Disjunction Rule): If A and B are mutually exclusive, then P(A or B) = P(A) + P(B)

For example, what is the probability of drawing either a spade or a heart? Since drawing a spade and drawing a heart are mutually exclusive, Rule 6 can be applied. The probability of drawing a spade is 13/52 (i.e., 1/4) and the probability of drawing a heart is 13/52. The probability of drawing either one or the other is: 13/52 + 13/52 = 26/52 = 1/2.

What is the probability of getting *at least* one head on two tosses of a coin? Are these events mutually exclusive? No. Hence, Rule 6 does not apply. In fact, if we were to use it, we would get a value of 1.0 (½

+ ½), which is clearly wrong. However, we can use Rule 3 to obtain an answer. Getting at least one Heads in two tosses of a coin is equivalent to either one Head on the first toss or one Head on the second toss or Heads on both tosses. The various possibilities are:

1st Toss	2nd Toss
H	H
H	T
T	H
T	T

This in turn is equivalent to not getting two tails (the last case). Hence, by Rule 3, this is P ~(T & T). By Rule 8, P(T & T) = P (½ x ½) = ¼. So, P ~(T & T) = 1 - P(T & T) = 1 - ¼ = ¾.

Suppose someone asks the question, What is the probability of drawing an ace or a spade? Here we obviously do not have mutually exclusive events, since the ace drawn could also be the spade drawn. Since these two events are not mutually exclusive, Rule 6 cannot be applied. However, a more general rule can be applied:

Rule 7. (General Disjunction Rule): P(A or B) = P(A)+ P(B) - P(A & B)

The difference between the two forms of the disjunctive rule concerns the last term, which is being subtracted from the sum of the first two probabilities. In this example, P(A & B), the probability of something being both an ace and a spade (i.e., the ace of spades) is 1/52 and this element in common must be subtracted. Since the probability of an ace = 4/52 and the probability of a spade is 13/52,

P(Ace or Spade) = P(Ace) + P(Spade) - P(Ace & Spade)
= 4/52 + 13/52 - 1/52 = 17/52 - 1/52 = 16/52 = .3

Of course, if two statements have no elements in common and hence are mutually exclusive, then the general disjunction rule reduces to the special form. For example, what is the probability of drawing an ace or a king? Using the general form we have:

P(Ace or King) = P(Ace) + P(King) - P(Ace & King)

Since nothing can be both an ace and a king, the last component is equivalent to 0.

Exercise 6.1

1. Suppose you have an ordinary deck of cards with 52 members. Each card has a probability of 1/52 of being drawn. What is the probability of drawing:

a. The ace of spades
b. The ace of spades or the queen of hearts
c. An ace
d. A face card (king, queen or jack)
e. A card that is not a face card
f. An ace or a spade
g. A queen or a heart
h. A queen or a non-spade

2. What is the probability of drawing a seven or an eight in a single draw from a standard deck? What is the probability of drawing a seven or a spade? What is the probability of drawing a seven or a black card?

3. The probability of getting a cold is .2; the probability of getting the flu is .15. If these events are mutually exclusive, what is the probability of getting one or the other? Suppose they are not mutually exclusive and the probability of a cold and the flu is .1?

4. The chance of the stock market going up is .58; the chance of it going down is .41. What is the chance of it going either up or down?

5. Forty percent of all citizens in the town are college graduates. Twenty percent are mechanics. The probability that a citizen is both a college graduate and a mechanic is .05. What is the probability that any citizen encountered is either a college graduate or a mechanic?

6. Fill in the blanks.

P(A) + P(~A) = ____
P(A v B) = P(___) + P(___)
P(~A) = ____

7. What is the probability of getting at least one tail in three tosses of a fair coin?

8. What is the probability of getting at most three heads in four tosses of a coin?

Conjunction Rules and Conditional Probability

Before discussing Rules 8 and 9, we must first introduce the notion of two events being independent of each other.

Two events are ***independent*** *of each other* $=_{df}$ *the occurrence of one of them does not influence the probability of the other event occurring.*

If we toss two coins, for example, what face turns up on the one coin does not have any influence on what face turns up on the other coin; these events are thus independent. Likewise, if one draws a card from a standard deck, replaces it, shuffles the deck, and then draws another card, two draws are independent because the first draw had no influence on the second draw. But suppose we had not thoroughly shuffled the cards; then the first event might have had some influence on the second. Likewise, the event of drawing an ace (and not replacing it) is not independent of the second draw of a suit, because the ace had to be of some suit, say, spades, and, without replacement, one would have one less spade to draw, thus decreasing the probability from 12/52 to 11/52.

Rule 8. (Special Conjunctive Rule): if A and B are independent of each other, then

$$\mathbf{P(A \& B) = P(A) \times P(B)}$$

For example, what is the probability of getting a head on each of two tosses of a coin? The probability of a heads (H) on the first toss is .5 and the probability of a heads (H) on the second toss is .5. Assuming the two tosses are independent, the probability of (Heads and Heads) is .5 x .5 = .25.

The *conditional probability* of a proposition B, given A (we will recall) is

$$P(B/A) = \frac{P(A\&B)}{P(A)} \text{ if } P(A) \neq 0$$

According to this rule, the conditional probability that event B occurs, given that event A occurs, is equal to the joint probability of events A and B divided by the probability of event A.

For example, what is the probability of a die coming up 2 *given* that the die shows an even number? Since an even number can occur if the die shows 2, 4, or 6, the question is: what is the probability that the die is 2 given that it was 2, 4, or 6:

$$(2/\text{even}) = \frac{P(2\&\text{even})}{P(\text{even})} = \frac{P(2)}{P(\text{even})} = \frac{P(2)}{P(2,4,6)} = \frac{2}{2,4,6} = \frac{1}{3}$$

Suppose we ask, What is the probability that a die is even, given that it was either 1 or 2 or 3 or 4 or 6? Obviously there are only three ways a die can be even: if it is a 2, 4 or 6, and P (even & 1 or 2 or 3 or 4 or 6) = P (2 or 4 or 6 and 1 or 2 or 3 or 4 or 6) = P (2 or 4 or 6). Thus

$$P(\text{even}/1 \text{ or } 2 \text{ or } 3 \text{ or } 4 \text{ or } 6) = \frac{P(\text{even} \,\&\, 1 \text{ or } 2 \text{ or } 3 \text{ or } 4 \text{ or } 6)}{P(1 \text{ or } 2 \text{ or } 3 \text{ or } 4 \text{ or } 6)} = \frac{2,4,6}{1,2,3,4,6} = \frac{3}{5}$$

Another example: suppose the probability of dying next year is .3, which means (roughly put) that the relative frequency of dying next year will be 3 out of 10. We can expect that the probability of dying given that one smokes would be considerably higher than .3, since the fact that one smokes increases the likelihood of dying. This is represented in the following **contingency table**.

	Smoke	Not-smoke	Total
Not-Die	15	55	70
Die	20	10	30
Total	35	65	100

The probability of dying is .3 (30/100), the probability of smoking is .35 (35/100), the probability of dying and smoking is .2 (20/100), and so forth. What is the probability of dying, given that one smokes? One can calculate this in two ways. First, by the formula P(B/A) = P(A & B)/P(A) we have

$$P(\text{Dying}/\text{Smoking}) = \frac{P(\text{Dying} \,\&\, \text{Smoke})}{P(\text{Smoking})} = \frac{.2}{.35} = .57$$

Looking at the table, one can also calculate P(Dying/Smoking) in the following way. The probability of smoking is .35. How does this make a difference to the frequency of dying? If one smokes, then only the first column is relevant. What is the frequency of dying in this column? It is the ratio: 20/35: the probability of dying given that one smokes.

Using this notion of conditional probability, we can calculate the probability of events which are *not independent*, events in which the probability of the one has some influence on the probability of the other. Clearly, this is the case in the above example of smoking, where smoking and dying are related (i.e., not independent). In such cases we use the following rule:

Rule 9. (General Conjunctive Rule):
P(A & B) = P(A) x P(B/A)

Exercise 6.2

1. What is the probability that, given a die is even, the face is a 2? What is the probability that it is even given that it is either 1 or 2 or 3 or 4 or 6?

2. What is the probability of an ace on three draws from a deck of cards with replacement? Without replacement (and assuming one doesn't actually draw an ace on any trial)? Without replacement (and assuming one draws an ace on each trial)?

3. What is the probability that two coins tossed will both land 'heads'?

4. Express the following in symbolic notation?

a. The probability of C
b. The probability of C given A
c. The probability of C and B
d. The probability of A and B given C
e. The probability of C given A and B

5. Suppose there are just three elements in the world (A, B, C). Fill in the following blanks

a. P(A & ___) = P(___)P(A/C)
b. P(C & B) = P(C)P(__/__)
c P(A/C) = P(___ & A)

P(___)

6. Suppose we divide scientists into natural scientists and social scientists and suppose we divide their incomes into 'wealthy' and 'poor.' Suppose also we find the following:

	Nat. Science	Soc. Science	Total
Wealthy	250	200	450
Poor	150	400	550
Total	400	600	1000

Calculate the following probabilities.
a. Of being a social scientist
b. Of being a natural scientist
c. Of being wealthy
d. Of being poor
e. Of being wealthy given that one is a social scientist
f. Of being poor given that one is a natural scientist
g. Of being wealthy given that one is a natural scientist

h. Of being poor given that one is a social scientist
Are these two classes (income, scientific status) independent or dependent? Why? Calculate the probability of being wealthy and a natural scientist? Poor and a social scientist?

Bayes' Theorem

We have already introduced the concept of *conditional probability*, the probability of a proposition given another proposition. In the context of science, conditional probability obviously applies to a hypothesis (H) and the evidence (E) supporting it, so that the probability of a hypothesis given some evidence is a certain amount: P(H/E) = r. Here, we are obviously concerned with the question, given the evidence, how likely is the hypothesis? This can sometimes be determined by what is called *inverse probability*, by inferring P(H/E) from the probability of E given H [P(E/H)] (called the *likelihood*). This can be done by means of *Bayes' Theorem*. There are several different forms of Bayes' Theorem.

The simplest version of Bayes' Theorem is the following:

$$\mathbf{P(H/E) = \frac{P(H) \times P(E/H)}{P(E)}}$$

where **P(H/E)** is the **posterior probability** of H given the evidence E, **P(H)** is the **prior probability** of the hypothesis (i.e., its probability prior to the evidence), **P(E/H)** is the **likelihood** of the evidence given the hypothesis, and **P(E)** is the probability of the evidence independent of the truth of the hypothesis.

Suppose, for example, one is interested in knowing the probability of a particular hypothesis (H) that drunk driving is responsible for traffic accidents (E). Suppose we know that 2% of the general population are drunk drivers so that P(H), the probability of a drunk driver, is .02. Suppose the probability of an accident is 15%, so that P(E) = .15, and that 25% of the drunk drivers have accidents, i.e., P(E/H) = .25. What is the probability of our hypothesis *given* our evidence about accidents, i.e., how does the additional knowledge contained in the evidence E affect the probability of our hypothesis H? By Bayes' Theorem:

$$P(H/E) = \frac{P(H)xP(E/H)}{P(E)} = \frac{.02x.25}{.15} = .033$$

What is the effect of the evidence E? It has raised the probability of H from .02 (its prior probability) to .033 (its posterior probability), thus increasing the chances that it is true. Suppose, however, P(E/H) was only .15. Then

$$P(H/E) = \frac{P(H)xP(E/H)}{P(E)} = \frac{.02x.15}{.15} = .02$$

Since the posterior probability equals the prior probability, the evidence E made no difference to the likelihood of H. This is because P(E/H) = P(E), i.e., the presence of H makes no difference to the probability of the evidence.

Exercise 6.3

1. If 1% of the population has respiratory problems, and 40% of coal miners do, and .5% of the population works in coal mines, using Bayes' Theorem, what is the probability of being a coal miner given that you have respiratory disease?

2. Suppose 1% of the general population develops lung cancer, 2% of smokers develop lung cancer, and 40% of all people smoke. Given that a person has lung cancer, what is his chance of being a smoker?

3. Suppose 90% of skin cancer patients work in the sun. What is the probability of developing skin cancer if you work in the sun (supposing 40% of the population works in the sun and only 1% of the whole population develops skin cancer)?

Other, More Complex Forms of Bayes' Theorem

Bayes' Theorem can be used in a variety of contexts but clearly it is of special value when it comes to assessing the epistemic status of hypotheses, often causal hypotheses. Ordinarily, a scientist will have a hypothesis H about the explanation of some phenomenon (E), the latter constituting evidence for the former. For example, a physician encounters a patient who has a particular set of symptoms, e.g., inability to urinate but extreme pressure on the bladder. Call this evidence E. The physician may have a hypothesis H for such things, say, prostatitis, (a swelling of the prostate gland), which constricts the urethra. Ordinarily, H has an antecedent plausibility--its *prior probability* (P(H)). How this prior probability is to be determined is a complex question but, for the Bayesian, it is determined by the individual's personal probability—her own belief about how likely H is—see below). (Often, scientific background knowledge will determine the prior probability, for example, how frequently in the past such a hypothesis has been found to be related to urinary problems, how likely such a connection is to be expected on theoretical grounds, what is the degree of belief in H by the medical community?, etc.)

The probability of the evidence given the hypothesis (P[E/H]) is the likelihood that the evidence E would occur if H were true. Sometimes H *entails* E and sometimes H makes E *probable*. When H entails E, we have a universal deterministic hypothesis, and P(E/H) will be equal to 1.0. For example, suppose there was a universal generalization that tuberculosis is caused by a tubercle bacillus organism. Then the likelihood of tuberculosis (E) given the presence of this organism (H) is 1.0. In other cases, P(E/H) may be less than 1.0, namely, where H makes E likely but does not entail it, This will happen, for example, where the causal hypothesis is a statistical regularity. Suppose the relative frequency of lung cancer in smokers is .9. Then the probability of an individual (S) getting lung cancer (E) is (roughly) .9 since 90% of smokers get lung cancer. This can be represented as a statistical syllogism:

Probability (lung cancer/smoker) = .9
S is a smoker
========================= [makes likely to degree .9]
S will have cancer

Here the conclusion (E) is rendered likely by the premises (H) to degree .9. H doesn't entail E, but makes it very likely.

Thirdly, there is the concept of the *probability* of the evidence: P(E)--how probable the evidence would be independent of the truth of the hypothesis (its degree of expectation). Sometimes we know P(E) because of our background knowledge concerning the frequency of this phenomenon; at other times, its expected value can be determined on theoretical grounds. Although we may not know P(E) directly, we can sometimes calculate indirectly. It can be shown that:

$$P(E) = [P(H) \times P(E/H)] + [P(\sim H) \times P(E/\sim H)]$$

Hence, if we know the prior probabilities of the two incompatible hypotheses and their respective likelihoods, we can determine P(E). This leads to the *second form* of Bayes' Theorem:

$$P(H/E) = \frac{P(H) \times P(E/H)}{[P(H) \times P(E/H)] + [P(\sim H) \times P(E/\sim H)]}$$

Example: a physician is concerned to evaluate the hypothesis (H) that a patient of hers has eaten a food--strawberries--to which he is extremely allergic. Suppose the prior probability of H is .2 (since the patient knows of his allergy and systematically tries not to eat his favorite fruit). Suppose, furthermore, if her patient eats strawberries, the probability of a red rash (P(E/H) is .7) and if he doesn't eat strawberries, the probability of a red rash P(E/~H) = .1. Upon seeing her patient, the physician notes a profuse red rash. What is the probability that he ate strawberries? Since, p(H) = .2, p (~ H) = .8

$$P(H/E) = \frac{P(H) \times P(E/H)}{[P(H) \times P(E/H)] + [P(\sim H) \times P(E/\sim H)]}$$

$$\frac{.2 \times .7}{[.2 \times .7] + [.8 \times .1]} = \frac{.14}{.14 + .08} = .64$$

P(H/E), the *posterior probability* of the hypothesis given the evidence, is the probability of the hypothesis after taking into account the evidence. If the posterior probability of H is higher than its prior probability (P(E/H) > P(H)), the hypothesis is confirmed; if it is lower (P(E/H) < P(H)), the hypothesis is disconfirmed. This is to be expected since the function of an empirical test (new evidence) is to subject the hypothesis to additional evidence; if the evidence supports the hypothesis, its probability should be increased, whereas if the evidence does not support the hypothesis, its probability should be lowered.

We can generalize the second form of Bayes' Theorem to consider not just two theories (H vs. ~H) but several.[2] Suppose, then, we have several hypotheses: H_1, H_2, . . .,H_n such that these hypotheses are mutually exclusive and exhaustive. Then

$$P(Hi/E) = \frac{P(Hi) \times P(E/Hi)}{\sum P(Hi) \times P(E/Hi)}$$

where Σ stands for the summation of all the hypotheses, and i stands for any arbitrarily selected hypothesis.

Suppose, for example, a scientist is considering three hypotheses about the origin of the universe: H_1 (the steady state theory), H_2 (the original big bang theory), and H_3 (the inflationary version of the big bang theory). Suppose he is considering the evidence E, which is the existence of background radiation in the universe. Suppose furthermore:

$P(H_1) = .5 \qquad P(E/H_1) = .1$
$P(H_2) = .3 \qquad P(E/H_2) = .4$
$P(H_3) = .2 \qquad P(E/H_3) = .4$

Find $P(H_1/E)$:

$$P(H_1/E) = \frac{P(H_1) \times P(E/H_1)}{[P(H_1) \times P(E/H_1)] + [P(H_2) \times P(E/H_2)] + [P(H_3) \times P(E/H_3)}$$

$$\frac{.5 \times .1}{[.5 \times .1] + [.3 \times .4] + [.2 \times .4]} = \frac{.05}{.05 + .12 + .08} = .2$$

Since $.2 < .5$, the new evidence disconfirms H_1.

Exercise 6.4

1. Suppose we observe 4 tosses in a row of a coin and the coin turns up heads each time. Is the coin fair (H) or not fair(~H)? If the coin is fair, the probability of 4 heads in a row is .0625, whereas if it is not fair but weighted so as to show only heads, the probability is 1.0. Suppose the prior probability of its being fair is only .2 (because the coin is owned by one of your slick gambling friends). What is the probability that it is fair, given the evidence of 4 tosses in a row?

2. Darwin's hypothesis of evolutionary descent by natural selection is judged by the scientific community to have a prior probability of .7. Let the relevant evidence be the fossil record as contained in geological strata and suppose the likelihood of this fossil record on the assumption of Darwin's hypothesis is .6; on the hypothesis of creationism, the likelihood of the evidence .1. Assume the two hypotheses are mutually exclusive and jointly exhaustive. How is the evaluation of Darwin's hypothesis affected by the fossil evidence?

3. Now assume we have three evolutionary hypotheses: H_1: Darwinism, H_2: creationism: H_3: Lamarckianism. Assume the prior probability of H_1 is the same as in problem #2, the prior probability of H_2 = .1 and the prior probability of H_3 = .2. The likelihood of the fossil record assuming Lamarckianism is the same as that of Darwinism, which is .6, whereas the likelihood of Creationism is .2. Calculate the posterior probability of Darwinism. How do you explain the difference between the answers to problems #2 and #3?

Summary of the Rules of the Probability Calculus

Rule 1. $0 \leq P(A) \leq 1$
Rule 2. If a proposition A is a tautology, then $P(A) = 1$.
$P(A \vee \sim A) = 1$
$P(A) + P(\sim A) = 1$
Rule 3 (Negation): $P(\sim A) = 1 - P(A)$
Rule 4. If a statement A is a contradiction, then $P(A) = 0$
$P(A \,\&\, \sim A) = 0$
Rule 5. If two statements A and B are logically equivalent, then $P(A) = P(B)$
Rule 6 (Special Disjunction Rule): If A and B are mutually exclusive, then $P(A \text{ or } B) = P(A) + P(B)$
Rule 7 (General Disjunction Rule): $P(A \text{ or } B) = P(A) + P(B) - P(A \,\&\, B)$
Rule 8 (Special Conjunctive Rule): if A and B are independent of each other, then $P(A \,\&\, B) = P(A) \times P(B)$
The conditional probability of a proposition A, given B

$$[P(A/B)] = \frac{P(A\&B)}{P(B)} \quad \text{if } P(A) \neq 0$$

Rule 9 (General Conjunctive Rule): $P(A \,\&\, B) = P(A) \times P(B/A)$

Bayes' Theorem:

$$(1)\ P(H/E) = \frac{P(H) \times P(E/H)}{P(E)}$$

$$(2)\ P(H/E) = \frac{P(H) \times P(E/H)}{[P(H) \times P(E/H)] + [P(\sim H) \times P(E/\sim H)}$$

$$(3)\ P(Hi/E) = \frac{P(Hi) \times P(E/Hi)}{\sum P(Hi) \times P(E/Hi)}$$

Interpretations of Probability

The probability calculus consists of a set of rules which licenses us to infer probability values from other probability values. As such, it is like the rules of formal inference for deductive logic, which allow us to infer truth value from truth value. The rules of deductive logic are truth-preserving, passing on truth values to the conclusion. The rules of the probability calculus allow us to deductively infer probability value from probability value. The rules of the probability calculus are probability-preserving and pass on probability value to the conclusion.

The key question that remains, however, is: From where comes the original probability figures, i.e., the probability assigned to one's original premises? We can call these the *initial probabilities*. For an answer to this question, we need to consult the different theories of probability. These theories propose answers to the questions: What is the meaning of 'probability;' and How are these initial probabilities to be determined?[3]

I. The Classical Theory

The first theory to be considered is the classical theory of probability. This theory originated in the 17th century with the work of Blaise Pascal (1623-1662), Pierre de Fermat (1608-1685), and especially **Pierre Simon de Laplace**.

Laplace was a staunch defender of determinism with respect to the course of events in the natural world and believed that every event is so rigidly determined that if one knew the underlying laws of nature and the set of initial conditions, one could predict the entire future course of events. But unfortunately, human beings do not have such all-extensive knowledge and hence cannot have perfect knowledge. In such cases, we speak of *chance*, which is a reflection of our ignorance. Probability, therefore, is relative to our knowledge and to our ability to decide which event will occur. "The theory of chance," Laplace said

> consists in reducing all the events of the same kind to a certain number of cases equally possible, that is to say, to such as we may be equally undecided about in regard to their existence, and in determining the number of cases favorable to the event whose probability is sought. The ratio of this number to that of all the cases possible is the measure of this probability which is thus simply a fraction whose numerator is the number of favorable cases and whose denominator is the number of all the cases possible. (1795/ 1951, p. 28)

But although probability is a lack of knowledge, we can measure it. According to this view, therefore, probability is the ratio

$$\frac{\text{The number of favorable alternatives}}{\text{The total number of alternatives}}$$

For example, the probability of obtaining a 2 on a throw of a die is 1/6 because there is just one way to obtain a 2 and six possible ways in which a die may turn up: 1, 2, 3, 4, 5, 6. Likewise, the probability of getting a 1 or 2 or 3 on a throw of a die is 1/2 since there are six possible outcomes and three of these--1, 2, 3--are the favorable ones. Laplace thus provided an important formulation of several rules of the probability calculus.

Principle of Indifference. According to the classical theory, probability is to be determined by comparing favorable alternatives to the total number of alternatives. It is this calculation which provides one with the initial probabilities. But how do we decide which alternative should be counted? It doesn't seem reasonable to argue, for example, that the probability of throwing a 6 on a die is 1/2 since there are just two alternatives: 6 and not-6.

One answer provided by classical thinkers is to invoke the notion of *equipossible*: just count those alternatives that are equipossible. Not-6 (i.e.,1,2,3,4,5) is not equipossible with 6 since there are more alternatives in "not-6" than in "6", but 1 is equipossible with 6. Hence in determining the total number of alternatives, count those alternatives that are equipossible. But what does 'equipossible' mean here? One answer is that it means 'equiprobable': count those alternatives that are equiprobable. But to avoid the apparent circularity involved here, equiprobable must be analyzed further.

The classical answer to this problem is that something is equipossible if we can't decide if it will occur or not. This has been called by later individuals the **principle of indifference:**: *two possibilities are equipossible if and only if there is no ground for choosing between them.* As Laplace (1795/1951) said: "if we do not know which of the three urns contains black balls only, so that there is no reason to believe that it is C rather than B or A, these three hypotheses will appear equally possible" (p. 29).

The principle of indifference operates so that, in the case of a die, we expect each side to occur an equal number of times. This is because we have no grounds for choosing one over the other. Likewise, in two consecutive tosses of a coin, the second toss is independent of the first toss because we have no grounds for believing otherwise.

Of all the parts of the classical theory of probability, the principle of indifference has been subject to the most criticism. The basic objection is that "it turns ignorance into knowledge." For example, suppose you know a die is biased but you don't know which side is biased. According to the usual interpretation of the principle of indifference, this means that you would have to treat each side as being equipossible (since you don't know which side is biased), even though this seems wrong.

A second issue centers on the question of whether, according to the classical theory, 'probability' refers to something subjective--a psychological state of knowledge or ignorance--or something more objective like the laws of logic. Although Laplace seems to have wanted to argue that probability is an objective measure, he also seems to have equated it with our internal states of knowledge. Moreover, the classical theory seems to be inadequate as a complete account of probability because many of the probability statements of science seem to refer to the objective, empirical world of natural events, to how frequently, for example, one set of events follows another set. But the classical theory seems to be silent on such matters.

A final objection to the classical theory of probability, therefore, is that it appears to be limited to rather restricted and unusual contexts, namely, situations that are like games of chance. Although probability has a clear sense in these contexts, it is not the only realm, nor the most relevant one, for the assessment of scientific theories. When there is a body of evidence E and a scientific hypothesis H, we want to determine how E makes H more (or less) probable. Here the classical theory of probability offers us little help. For these (and other) reasons, most individuals do not consider the classical theory of probability to be an adequate theory. The parts of it which appear to be most useful have been incorporated into later theories of probability.

Summary: We may summarize the basic tenets of the classical theory in the following way (Weatherford, 1982):

1. Probability is the ratio of the number of favorable cases to the total number of cases.
2. The Principle of Indifference
3. There is no objective chance (indeterminacy) in the universe. Probability is an expression of our ignorance.
4. Initial probability statements are objective statements, since there are objective rules for generating initial probabilities. But this objectivity is not a part of an independently existing reality.
5. Individual (initial) probability statements are rationally (intuitively) known; hence, they are a priori, i.e., known prior to sensory experience.

Exercise 6.5

1. Suppose an ideal die is tossed. What is the probability that

a. a 6 turns up
b. a number less than 6 turns up
c. an odd number turns up
d. a number greater than 2 but less than 6 turns up.

2. Does the classical interpretation of probability justify the following argument? If so, how? If not, why not?

A pair of coins is tossed. There are three possible outcomes: a pair of heads, a pair of tails, and one of each. There is no reason to suppose one of these outcomes to be more probable than any other; hence, the probability of a pair of heads is 1/3.

3. Use the principle of indifference to calculate the probability that Longshot will win a horse race against seven competitors. Why is or isn't the principle of indifference appropriate here?

II. Relative Frequency Interpretation

A widely accepted interpretation of probability does not interpret it in an *a priori* fashion, as a ratio of favorable cases to the total number of cases, but in an *a posteriori* way, as something to be determined empirically. On this view, called the relative frequency interpretation, probability is a term that refers, first and foremost, to *events*--classes of events. The individuals who have been most closely associated with this interpretation of probability in this century are Richard von Mises (1937) and **Hans Reichenbach** (1949). Reichenbach's account has been especially influential and it will be the focus of our discussion.

On this view, the probability of something is the relative frequency of that kind of event--the frequency of its occurrence--in relation to a larger class of events, called *the reference class.* Suppose, for example, one wanted to know the probability that the child you are about to bear will be a girl. Since the relative frequency of girl births is .49 (49%), the probability is .49 that your child will be a girl. Such a probability is calculated in the following way:

$$\frac{\text{Number of female births}}{\text{Total number of births}} = 49\%$$

In general, the probability of an event is equal to the number of observed favorable events divided by the total number of observed events. So, what is the probability that a 16 year old driver will have an accident? It is the relative frequency:

$$\frac{\text{Number of accidents of 16 year old drivers}}{\text{Total number of accidents}}$$

Reference Class. A crucial concept in the relative frequency interpretation of probability is that of the *reference class*, that class to which the class of events in question belong. An individual event may belong to many different reference classes: a human may belong to the class of females, female students, female students who ski, etc., just as an illness may belong to the class of cases of tuberculosis, serious cases of tuberculosis, serious cases of tuberculosis leading to death, etc. Different classes of references will give us different probabilities. Which of these should we choose? For example, the probability of skin cancer given that an individual is a woman may be .2, the probability of skin cancer given that she is a woman surfer may be .4, and the probability of skin cancer given that she is a woman surfer who uses sun screen of 15 may be .3 Which reference class should we employ in determining the probability of skin cancer? Reichenbach's answer is: use the "narrowest class for which reliable statistics can be compiled." Since the class of women surfers is narrower than the class of women, we should use that reference class. Suppose we compare the probability of skin cancer, given that one is a female skier with the probability of skin cancer given that one is an individual who uses sun screen of 15. What should we do here? Reichenbach's answer is: choose the common class e.g., the class of humans who spend time in the sun (assuming we have reliable statistics for this common class).

There are problematic cases, however, with this reference class. Consider, for example, the following. Out of a sample of 128,000 Coloradoan skiers, 35,718 developed skin cancer. What is the probability that Jane Smith, a Coloradoan skier, will develop skin cancer? The probability of developing skin cancer given that you are a woman is .24. Here we are using a narrower reference class. In a sample of 7,418 people, only 67 got skin cancer when they used sun screen with a rating of 15. So, the probability of developing skin cancer given that you use sun screen with a rating of 15 is .009. What is the probability that Jane Smith, a woman Coloradoan skier who uses a sun screen with a 15 rating, will get skin cancer? To answer this question, we would need to have the relative frequency of skin cancer in this specific reference class. Suppose we don't have such statistics? Then we cannot calculate the probability.

As the last exercise illustrates, on the relative frequency interpretation, the probability of an event can change depending on the reference class that one selects. This means that *probability is relative to a reference class* even though this probability is an objective and real feature of the world. The probability of an event might be one value given a particular reference class and another value given a different reference class. So if someone belongs to both reference classes (e.g., a woman who is a skier and who also uses sun screen) and we do not have these statistics, we cannot calculate the probability. The probability calculus is of little assistance here, since what we need are additional empirical data.

Exercise 6.6

Using the relative frequency theory, determine the probability of the following events:

1. Suppose the average number of male births in the United States per year is 2,180,000 out of 4,258,000 births. What is the probability that the next baby born will be a boy?

2. Examine the appropriateness, pro and con, of the empirical interpretation of 'probability' in the following statements.

a. The probability is 1/6 that a roll of this die will result in a six

b. The probability of rain tomorrow is .4.

c. The probability of snow this Christmas is high.

d. The probability of getting at least one head on the next ten tosses of a coin is about 99 out of 1000.

e. The probability of getting about 1/6 sixes on a thousand rolls of this die is very high.

f. The probability that the train will be on time is high.

g. The probability that a strong and warlike nation will be invaded by one of its weaker neighbors is rather small.

3. Of 100 juniors who took IQ texts at Rocky Mountain High School last year, 27 had scores between 110 and 118. The scores of the whole group ranged between 102 and 167. If next year's class is similar to this year's class, what is the probability that a randomly selected junior next year will have an IQ between 110 and 118?

Limits. The relative frequency theory of probability appears to be a very empirical theory of probability--one simply collects the relevant empirical evidence and then determines the frequency of a certain kind of event. It is also a theory that claims a great deal of objectivity, for on this view the probability of an event doesn't depend on one's subjective beliefs, nor upon a priori logical relations. The probability of an event is a real objective physical property of the world, e.g., the actual frequency of a certain kind of event in a sequence of events.

Some sequences are finite and some are not. Imagine, for example, an infinite number of tosses of a coin as compared to the number of times in the past the sun has risen. A theory of probability should be able to deal not only with finite sequences of events, but also with those unlimited sequences. One can count the actual number of events in a finite sequence and determine its relative frequency, but one cannot actually count the number of events in an infinite sequence. What, then, does probability mean in this latter case?

Fortunately, there are well-known mathematical properties of such infinite sequences that assist us in answering this question. Crucial here is the mathematical notion of a **limit**--an ideal mathematical point in a series never actually reached but approximated as the number of cases reaches infinity. Consider, for example, the following infinite sequence of tosses of a coin:

H T H T H T H T H T H T H T H T H T H T H T H T H T H T . . .∞

Although the series goes on forever, it does appear that the relative frequency of heads is 1/2. But since there is no numerical ratio here--because we would have an infinite number in the numerator and in the denominator--what can such a notion mean?

Consider any finite initial segment of this sequence: H, HT, HTH, HTHT, etc. For each segment there is a corresponding relative frequency

Initial Segment	*Relative Frequency of Heads*
H	1/1
HT	1/2
HTH	2/3
HTHT	1/2
HTHTH	3/5
HTHTHT	1/2
HTHTHTH	4/7

Here we would have the following sequence of relative frequencies of heads:

1 1/2 2/3 3/5 1/2 4/7 . . .

If we ignore the cases of 1/2, the progression of fractions is getting closer and closer to 1/2: 1, 2/3, 3/5, 4/7, 5/9, 6/11, 7/13, 8/15, etc., i.e., it is approaching 1/2. Will it ever reach 1/2? Only as an ideal limit as the numbers continue on to infinity. According to the famous mathematical theorem of Bernoulli--**the Law of Large Numbers**--this sequence has a limit L as the numbers go to infinity just in case for every positive number e, there is a number N such that if $n > N$, then $|f_n - L| < e$ (where f_n is the relative frequency of an event among n trials, L is the limit and e is any arbitrarily small number). In other words, *for any finite error tolerance e chosen, no matter how small, there is a position N in the sequence such that every subsequent member of that sequence is within e of L.* With regard to infinite sequences, therefore, we can define the probability of an event as *the limit of the sequence of relative frequencies of a certain kind of event.* ***Example.*** Suppose in an actual toss of a coin, we found the following:

Initial Segment	*Relative Frequency of Heads*
H	1/1
HT	1/2
HTH	2/3
HTHH	3/4
HTHHH	4/5
HTHHHT	4/6
HTHTHTH	4/7
HTHHHTH	5/7
HTHHHTHT	5/8
HTHHHTHTH	6/9
HTHHHTHTHH	7/10
HTHHHTHTHHT	7/11
HTHHHTHTHHTT	7/12
HTHHHTHTHHTTT	7/13

Bernoulli's Theorem (the law of large numbers) applies to this sequence (and there is a limit L) just in case, as the numbers continue, the series approaches L as a limit. Suppose we let e = .25, and L = .5. According to Bernoulli's theorem, this series has the limit L if there is a number N such that for all numbers greater than N, the quantity f_n - L is less than .25 (where f_n is the relative frequency of the event). Suppose we take the third entry, in which f_n is 2/3. Then | f_n - L | is 2/3 - 1/2 = | .667 - .5 | = .167 which is less than e (= .25). But is that true of all subsequent relative frequencies? What about entry number 4, where f_n = 3/4? Entry number 7 is a better candidate since | 5/7 - 1/2 | < .25 and all subsequent entries have relative frequencies that deviate less than .25 from 1/2. Hence, there is a limit L (of 1/2) to this series.

For finite sequences, therefore, the relative frequency of the class of events is the ratio of the number N of As and Bs in the first n members divided by the number N of As in the first n members. For infinite sequences, however, the relative frequency of the class of events B. As is the limit of the ratio B/A as n approaches infinity.

Source of Initial Probabilities. How do we determine the relative frequency of a series, especially an infinite series? Although in a finite sequence, one may actually count the entire sequence, this is not possible for an infinite sequence. What should we do? Reichenbach's answer is that we should proceed empirically (a posteriori) via *induction by simple enumeration*: we simply count the relative frequency of an initial segment of our sequence and then infer that the relative frequency will continue to persist for the rest of the series. Here we are inferring from all observed cases (which is n) to all cases. This solution--of taking some small initial part of the sequence and using it as a basis for inferring the relative frequency of the entire sequence--is called the *straight rule* (induction by simple enumeration). According to Reichenbach, this will provide one with the statistical evidence necessary to calculate the relative frequency of a series, and hence, to determine its probability. In short, one must do empirical research in order to decide what the probability is.

Induction by simple enumeration provides one with an estimate, guess, or posit concerning the relative frequency of the series. It also provides one with a means for determining what the probability of a *particular* individual event is.

On the relative frequency view, the probability of an event (e.g., snow tomorrow) is the relative frequency of a class of events, namely, a ratio of the favorable class of events (snowing) to the total number of events (all days). But what is the probability that a single event will occur? On the relative frequency view, there appears to be no answer, since probability concerns classes, not individuals.

Reichenbach tried to accommodate the probability of a single event by introducing the concept of a posit. We assert that a single event will probably happen (e.g., it will probably snow tomorrow) because the relative frequency of such events is relatively high (say .9). When we reason this way, we are making a posit (a bet): we behave as if this statement were true. "A posit is a statement with which we deal as true, although the truth value is unknown" (Reichenbach, 1949, p. 19). Because of the high relative frequency (probability) of this kind of event, we will be correct (or successful in our action) about 90% of the time, i.e., the relative frequency of the success of our posit would be .9. This relative frequency is called the *weight* of the posit, which is its truth value--how often it is true. "We shall use the term *weight* for the truth value of the posit. The probability of a single case therefore is the weight of the posit" (pp. 378-9). Posits can be improved upon (their weights adjusted) as a result of new evidence. We can therefore calculate the relative frequency of the truth of certain kinds of posits (e.g., positing snow when its relative frequency is high) and can thus determine we have been successful a certain percentage of the time.

The concepts of posit and weight can also be developed into a *truth-frequency theory* account of scientific theories. What, for example, is the probability of a particular scientific theory, say, Darwin's theory of evolution? The probability of an individual scientific theory does not appear to be analyzable as a relative frequency of two classes of events. But according to Reichenbach, the probability of a theory can be analyzed as a weight. According to this account, the success of a particular kind of theory is to be measured by the relative frequency of the truth of these kinds of theories, given the total number of theories. The probability of a singular hypothesis (belonging to the class of theories) is thus its weight. Assume we have available the relative frequency of a certain class of theories T, i.e., T-type theories have been true a certain portion of the time, say,

$$\frac{\text{Truth of T-type theories}}{\text{Truth of all theories}} = .4$$

Suppose we are confronted by a new instance of T, say, T_n. What do we say about the probability of T? We posit that T_n is true (i.e., predict it is true). Such a posit has a certain weight in this case .4. Hence the probability of T_n = weight of posit of T_n = relative frequency of T.

The relative frequency theory thus appears to be able to handle two of the objections raised against it: the problem of the single case and the probability of a hypothesis. There are, of course, other objections that have been raised against this theory. One concerns the problem of the reference class. As we have seen, an individual event can belong to more than one reference class. So, in determining the relative frequency of an event, e.g., the probability of getting cancer, it is important that we specify the reference class. But since there are several possible reference classes, which one should we use? Reichenbach's answer was: select the narrowest reference class, i.e., that reference class that contains the most detailed information. However, the problem is that we may not always have such a reference class. In addition, some have suggested that we should select the most relevant reference class, i.e., that class that has the most relevant information about the subclass in question. But although this is obviously valuable advice, it is difficult to specify what 'relevant' means.

Relative frequency, on the standard interpretation, involves the notion of a limit to the (infinite) sequence. But since the notion of a limit involves the notion of approximation to a particular numerical value as the number of instances approaches infinity, it will be difficult to determine what and whether a series has a limit or not. In other words, the claim that a series has a limit of, say, L is a claim that will be difficult to test empirically because all that we will ever have in our knowledge base will be a finite number of instances, and such a finite number is compatible with the sequence having this limit L as well as its not having L. How will it be possible, then to falsify or disconfirm a claim that a series has a limit? In short, it is difficult to know whether a series has a limit and if it has one, what this limit is.

Summary: According to the relative frequency view (Weatherford, 1982):

1. Probability means either:
   a. The relative frequency of a property within a finite sequence; or
   b. The limit of a relative frequency in an infinite series.
2. Probability statements are objective features of the real world.
3. Individual probability statements are synthetic, empirical and factual.

III. The Logical Entailment View

In an earlier chapter, we characterized an inductive argument as one in which the premises give a certain amount of evidential support for the conclusion but do not (as in deductive logic) entail it. The reader will recall that deductive arguments possess the logical property of *validity*. The corresponding weaker logical relationship possessed by inductive arguments is *inductive strength*. We characterized this concept in the following way:

> An argument is *inductively strong* if
> (1) The premises *inductively support* the conclusion.
> (2) If the premises were true, the conclusion would *probably* also be true.
> (3) It is *improbable* that the premises are true and the conclusion false.

Clearly, because 'strength' comes in degrees, so does the adequacy of an inductive argument. Hence, there are degrees of inductive strength. But if so, one should be able to assign numbers to the degree of inductive strength. Furthermore, the three definitions of 'inductive strength' seem to be referring to the same underlying notion.

The logical interpretation of probability (also called the range interpretation of probability, and the degree of entailment theory of probability), associated with the names of John M. Keynes and especially **Rudolph Carnap**, can be seen as an attempt to cash in on the above line of argument by suggesting that probability can be interpreted in terms of inductive strength.

According to this interpretation, to say that something is *probable* is to make a claim about the logical relation between propositions, not events. This logical relation is conceptualized to be analogous to validity, but only weaker, namely, *partial entailment:* in a probabilistic argument, the premises *partially* entail the conclusion. Figure 1 represents the relation between premises and conclusion in a deductive argument.

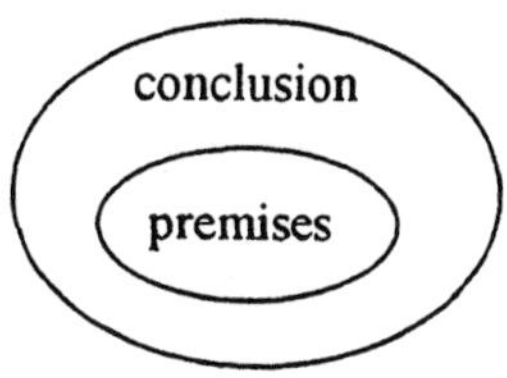

Figure 6.1. The Relation of Deduction

If one statement entails another, the range of the second is included within the range of the fist. This means that every possible state of affairs of the universe in which the first is true is a possible state of affairs of the universe in which the second is true. Figure 6.2 represents the partial entailment view.

If one statement partially entails another, the range of the second is partially included in the range of the first. In every possible state of the universe in which the first is true, there are some of these in which the second is true. Figure 6.3 shows degree of partial entailment

In Figure 6.3A, we can say: If the premises were true, the conclusion would probably be true to a very weak degree since the intersection of Premises and Conclusion is very small. But in Figure 6.3C, if the premises were true, then the conclusion would very likely be true because so much of the premises are contained in the conclusion. Hence, the greater the degree of overlap between the circle representing the premises and the circle representing the conclusion, the greater the degree of entailment.

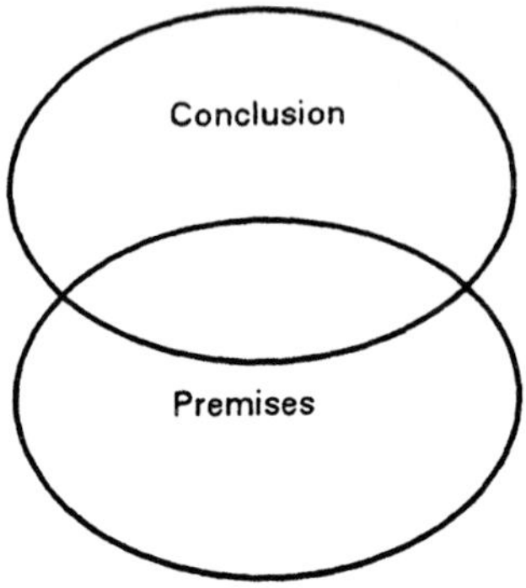

Figure 6.2. Partial Entailment

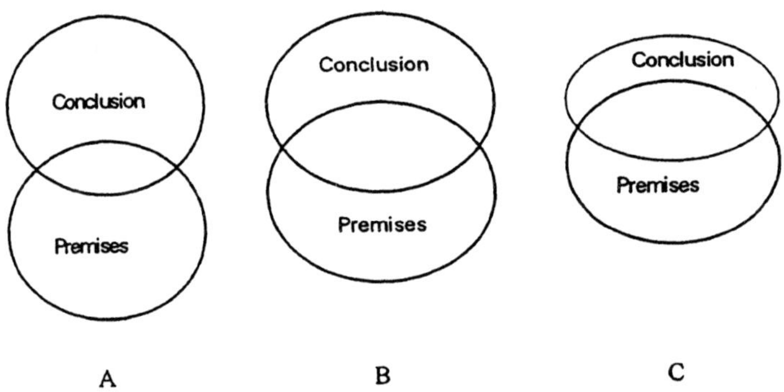

Figure 6.3. Degrees of Partial Entailment

Unlike the relative frequency interpretation of probability, which is empirical and contingent, the logical interpretation is analytic and necessary. On the relative frequency view, the truth of a probability statement depends upon actual facts in the empirical world, but, on the logical interpretation view, the truth of a probability statement, such as P(H/E), does not depend on the empirical facts but only on the logical relations between the premises and the conclusion.

The Range Interpretation. As a logician, Carnap was concerned with setting up an ideal, logical language by means of which one could express what has been called the range interpretation of probability. Let us assume, therefore, that our ideal language has a single predicate P describing some property and that there are just three individuals in the world denoted by a, b, and c. If 'P' stood for *apple*, then one could express the proposition that 'a is an apple' by means of the logical symbolism *Pa* (a has the property appleness); likewise for *Pb* and *Pc*. We might imagine the case in which a scientist observes an apple at a particular time and in a particular region of space and asserts 'this is an apple' (Pa). Later, the scientist might observe another example of an apple (Pb) and still later another example (Pc).

Suppose in our ideal world, we wanted to give a complete description of the state of this world--a *state description*. Such a state description would consist of a listing of all the facts in the world. But suppose that we want more than a description of our actual world; we want a description of all possible worlds, of ways our world might be. It is clear that the following would be an exhaustive listing of all possible state descriptions of our world:

(1)	Pa	&	Pb	&	Pc
(2)	Pa	&	Pb	&	~Pc
(3)	Pa	&	~Pb	&	Pc
(4)	~Pa	&	Pb	&	Pc
(5)	Pa	&	~Pb	&	~Pc
(6)	~Pa	&	Pb	&	~Pc
(7)	~Pa	&	~Pb	&	Pc
(8)	~Pa	&	~Pb	&	~Pc

Figure 6.4 Eight State Descriptions

Suppose one had a hypothesis (H), that c is P, and the evidence (E) consisted of two facts: Pa and Pb. What would be the probability that H is true given the evidence E, i.e., what is p(H/E)? To put it slightly differently, what is the degree of confirmation that evidence E provides for H, represented as *c(H/E)*?

On the simplest interpretation of degree of entailment (the range interpretation), one would simply calculate the range of E (those state descriptions in which E is true)--our premises--and the range of H (those state descriptions in which H is true), which constitutes our conclusion. The degree of evidential support that E offers to H is the degree to which E entails H, the degree of overlap between the ranges of E and H. Clearly, if there is some overlap, H and E will have something in common and this will simply be E itself. Hence, in comparing the ranges of H and E, we are really comparing the ranges of E and (H & E). In fact, what we have is the following familiar formula:

$$p(H/E) = \frac{p(H \& E)}{p(E)}$$

but interpreted in the following way:

$$c(H/E) = \frac{m(H \& E)}{m(E)}$$

where c is the confirmation that evidence E gives to a hypothesis H and m is a *measure* of the degree of confirmation. The *confirmation* of H given E is equal to the *measure* of the range of (H & E) divided by a measure of the range of E. But how should their ranges be measured, i.e., what probability values should be assigned to them? Carnap discussed two main possibilities: c+ and c*.

One possibility suggested by Carnap is to assign each state description an equal probability. (Let us call this particular confirmation function c+). Since we have 8 such state descriptions in our world, (see Figure 6.4), each would receive a value of 1/8 as shown in Figure 6.4.

(1)	Pa	&	Pb	&	Pc	1/8
(2)	Pa	&	Pb	&	~Pc	1/8
(3)	Pa	&	~Pb	&	Pc	1/8
(4)	~Pa	&	Pb	&	Pc	1/8
(5)	Pa	&	~Pb	&	~Pc	1/8
(6)	~Pa	&	Pb	&	~Pc	1/8
(7)	~Pa	&	~Pb	&	Pc	1/8
(8)	~Pa	&	~Pb	&	~Pc	1/8

Figure 6.5. Equal Probability Assigned to Each State Description

In determining the c+(H/E), therefore, we would have:

$$c+(H/E) = c+(Pc/Pa \& Pb) = \frac{m+(H \& E)}{m+(E)} = \frac{m+(Pa \& Pb \& Pc)}{m+(Pa \& Pb)} =$$

i.e., one would first find that particular state description in which both H and E are present, which is state description #1. This has a

$$\frac{\text{state description \#1}}{\text{state description \#1 \& \#2}} = \frac{1/8}{1/8 + 1/8} = .5$$

probability 1/8. One would then determine the state descriptions in which Pa & Pb occur--#1 and #2--each of which has a probability of 1/8. On this account, the degree to which E confirms H is .5.

Note that this degree of confirmation is the same as that of the hypothesis Pc just on the evidence Pb alone, namely, .5. Hence, the confirmation of our hypothesis (Pc) was not increased by the addition of empirical evidence Pa! Something has gone wrong here.

To handle the above problem, Carnap suggested a second way of measuring confirmation, one in which we assign equal probabilities not to state descriptions but rather to what he called the underlying structure of state description--its *structure description.* Two state descriptions have the same structure just in case the one is obtained from the other by permutation of individual constants, e.g., *Pb & ~Qb* can be obtained from *Pa & ~Qa* by permutating 'a' and 'b.' This structure can be read as: *one individual is P and one individual is not-Q*. This will produce a different c-function: c*. For example, Pa and Pb have the same structure, namely one individual has the property P. Similarly, Pa & ~Pb has the same structure as that of ~Pa & Pb--one individual has P and one individual doesn't have P.

If we look at our state descriptions in Figure 6.4, we will notice that numbers (2), (3) and (4) have the same structure, and that (5), (6) and (7) have the same structure. Together with (1) and (8), therefore, we have four structures. Carnap suggested, therefore, that we assign equal probabilities to the same structure; hence, each structure would get a value of ¼. See Figure 6.6

(1)	Pa	&	Pb	&	Pc	¼
(2)	Pa	&	Pb	&	~Pc	}
(3)	Pa	&	~Pb	&	Pc	} ¼
(4)	~Pa	&	Pb	&	Pc	}
(5)	Pa	&	~Pb	&	~Pc	}
(6)	~Pa	&	Pb	&	~Pc	} ¼
(7)	~Pa	&	~Pb	&	Pc	}
(8)	~Pa	&	~Pb	&	~Pc	¼

However, since there are three examples of the second structure, we must divide the probability of 1/4 over these three, so that each gets a value of 1/12. This would give us the following

(1)	Pa	&	Pb	&	Pc			¼
(2)	Pa	&	Pb	&	~Pc	1/12	}	
(3)	Pa	&	~Pb	&	Pc	1/12	}	¼
(4)	~Pa	&	Pb	&	Pc	1/12	}	
(5)	Pa	&	~Pb	&	~Pc	1/12	}	
(6)	~Pa	&	Pb	&	~Pc	1/12	}	¼
(7)	~Pa	&	~Pb	&	Pc	1/12	}	
(8)	~Pa	&	~Pb	&	~Pc			¼

Now let us recalculate the new degree of confirmation (c*) of Pc given Pa and Pb:

$$c+(H/E) = c+(Pc/Pa \,\&\, Pb) = \frac{m+(H \,\&\, E)}{m+(E)} = \frac{m+(Pa \,\&\, Pb \,\&\, Pc)}{m+(Pa \,\&\, Pb)} =$$

$$\frac{\text{state description \#1}}{\text{state description \#1 \& \#2}} = \frac{1/4}{1/4+1/12} = .75$$

Given the confirmation function c*, E confirms H to degree .75 (instead of .5). What about the probability of Pc given Pb? It is .67 (This is left as a homework problem.) On the new system c*, therefore, the degree of confirmation of Pc changes as a result of new evidence from .67 to .75, which seems much more reasonable.

The above language was introduced to illustrate how Carnap's system works. Things get much more complicated, however, when we add more individuals and properties. For example, if we have two predicates P and Q and two individuals a and b, we would have 16 state descriptions. Determining the underlying structure here is much more complicated. However, the basic process of assigning numerical values is the same as in the above example.

Exercise 6.7

1. Assume language L has one predicate (P) and three individuals (a, b, c). Show that c+(Pc/Pb) = .5 *(see page 33). Calculate the probability of Pc given Pb using the c*-confirmation function.

2. Suppose we have in our language two properties--Black (B) and White (W)--and two individuals--a, b. How many state descriptions are there? Construct a complete list of them.

3. Suppose we have the language specified in problem 2. Suppose our hypothesis is Ba and Bb and Wa and Wb and our evidence is Ba and Bb. What are the corresponding c+ and c* values?

Carnap spent a great deal of his life attempting to lay down the basic principles of a logic of confirmation, an inductive logic that would result in numerical confirmation functions. Given that we say things such as "given all the evidence, the probability of that theory is .7," Carnap tried to provide a system of rules that would explain how we could say such things.

Carnap began with a simple logic and with simple logical languages, languages which were not rich enough to handle the propositions of science. The predicates of these languages, for example, were too simple since they did not contain relations. Furthermore, a problem Carnap encountered early on was that, on his system of inductive logic, all scientific laws would have a value of 0 (if they ranged over an infinite number of individuals). Carnap tried to answer this objection but without much success.

Another problem with Carnap's account concerns the choice of a confirmation function. In addition to c+ and c* , Carnap argued there was an indefinite number of other confirmation functions and that there was no compelling argument for one over another. This has seemed to be a problem to many individuals, since we would like to be able to provide a rational basis for our choice of a confirmation function.

The verdict of history seems to be that Carnap produced a magnificent system of inductive logic--an ideal and abstract model of a quantitative system of inductive logic, but one that is not very closely tied to actual science and one that has several flaws. As a result, many individuals have turned to other accounts of probability, one of the most popular approaches being that of subjectivism (personalism).

Summary. In summary, the following set of theses characterize the logical (a priori) interpretation of probability (Weatherford, 1982):

1. Probability is a logical relation between statements.
2. The logical relation is completely determined by means of the rules of logic and the rules of the probability calculus.
3. Every probability statement is relative to the evidence.
4. Every probability statement is analytic (logically true) and a priori (independent of empirical evidence).

.ıjectivism

.ne final interpretation of probability we will mention is ,ubjectivism (sometimes called **personalism**), a view associated with the thinkers Frank Ramsey (1926/1960), Bruno de Finetti (1937/1964), and Leonard Savage (1954). Although subjectivism was not a serious contender when it was first advanced in the '30s (and received bitter criticism by famous statisticians), it has become increasingly popular and influential since the 1960's (largely due to the work of Savage [1954]). Today it is closely associated with the position known as **Bayesianism.** Even the arch-rationalist Carnap, who championed a logicist interpretation of inductive logic, moved toward subjectivism in his later writings. Today, the subjective interpretation of probability has become one of the dominant positions regarding the nature of probability.

According to the subjectivist, probability is not to be identified with an objective, empirical property (as in the relative frequency view), nor an objective, logical property (as in the logical entailment view), but simply as the degree (intensity) of one's subjective belief. Hence, to say the probability of rain is .7 is simply to say that for a particular individual S, his personal (subjective) degree of belief in rain tomorrow is .7.

In claiming that the meaning of probability is just one's personal degree of belief, probability is relative to the individual, since probability is the personal belief of a particular individual. Hence, two individuals can have different personal beliefs about the same thing in the sense of attaching different probability values to them. Since probability is just the actual degree of belief, it makes no sense to say one belief is correct/incorrect, or that one probability value is correct/ incorrect.

All of this makes it sound as if subjectivism is indistinguishable from an investigation into the psychology of individual belief, and that it is relativistic, irrational and "subjective." After all, it is notorious that individuals can believe the most outrageous things and feel extremely confident about them. Hence, if subjectivism interpreted 'probability' to refer just to the strength of one's actual beliefs, this would seem to be an important objection against it. Fortunately, the subjectivist has a reply, one showing that things are not as bad as they seem. But this reply depends upon understanding the subjectivist's answer to the question: How does one determine how strong one's personal beliefs are? How, in short, do you measure one's degree of belief?

To answer this question, one could simply ask the person how strongly they believed something and record their answers. But subjectivists have preferred to follow the old adage: actions speak louder than words. Hence, they have looked to one's public *behavior,* in particular to the bets (wagers) one is willing to make, in order to determine how strongly one believes something.

Everyone is familiar with betting odds--the odds one is willing to give that a particular event will occur. If Jimmy the Greek is convinced that Dallas will win the Superbowl and gives odds of 2:1, he in effect is saying: I am so confident that Dallas will win, that I am willing to bet money on it in the following way: if Dallas wins, you give me $1; and if Dallas loses, I will give you $2. In this case, Jimmy the Greek's subjective probability is 2/3, calculated in the following way: 2/2+1. Similarly, if Francine believes the bet (wager) is fair and agrees to bet with Jimmy the Greek, the odds she is willing to give on Dallas winning is 1:2, or 1/1+2 = 1/3. Her subjective probability is thus .33. Suppose Francine believes the chances are even worse than 1:2 that Dallas will win, say, 1:4. Then her subjective probability is 1/1+4 = .2

Individuals can also bet on scientific theories. For example, Donna believes creationism is the correct (or best) theory about the origins of life. We can calculate her subjective probability by asking for her betting odds: What odds would she give that creationism is correct? Suppose her confidence is so great, she would give odds of 1000:1. Then her subjective probability is 1000/1000+1 = .999. What odds would Donna give that a contradictory or contrary theory, say, the theory of evolution is correct? 1/1000+1 = .0009.

Although we have a way of measuring one's degree of belief, we have yet to answer the objection that subjectivism is irrational. How does the personalist reply to this objection? By showing that if one's personal beliefs are not to be irrational, they must be in accord with the probability calculus; otherwise, one can give a Dutch book argument against them.

Exercise 6.8

Erasmus believes the theory of evolution is correct and is willing to give odds of 2:3. Calculate his subjective probability. Suppose his odds are: 6:7; 7:7; 8:2?

Dutch Book Arguments

Many individuals believe very implausible theories with high degrees of subjective probability, e.g., some individuals still believe the earth is flat. They may even be willing to bet their entire life savings on the truth of such a theory and give odds of, say, 5:1. In such a situation, it seems that an individual's subjective degree of belief in the earth being flat would be irrational, just as it would be irrational to believe offspring are produced by whispering words into the wind. But according to subjectivism, 'probability' means no more than that an individual has an internal psychological state of belief (confidence) such that they are willing to bet on such a belief. Accordingly, if this were all there was to the subjectivist theory, it would have few followers, for such a view is patently inadequate as a theory of probability (and rationality).

But fortunately, this is not all there is to the story. Instead, some individuals (e.g., Savage and Carnap) have claimed that instead of defining probability as one's *actual* degree of personal belief, one should define it as one's rational degree of belief. What then does 'rational' mean here?

The answer most personalists give is simple: for an actual belief to be a rational one, it must be *coherent* (consistent), and this means that it must satisfy the rules of the probability calculus. If a belief does satisfy these axioms, it is coherent and rational; and if a belief does not satisfy them, it is not coherent (not rational). According to most personalists, this is all that a belief set need be in order to be rational.

What is the argument to show that a necessary and sufficient condition for rationality is that a belief set satisfy the probability calculus? The answer is known as the *Dutch Book argument*, which is designed to show that if one's beliefs do not satisfy the probability calculus, they are not rational; they are not rational because, in such a case, a clever gambler can construct a "Dutch book" against you.

A Dutch book is a scheme of wagers such that you will lose no matter how the event comes out. Suppose, for example, I am willing to bet that a certain event E will occur, say, the Dallas Cowboys will win the Superbowl, and I offer betting odds of 3:1 (.75) in favor of this event. Suppose I also am willing to bet 2:1 (.67) that Buffalo will win the Superbowl (~E). Now, either Dallas will win or Buffalo will win. But no matter what happens, I am assured of losing money on the outcome. Suppose, first of all, that Dallas wins. Then I collect $1, but will have to pay out $2 because Buffalo didn't win. Net loss: $1. Suppose, on the other hand, Buffalo wins. Then I collect $1, but have to pay out $3 because Dallas didn't win. Net loss: $2. Either way, I will lose money.

This Dutch Book argument shows there is something wrong with my betting odds and hence that I am irrational. What is wrong is that I have inconsistent beliefs. This is shown by the fact that I believe event E with a probability of .75 and I believe event ~E with a probability of .67. But since these outcomes are mutually exclusive and exhaustive, according to the probability calculus, these two states (P(E) + P (~E)) should only equal 1.0. But in the case of my betting odds, they equal 1.32. Hence my beliefs don't satisfy the probability calculus. Suppose my beliefs were changed so that they did satisfy the probability calculus. Suppose, P(E) = .75 and P(~E) = .25. If P(~E) = .25, then my betting odds might be 1:3. It's very easy to show that on this revised system, a Dutch Book can't be made against me. In short, if my betting odds satisfy the probability calculus, then one can't make a Dutch Book against me.

Bayesian Conditionalization

An individual's subjective probabilities are rational if they satisfy the probability calculus. If they do, no Dutch Book can be made against them. But if they don't, then a Dutch Book can be made, thus showing one's personal probability to be irrational. From this it follows that to be rational, one's subjective probabilities must be coherent.

No doubt, one sense in which a belief is rational would be if it were in keeping with the probability calculus. But are there any other senses

$$P(H/E) = \frac{P(H)P(E/H)}{P(E)}$$

of 'rationality' relevant here? If so, how would the subjectivist handle these?

Rationality can be manifested over time (*diachronic rationality*) as well as at a single instant of time (*synchronic rationality*). Suppose, for example, I have a personal probability value P(H) at time t_1; i.e., I believe at time t_1 that H is true with a certain degree of confidence. What if, at a later time t_2, I change this value of P(H) to P'(H)? Is such a change a rational one? Suppose such a change does not open me up to a Dutch Book argument. Still, we may ask, was the change from P(H) to P'(H) a rational one? If one is being consistent over time, then P'(H) should equal P(H) unless there was some reason to change the value. If P(H) is one's degree of confidence in H, why would one change this degree of confidence? One obvious reason would be if one happened to have new evidence E that was relevant to H. Suppose between t_1 and t_2, one acquires such evidence E. In this case, one's degree of belief in E is certain, i.e., P(E) = 1. How, then, should P(H) change to P'(H)? The answer personalists give is, (H) should change to P'(H) by *conditionalization*--by conditionalizing on the evidence E. This means that the new probability of the hypothesis P'(H) should be equal to the probability of the hypothesis given the new evidence. P'(H) = P(H/E).One way of interpreting conditionalization would be to say: The new probability of H given the acquisition of new evidence E = the probability of H, given the evidence E. Read this way, the claim makes obvious sense. It can be interpreted as saying: at time t_1, an individual proclaims what she believes the probability of a hypothesis would be if she were to acquire new evidence (i.e., P(H/E)). Now, if she does acquire this new evidence E, the revised probability of H should be in keeping with what she said she believed it to be; otherwise, she would be inconsistent. For example, she says: if I were to observe a certain pattern of clouds in the sky in the morning (E), I would believe the probability of it raining today (H) would be .7. (P(H) = .7). Suppose she does observe that pattern of clouds in the sky (E). What then is the probability of her hypothesis H? To be consistent, she must say .7. Hence P'(H) = P(H/E). In short, one should revise one's probabilities by conditionalizing on the evidence.

Most personalists believe not merely in updating your beliefs by conditionalization, but they also believe that you should conditionalize by using Bayes' Theorem. According to the first version of Bayes' Theorem, one should conditionalize in the following way:

$$P'(H) = P(H/E) = \frac{P(H)P(E/H)}{P(E)}$$

To be rational over time, therefore, is to revise (update) your beliefs in a certain way, namely, to conditionalize via Bayes' Theorem and hence to have a dynamically coherent set of beliefs.

A widespread criticism of personalism is that is it too subjective and arbitrary since any one can assign any prior probability to a belief (H) one wishes. For example, I can attach a prior probability of .999 to Darwinism and .001 to Creationism, but a creationist can attach a probability of .999 to Creationism and .001 to Darwinism.

Although this criticism has some substance to it, it is mitigated by a surprising result of Bayesianism termed "the swamping (washing out) of the priors." Suppose we use the above two sets of prior probabilities. Suppose furthermore that E_1 is fossil evidence such that the likelihood of E on H_1 (Darwinism) is .6 and .1 on H_2 (Creationism). Calculating the posterior probability of Darwinism results in two different values for the two individuals. For the Darwinist, $P(H_1/E)$ = .99, $P(H_2/E)$ = .002; for the creationist, $P(H_1/E)$ = .057, $P(H_2/E)$ = .94. Suppose, however, additional evidence E_2 becomes available, say, a particular kind of skeleton remains intermediate between that of the primates and humans. Suppose $P(E_2/H_1) = .8$ and $P(E_2/H_2) = .01$. Now suppose the Darwinist and Creationist calculate the posterior probability of Darwinism using as its prior probability the results of the earlier calculation.

Darwinist	*Creationist*
$P(H_1) = .99$ $P(H_2) = .002$	$P(H_1) = .057$ $P(H_2) = .94$
$P(E_2/H_1) = .8$	$P(E_2/H_2) = .01$
$P(H_1/E_2) = .99$	$P(H_1/E_2) = .83$
$P(H_2/E_2) = .000025$	$P(H_2/E_2) = .17$

The two (respective) posterior probabilities of H_1/E_2 are much closer; as a result of the two items of evidence E_1 and E_2, the Darwinist has not changed his evaluation of Darwinism very much, but the Creationist has changed his evaluation of Darwinism from .01 to .83--much more in keeping with that of the Darwinist. The Darwinist has changed his evaluation of Creationism from .002 to .000025; the Creationist has changed his evaluation of Creationism from .94 to .17. Hence, as more evidence accumulates, the judgments of the two individuals become closer. Assuming they agree on the respective likelihoods and have the same observational evidence, their posterior probabilities will approach that of each other. Hence, the influence of the initial prior probabilities become less and less influential as evidence accumulates.

Exercise 6.9

1. Suppose your hypothesis H is Darwin's theory of evolution, which you believe to be likely to a small degree (P(H))= .3. Later, after observing new fossil evidence (E), you update the probablilty of H (P'(H). What should P'(H) be? Suppose the original probability of E was .4 and P(E/H) = 1.0. Calculate P'(H).

2. You are considering two mutually exclusive and jointly exhaustive hypotheses H_1 and H_2. You assess $P(H_1) = .3$, $P(E/H_1) = .8$ and $P(E/H_2) = .6$. After observing the evidence E, what should your revised probability be?

3. Suppose there are two hypotheses regarding who will win the Superbowl: H_1: Dallas wins; H_2: Buffalo wins, Jimmy the Greek gives odds

of 5:4 that Dallas will win. The key question concerns whether the Dallas quarterback is well and will play. He has been sick the last two weeks, so P(E) = .3. After observing that the Dallas quarterback is well, Jimmy the Greek updates his probability assessment of H_1 to .65. Suppose he has been coherent. What was the original value of P(E/H)?

Subjectivism as a theory of probability has had a stormy career. Originally subjected to widespread criticism (and misunderstanding), it launched a comeback in the late 50s and 60s. In part, this was due to clarification concerning the meaning of 'subjective' so as to remove certain objectionable features from it. In claiming that probability was just a subjective degree of belief, it became clear that one needed to add: consistent with the probability calculus. The issue then became one of whether this coherence requirement was adequate or not as a theory of rationality, i.e., whether it is sufficient for rationality that one's beliefs just be consistent or coherent (as defined by the probability calculus). Although some would argue it is not, many would reply that it is, especially if one allows for the possibility that new observational knowledge (evidence) be subject to Bayesian conditionalization. In any case, as a theory of probability, subjectivism (coupled with Bayesianism) currently enjoys wide acceptance among many inductive logicians and philosophers of science (although most scientists and many statisticians continue to be somewhat skeptical of it).

Summary. In summary, subjectivism (personalism) can be interpreted as maintaining the following theses (Weatherford, 1982):

1. The probability of a proposition is the degree of belief of a person S.
2. Probabilities are determined (measured) in terms of their behavior, e.g., their betting behavior.
3. There are no objective probabilities.
4. An event or hypothesis has no unique probability. Since each individual is free to set his own value, probabilities are relative to the individual.
5. To be rational, the personal beliefs of an individual must be consistent with the Probability Calculus.

For Further Reading

Most standard accounts of the probability calculus require some knowledge of calculus. The most elementary account is Cramer (1955).

Bayes' Theorem (Bayes 1783/1958) was first published in the 18th century and is reproduced in several collections. For a thorough treatment of Bayesianism, see Howson & Urbach (1993). For a history of theories of probability, several works are available, including Hacking (1975).

An excellent and complete introductory coverage of the major interpretations of probability can be found in Weatherford (1982). Laplace's views can be found in Laplace (1795 /1951). Reichenbach's major work (Reichenbach, 1949) is not easy reading for the beginner. A good selection from this work can be found in Luckenbach (1972). The simplest introduction to Carnap is Carnap (1960). His major work (Carnap, 1962) is for the advanced student and presupposes a basic knowledge of symbolic logic. Like Reichenbach (1949), Carnap (1962) has been an extremely influential and important work. For selections from the subjectivist point of view, see Kyburg & Smokler (1964). Another excellent, general anthology is Foster & Martin (1966).

Discussions of the Dutch book argument and Bayesian kinematics tends to be at the intermediate level of difficulty. For a beginning see Howson & Urbach (1993, ch. 6) and then Jeffrey's important (1983, ch. 11). For Carnap's later turn towards personalism, see his 1962/1972 . Finally, an excellent discussion of all these issues can be found in Salmon (1967), which is highly recommended, as is Skyrms (1986).

Notes

[1] Although it is customary to present an axiomatic presentation of the probability calculus and to identify a set of axioms sufficient to generate an appropriate set of theorems, such an approach is not followed here.

[2] We have been assuming that there are two mutually exclusive and jointly exhaustive hypotheses: H vs. ~H, which we could have abbreviated H_1 vs H_2. Because most sets of hypotheses are not jointly exhaustive (but are mutually exclusive, i.e., they are contraries but not contradictories), this second form of Bayes' Theorem has limited application.

[3] It should be pointed out that there are other theories of probability, e.g., the propensity theory, the epistemic theory, etc. But these are not sufficiently different from the four discussed to warrant separate treatment.

PART THREE

Scientific Method

PART THREE

Scientific Method: Introduction

Introduction

Our discussion in Part II of this book fundamentally concerned questions of logic. Scientific methodology—the theory of scientific method—contains several parts, in particular, an account of the procedures used in the acquisition of scientific knowledge and an account of the logic of scientific inference. The latter concerns the question of what conclusions may legitimately be drawn from premises (the evidence). This is the domain of logic, in particular, formal logic. Formal logic is concerned with the question of the formal relations between premises and conclusions and what inferences are formally valid--valid in virtue of their logical form.

Formal logic is concerned with establishing the principles for correct (valid) inference, principles, which belong to deductive logic. By contrast, inductive logic is a much less established field of logic and is also concerned with establishing comparable inductive rules of inference. But it is also concerned with non-formal relations between evidence and statements about the evidence. Hence, inductive logic is broader in scope than deductive logic and, not surprisingly, much less precise.

Probability is also concerned with the formal relations between premises (specifying the probability of a set of statements) and a conclusion (specifying the probability of a statement). These are codified in the rules of the probability calculus. Probability theory deals with other matters, of course, for example the meaning of 'probability' and the determination of initial probability values, and these are not purely formal. But it is convenient to discuss both of these aspects of probability theory together.

Formal vs. Empirical Science

The subject matter of Part II—logic and formal probability—concerns what may be called the *formal* aspects of scientific methodology, with the questions about the formal (structural) relations between statements. Mathematics is another discipline concerned with these formal relations and so is statistics, computer science, certain aspects of linguistics, etc. All of these sciences are traditionally labeled **formal sciences,** as opposed to the non-formal (empirical) sciences such as physics, astronomy, chemistry, biology, geology, psychology, economics, etc. which are concerned with questions about a particular subject matter that are not merely formal in nature. Because the formal sciences share certain features and follow similar methods, a discussion of scientific methodology should, therefore, include a brief discussion of the methodology of the formal sciences.

The Methodology of the Formal Sciences

Traditionally,the formal sciences have been distinguished from the non-formal sciences by their subject matter and methods. According to the traditional account, the formal sciences have a subject matter that is purely formal and this has usually been interpreted to mean not a part of the natural world. (Sometimes, therefore, the distinction is drawn between the formal realm and the material realm.) Biology, for example, may study the nature of cellular reproduction in plants—an aspect of the natural world—but mathematics, by contrast, at least pure mathematics,is not concerned about the addition of apples and oranges, in fact, not about any material objects; instead,its subject matter consists of the realm of ideal or abstract entities—numbers. On this account, a view championed in the fourth century B. C. by Plato, numbers do not exist in space and time; hey have no causal and material properties; they are not about particular, concrete objects. Instead,numbers are abstract entities—what Plato called Forms—existing in their own autonomous realm, transcendent of space and time; this realm was supposed to be a purely formal, ideal realm. On this standard "Platonistic" account, the subject matters of logic, linguistics, computer science, etc. also belong to a purely formal realm.

Secondly, the respective methods of the formal versus the non-formal sciences are alleged to be fundamentally different in kind and, as a result, the products of these respective methods—the kinds of knowledge obtained—are different (see Chapter 2). Several related distinctions are used to capture the difference between formal knowledge vs. non-formal knowledge:. (1) the formal sciences obtain *necessary* truth, whereas the non-formal sciences obtain only *contingent* truth, (2) the propositions of the formal sciences are *analytic* whereas the propositions of the non-formal sciences are *synthetic,* and (3) the formal sciences have **a priori knowledge,** whereas the non-formal sciences have **a posteriori knowledge.** The Latin term 'a priori' means "prior to"; as applied to knowledge, 'a priori knowledge' means or (more precisely): to know something a priori is to know it without having to subject it to subsequent empirical (observational, experimental) test involving the various senses. For example, one can know that 2+2=4 without having to subject this claim to experimental test by, say, counting two apples and two more apples. In earlier times, one was said to know this merely by thinking about it or performing certain mental operations. (Hence, 'a priori' is epistemologically aligned with rationalism.)

'A posteriori' means "posterior to"; a posteriori knowledge is thus knowledge one possesses only because one has tested it by subsequent empirical means. For example, one can only know "water boils at sea level at 100° C" or "ice is cold" by means of sensory observation, experiment, measuring operations, etc. ('A posteriori' is thus aligned with an epistemology of empiricism.)

The formal sciences produce a priori knowledge because they possess a method of knowing that does not involve subjecting certain claims to empirical check; because the non-formal sciences do test their claims by subsequent empirical examination, their knowledge is a posteriori. Hence, the methods of the formal versus the non-formal sciences are distinct. Part III will examine several of the methods of the non-formal, empirical sciences. But what about the methods and methodology of the formal sciences?

Our previous discussion (Part II) of the nature of logic will have provided the basic ingredients of an answer, for the primary method of the formal sciences consists in establishing formal proofs of a particular conclusion. Hence, the formal sciences are deductive in nature and employ deductive methods of proof, issues we have already briefly discussed.

The theory of constructing formal proofs is known as *proof theory,* which forms the core of the methodology of the formal sciences. Proof theory is concerned with the question of what are the appropriate methods to use in constructing proofs, what types of proofs are not appropriate or inappropriate, what kind of rational defense or justification can one give of such claims? etc. The latter in turn involve issues in the philosophy of logic and mathematics, issues of decidability, effective proof procedures, completeness vs. incompleteness, the specification of what an adequate axiomatic system is, etc. Clearly, we cannot discuss all of these issues here in any detail.

As we will see in more detail later, scientific method essentially involves not only the collection of data but also the organization and representation of the data. When one summarizes, organizes, categorizes, etc., one must focus on certain aspects of the data and ignore others; one must represent a part of the data by means of symbols. This abstract system of representation thus functions as a *model* of the data, with its symbols standing for certain aspects of the world.

It is impossible for a model to capture (represent) every aspect of the phenomenon under consideration; it must be selective in its representation. In a formal system of representation, the model focuses just on the formal aspects of the data and ignores everything else. For example, mathematics may represent certain properties of a group of individuals, e.g., the twelve Apostles. But it will only represent certain formal properties—their numerosity and mathematical relations—not their religious properties, the color of their robes, their psychological states, etc., properties that concern what is called *content* and not form. A different formal model might represent other formal properties of this group, but in so doing it would ignore other features. In short, *a formal model can only represent the formal properties of the data.*

Logic and mathematics are the most familiar examples of formal systems. Although originally they may have emerged as formal models, in time individuals began to treat them independently and to construct formal systems without regard to the question of what they were models of—what they were about. Individuals became interested in the question: what is the nature of a purely formal system? In a purely formal system one only has a system of signs or symbols—sometimes called a purely syntactical system—that needs to be given an *interpretation.* To give an interpretation to these signs is to indicate what these signs are about, what they mean. To do this is to leave the realm of syntax and enter the realm of semantics. (See Figure 1).

Historically, the best examples of formal systems came from Aristotle's logic and Euclidean geometry. Euclidean geometry took the form of an axiomatic system when Euclid axiomatized geometry. Because of its importance, Euclidean geometry became a model of

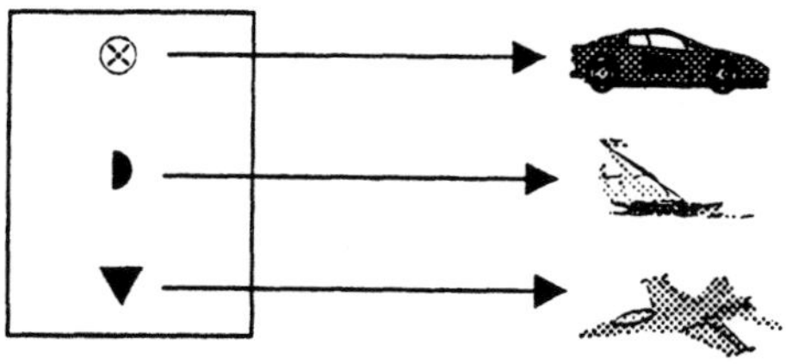

Formal system Interpretation

Figure 1. Formal system and Interpretation

what a formal model should be and hence the concept of an axiomatic system became central to the concept of a formal system.

In a formal, axiomatic system, there are several components. Although there is some disagreement concerning precisely what these components are, most individuals believe an axiomatic system should consist of the following: (1) a list of primitive (basic, undefined) terms—the vocabulary; (2) a list of formation rules rules indicating how the primitive terms may be combined to form larger units—formally acceptable larger units are called well-formed formulas; (3) a set of axioms (sometimes called postulates), which are the basic, foundational principles of the system; and (4) a set of rules of inferences indicating what types of conclusions (theorems) may be inferred from these axioms, such inferences constituting proofs.

The basic method of the formal sciences is the construction of such proofs—formally valid (deductive) proofs—several examples of which we have already discussed in part II. If one has constructed a proof of a conclusion C in a particular formal system, say, F, one has shown that C necessarily follows from the axioms and rules of inference (in F). One will have proven the proposition C and such a proof is clearly a priori; either the string of symbols is a valid deductive proof or it is not, and one can (with certain exceptions) determine by purely formal means whether the statement is the conclusion of a proof or not.[1] But even if this is too restrictive, it still appears to be true that one would not verify that it is a formally valid proof by examining concrete objects through one's senses, for it is not clear that any such empirical examination could show that something was or was not a formally valid proof. In addition, if something is a valid proof in F, then the proposition C is *necessarily true* in F, whereas it does not seem to be the case that one can establish that anything is necessarily true about the empirical world through the use of one's senses.

To produce an adequate formalization of a particular area of knowledge (e.g., linguistics, geometry, supply and demand), certain conditions must be met: the axiom system must be independent, consistent, and complete. An axiom system is *independent* if no axiom is implied by any other axioms; an axiom system is *consistent* if it is not possible to derive a contradiction from it; an axiom system is *complete* if it is possible to derive every true proposition of the subject matter from the set of axioms.

The above must suffice as a brief characterization of the methodology of the formal systems. Clearly, a great deal more needs to be said if we were to give an adequate characterization of a formal system. But perhaps enough has been said to give the student the rudiments of an understanding of the methodology of the formal sciences.

According to the standard account, there is a sharp distinction between the methodology of the formal and the empirical sciences. We have indicated what the distinction is supposed to be. It should be pointed out, however, that although the formal sciences do not use a posteriori methods, relying exclusively on formal methods, this is not true of the non-formal sciences, which use both a priori and a posteriori methods. Indeed, whenever a science is employing the hypothetico-deductive method, it is employing formal methods since it is engaged in the process of deduction, which is purely formal. Furthermore, whenever an empirical science is employing the use of formal models, it is also knee-deep in formal method. But the empirical sciences are not formal sciences because the former also employ non-formal methods. It is these non-formal methods, which have seemed to be particularly characteristic of scientific method, resulting in our knowledge of the natural world. Because of the importance attached to empirical knowledge, therefore, the remainder of the book (except for occasional exceptions) will be devoted to an examination of the methods of the empirical sciences. Unless otherwise indicated, the term 'scientific method' will from now on be used to refer to the method of the non-formal sciences.

The Nature of Scientific Method

Throughout the history of science, there has been considerable controversy over the question of the nature of scientific method. Historically, the major issue fought has been between *inductivism* and *deductivism*. We will distinguish two versions of each, a radical and inadequate form, and a more moderate and adequate form. This will result in four different views: deductivism, inductivism, hypothetico-deductivism, and inductive-deductivism.

Deductivism

The inspiration of deductivism comes from the methodology of the deductive sciences, in particular, the axiomatic method used in geometry. According to deductivism, one begins with a general proposition (hypothesis, law, theory) or set of general propositions, from which one deduces a lower-level proposition about, for example, a particular fact. In so far as one can deduce such a proposition, one has (according to this account) explained it since, on this account, .

The model here is the axiomatic method of geometry, where one begins with the basic components of an axiomatic system and from these deduces a theorem. This model, based upon Euclidean geometry, has been the standard or ideal of knowledge from the time of the Greeks until the 20th century. Although its natural home is the formal sciences, some philosophers and scientists have championed it as a method of the empirical sciences such as physics, where the axioms are expressed in terms of mathematical equations.

For example, given Newton's three laws of motion and the law of universal gravitation (the axioms), one can derive (approximately) Kepler's laws of planetary motion, Galileo's law of acceleration, the law of the pendulum, etc. Such an ideal can be found in the writings of Leibniz and Descartes, as well as in the writings of contemporary scientists who use mathematical (formal) models in their theorizing.

As stated in this form, deductivism clearly needs qualification and correction, for, otherwise, the empirical sciences would be indistinguishable from pure mathematics (e.g., pure geometry or algebra), which also follows the axiomatic method. Mathematical physics may be a kind of applied mathematics; as such it must come into contact with the natural world--it surely is not pure mathematics— in particular, the formal theorems deduced from the axioms require subsequent empirical testing by being matched against features of the actual world (as in the more adequate hypothetico-deductive method). Hence, most individuals have rejected deductivism as a model of the scientific method of the empirical sciences.

Inductivism

A second account of scientific method can be called radical inductivism. According to deductivism, science is a top-down affair, proceeding from the most general to the less general. According to inductivism, however, the scientific method is a bottom-up approach, proceeding from the level of the particular or less general to the most general. See Figure 2.

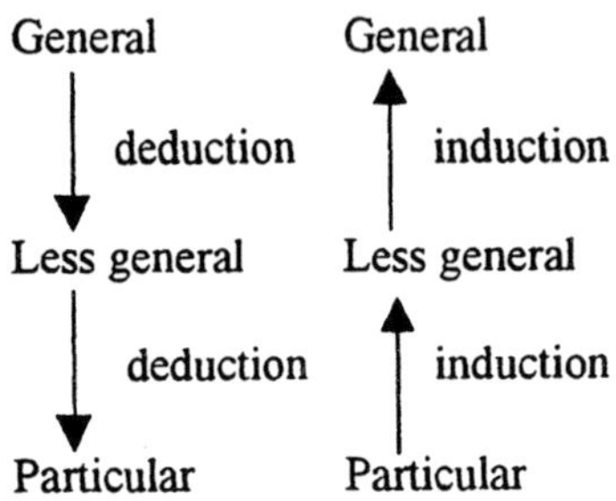

Figure 2. Deductivism vs. Inductivism

According to inductivism, science proceeds by first collecting particular facts and then inducing a generalization from them. This might occur, for example, by induction by simple enumeration (from "some" to "all"). Suppose several such low-level generalizations have been induced. The next step would then be to induce a higher-level generalization from these lower-level ones. An example sometimes given is this: from particular facts collected by Tycho Brahe, Kepler induced his three laws of planetary motion; and from particular facts collected by Galileo involving the inclined plane, Galileo induced his law of constant acceleration. From these two sets of laws, Newton then induced his three laws of motion (perhaps with his law of gravitation). Such an interpretation is supported by some of the remarks of Sir Isaac Newton himself. Writers such as Sir Francis Bacon and John Stuart Mill are also often cited as examples of inductivists. In the 20th century, examples of inductivism can be found in the work of behavioral and social scientists (e.g., B. F. Skinner).

Inductivism is a view actually championed by few individuals, since it has major conceptual problems. First, it does not permit the introduction of theoretical entities such as genes, electrons, black holes, etc. Second, as most inductivists have realized, once one has "induced" a high-level generalization, it is necessary often to test it further by deducing additional predictions from it. Third, as we will see in the next chapter, the inductivists are wrong in believing that scientific observations and facts are infallible, indubitable, and absolutely certain.

Hypothetico-Deductivism

According to a widely influential account, scientific method consists of the hypothetico-deductive method (see Chapter 12). The first part of the hypothetico-deductive method follows deductivism, namely, from a hypothesis H, one deduces a low-level generalization or prediction about some particular matter of fact. If the observation does not occur as predicted, H is disconfirmed or falsified (since one of its test implications has been proven false); on the other hand, and on this advocates of the hypothetico-deductive method disagree, if the observation occurs as predicted, then H has been confirmed (since one of its test implications has been proven to be correct). In short, on the most widely accepted interpretation of the hypothetico-deductive method, we have the following situation:

$H \supset O$	$H \supset O$
O	$\sim O$
=======	======
H is confirmed	H is disconfirmed

We must add several features to this simple account, however. For example, suppose our predicted outcome is false and hence H is falsified. Scientists typically do not stop here. They proceed to test H further or to modify it and then retest it. So the picture actually is more complicated (see Figure 3). Hence, we have something like a spiral account of theory testing.

What is thus important about the hypothetico-deductive method (over the deductivist model) is the requirement that one empirically test to determine whether the predicted observation actually occurs or

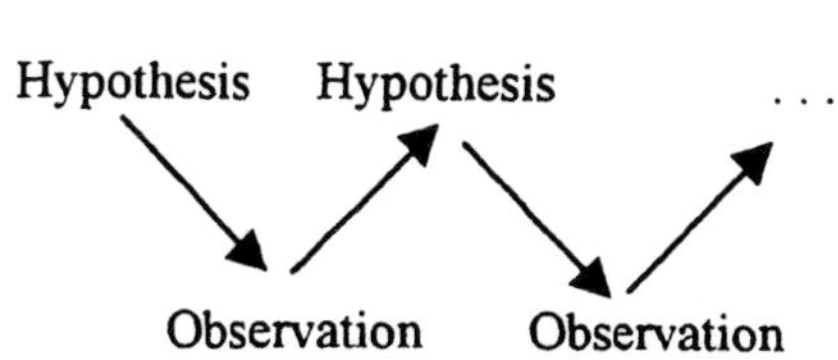

Figure 3. The Hypothetico-Deductive Account

not. For example, from a hypothesis about Halley's Comet returning at a particular time, one predicts that one should be able to see it through a telescope at a certain time of the year.

The strength of the hypothetico-deductive method is its ability to deal with theoretical hypotheses about unobservable entities and to test for their existence and nature indirectly by means of testing their implications. Although there are many critics of simple and even refined versions of hypothetico-deductivism, the basic insight of the hypothetico-deductive method is acceptable and we will incorporate it into our account of the scientific method.

Inductive-Deductivism

A final account of the nature of scientific method is an obvious modification of inductivism. According to inductivism, observations lead to hypotheses. But this is not the end of the matter since, as most inductivists have maintained, these hypotheses must then be tested by deducing test-implications from them (just as the hypothetico-deductive method suggests) and then either observing the prediction or not. See Figure 4.

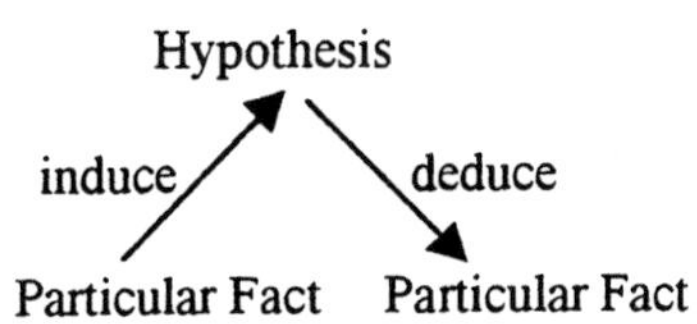

Figure 4. Inductive-Deductivism

If the predicted fact occurs, our hypothesis is confirmed and if it does not, our hypothesis is disconfirmed. In either case, our hypothesis (or modified hypothesis) is subject to additional empirical tests. In the meantime, we may have collected additional facts, which lead us to induce something hypothetical. We have, therefore, another example of a spiral:

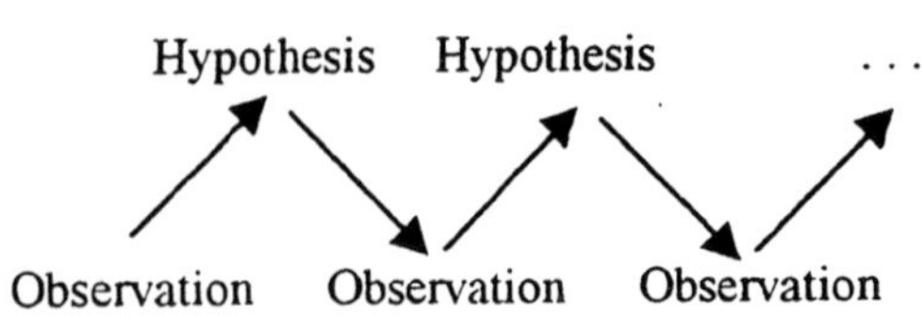

If one compares the two spirals, they appear to be very similar, with the crucial difference being the starting point: Was the initial hypothesis preceded by an observation or was the initial observation preceded by a hypothesis?

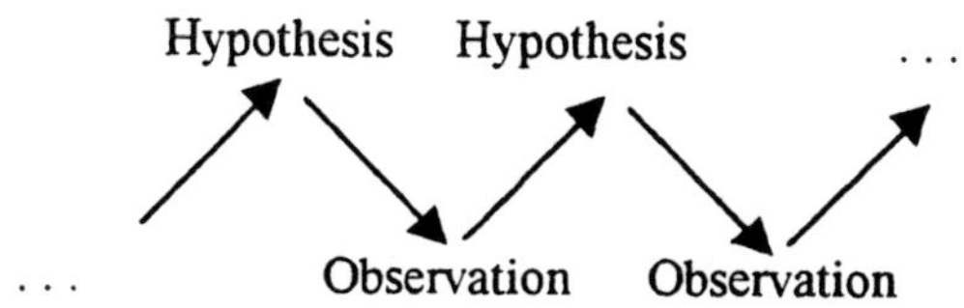

Suppose the scientific method has the characteristics of such a spiral proceeding in both directions. Then there is no terminus to the conduct of inquiry: hypothesis--observation--hypothesis--observation... But what about the beginning point?, where does the conduct of inquiry begin?

Can it go on forever backward, in a temporal sense? The answer is obviously no. What then is the actual starting point? On this question, the two sides are at loggerheads. The hypothetico-deductivists (e.g., Popper) claim that every observation is preceded by a theoretical conjecture; and as far back as we go in the history of the individual scientist, one always finds that a hypothesis precedes an observation. But how far back can we go with this line of reasoning? What about the newborn infant, who comes into the world and begins to observe things? According to Popper, we must conjecture that even with the infant's very first observation, it was preceded by a theory, in this case an According to Popper, therefore, we are born with certain theories.

Is this view correct? This is obviously an empirical question to be settled by scientists, in this case, developmental psychologists. What does the evidence show? Here the scientific answer is unclear and there are at least three schools of thought on this matter: Several developmental psychologists maintain that all concepts and hypotheses are acquired from experience and that nativism is false. Next, there are neo-nativists, who champion the notion that children are born with rich cognitive structures and "theories." A third school maintains that initial concepts/hypotheses are not innate but result from experience, which in turn is structured and directed by "ur-concepts" (theoretical precursors of concepts and hypotheses), with which we are born but which change and develop into regular concepts and hypotheses. The empirical evidence on this question remains unsettled.

The issue may be partly an empirical one about the temporal sequence in question, but it is clearly something more than this, something more epistemological (or logical). The hypothetico-deductivists might be claiming that every observation logically or epistemologically presupposes a hypothesis, since without a hypothesis it would be impossible to make an initial observation. There are reasons for thinking that such a claim is plausible, but before it can be evaluated, it must be made clear what a 'hypothesis' is supposed to mean. If it means something as generic as "something is connected to something else," a Columbus-hypothesis such as "if I sail west, I'll find land," or a physiological hypothesis (the act of eating presupposes the hypothesis that one has a stomach), then the notion of a hypothesis is being trivialized to the point of non-recognition. Certainly, hypotheses such as these are rarely the stuff of which scientific research is made. It also invites the inductivist retort: the hypothesis that "if I sail west, I'll find land" came from earlier experiences of sailing, just as one learns that something is connected to something else by virtue of a correlation between these two kinds of things. And so the issue remains open.

A third issue concerns the question of a **logic of discovery**: is the creation of a hypothesis something that is miraculous and incapable of explanation--an act of artistic intuition--or is there a logic to how hypotheses originate? If there is a "logic" to the discovery of hypotheses, is the underlying set of rules for generating a hypothesis mechanical in nature, something that a computer could follow--technically known as *algorithms*--or are there antecedent conditions that foster the development of plausible hypotheses but do not guarantee them—something known as a *heuristic*? The standard interpretation of hypothetico-deductivists such as Popper and Hempel is that there is no logic of discovery in the sense of algorithms; and if there were such a set of heuristics, such an inquiry would belong to psychology, not logic. The inductivist-deductivist, on the contrary, claims that in order to come up with an interesting and plausible hypothesis, it is first advisable to collect observational or experimental data and then generalize from these to a hypothesis. For after all, it is extremely unlikely on empirical grounds that someone who didn't know the relevant observational data could come up with a sensible hypothesis to explain them. If it were a matter of poetic insight and intuition, a layperson should be able to create as many plausible hypotheses as a trained scientist could, which is obviously false. Hence, the inductivist advice: before coming up with a hypothesis, collect the relevant observational data. Of course, the inductivist must then give an account of what 'relevant' means. And so the debate continues....

What are we to say of this debate? How are we to decide which model of scientific method is preferable? Assuming that the methodological spiral is correct as an account of the process of scientific inquiry, the answer seems to be that it is partly a matter of convention where one breaks into the spiral, stops the process of inquiry, and gives a description of the relevant set of procedures. The spiral is consistent either with hypothetico-deductivism or with inductive-deductivism. However, even though it is a matter of convention, it is not an arbitrary one. On balance, the arguments favoring the inductivist-deductivist views seem (to the author) to be more compelling than those favoring the hypothetico-deductivist's. But more than that, there are useful pedagogical reasons for adopting the inductive-deductive model, and such pedagogical reasons will motivate the remaining structure of the book. The inductive-deductive model will, thus, provide a useful (heuristic) model for organizing the various procedures of science into a coherent whole.

Following the inductivist-deductivist model, we will begin with a discussion of the nature of scientific observation (Chapter 7), together with related topics concerning the nature of data, evidence, and facts. Then we will discuss various procedural steps leading from the collection of data to the generation of scientific hypotheses; in particular, we will discuss the procedures useful for inferring a hypothesis from the data. This will include the topics of classification of data, the nature of definition, quantitative induction and measurement (Chapters 8 and 9). After this, we will proceed to discuss two traditional inductive procedures for discovering hypotheses--Mills' methods of experimental inquiry (Chapter 10) and the role of analogy and models in the scientific enterprise (Chapter 11). Next, we will discuss the nature of hypotheses and hypothesis-testing--the heart of the hypothetico-deductive method (Chapter 12), followed by an account of the nature of the experimental method (Chapter 13) and the nature of statistical inference as it relates to the experimental method (Chapter 14).

For Further Reading

For an excellent discussion of formal sciences, see Chapter 7 of Cohen and Nagel (1934). Few individuals have championed deductivism, a notable exception being perhaps Descartes. But some individuals are now disputing even this. Passages in Popper also suggest a commitment to deductivism. For the clearest area where deductivism is prevalent, one should consult the construction and use of mathematical models in science, e.g. Freudenthal (1961). For a discussion of the axiomatic method of Euclid see Losee (1993, ch. 3).

For a discussion of whether computer science is a formal science or not, see Newell & Simon (1976/1981).

It is difficult to find anyone championing a radical inductivism, although individuals often cite Bacon and Mill as proponents of such a view. Ernst Mach and Karl Pearson are also sometimes mentioned.

There have been several criticisms of all forms of inductivism by Popper (1965) and certain naive forms by Hempel (1966). Popper argues there is no such thing as induction and scientists never use it. Peter Medawar, a Nobel laureate scientist strongly influenced by Popper, attacks inductivism in his (1969).

The hypothetico-deductive method, at least one widely accepted version of it, is discussed in Chapter 12. Standard defenses of this model include, besides the earlier account of Whewell (1989), Jevons (1905), and Dewey (1938), the more recent writings of Hempel (1966)and Popper (1934/1965). Hypothetico-deductivists disagree over several fundamental issues. There are several extant criticisms of this model in the literature, an influential example of which is Glymour (1980), which is somewhat technical.

There are few defenders of the inductive-deductive model, at least among 20th century thinkers. Aristotle seems to hold something like this view and so did Sir Francis Bacon (1620/1980) and John Stuart Mill (1872)--a convenient selection of whose work is Nagel (1961). In the twentieth century, this view has been endorsed by a variety of people, e.g., the logician Kemeny (1959), the physicist Frank (1956), and the psychologist Cattell (1966), whose idea of a spiral I have adopted. A recent defense of a modified form of inductivism can be found in the influential Hacking (1983, ch. 9). The number of defenders of inductivism seems to be increasing.

Recent discussions of a "logic of discovery" are extensive. A good introduction can be found in Hung (1997, ch. 5). A good anthology is Nickles (1980); see especially his introduction. On the question of whether one can write computer programs to make scientific discoveries see, first, Hung (1997, ch. 32) and then one of the several recent books by Langley, et al., (1987), Holland, et al., (1986).

Notes

[1]Here a set of related questions and issues arise, e.g., the notion of an *effective decision procedure*. Fortunately we can ignore these questions.

7

Scientific Observation

Introduction

What is the first step in the scientific method? According to many individuals, it is observation. Before we can begin the work of discovering scientific laws, determining causal connections, and constructing hypotheses, we must first acquire knowledge of the empirical world; and this consists of making empirical observation, collecting data, discovering the relevant facts, in short, acquiring empirical evidence.

This is a widespread view. A biologist, for example, writes: "All science begins with observation, the first step of the scientific method" (Weisz, 1959, p. 4). In a chemistry textbook it is stated:

> Science starts with observation. *Observed facts*, be they qualitative or quantitative, are called *data*. . .Over the years the data accumulated has [sic] reached staggering proportion. (Sienko & Plane, 1961, p. 13)

A more recent chemistry textbook repeats this point:

> [In "the scientific method"] once the goal has been defined the next step involves performing experiments, making careful observations, and collecting bits of information about the system. The bits of information are called "data". (Chang, 1984, pp. 2-3)

Finally, a psychology textbook has this to say about "Scientific Methods in Psychology":

> Psychology as a science is first of all *empirical*. That is to say, data are gathered by experiment and observation; no reliance is placed on intuition, opinion, or belief. . . What make his science secure are the unarguable facts, the observations that he has made. . . Through research , psychology has already acquired a wealth of facts, and it grows richer every day. . . .
>
> . . .We make observations wherever we can, often without too much rhyme or reason to them, and then, after careful analysis, we attempt to formulate tentative principles that we think pretty well summarize our observations. (Morgan & King, 1975, p.14)

Comments such as these represent the *standard* or received interpretation of the first step in the scientific method. Although, as we will see, there is an important kernel of truth in this account of the acquisition of scientific knowledge, it tends to be somewhat misleading, especially in its interpretation of the nature of scientific observation. Although observation may well be the first step in the scientific method, the really important question concerns its epistemological nature. Historically, there have been two competing views concerning this question, the standard (received) view (prevalent throughout much of the history of philosophy) and the "theory-laden" view of observation (present in modern times but most forcibly argued in the twentieth century).

The Received View

Observation, Fact, Data, and Evidence

The concepts *observation*, *fact*, *data*, and *evidence* are closely related. The term *datum*, for example (whose plural form is *data*), comes from the Latin *datus* which means "giving." A datum, therefore, is something that is given, that is, something that is there, independent of the individual and separate from any interpretation. It is something which "gives itself" to the observer. More importantly, it is the basis of all subsequent reasoning. The term *data* in science thus typically has the sense of *factual information used as a basis for scientific reasoning and evaluation.*

This concept of *data* is closely related to another correlative notion, that of *evidence*. As everyone admits, scientific evidence is crucial to the scientific enterprise; according to many, it is the presence of such objective evidence that distinguishes science from non-science. Etymologically, *evidence* constitutes the proof of something and is thus related to the notion of something's being *evident*—something, which is clear, plain, and obvious to the understanding. The evident, therefore, is that which shines forth in a kind of direct immediacy. It is because of this that individuals often claim that evidence eliminates any doubt or uncertainty.

In addition to data and evidence, scientific *facts* also have an honorific status. *Fact* comes from the Latin *factum*, which means a deed or act. It is that which is *actual*, or true. Hence, a fact is *an item of information having objective reality*.

Finally, data, evidence, and facts are related to a fourth notion, that of *observation*: On the traditional view, one obtains facts, data, and evidence by means of observation, by observing something through one's senses. In so doing, one observes what is there independently of the observer's subjective wishes. This datum or fact, because it is clear and obvious, constitutes evidence, which in turn functions as the grounds for one's hypotheses.

The received view concerning the role of observation in scientific method is a view firmly entrenched in the standard way of thinking about scientific method. In fact, it can be said to form a fundamental part of what can be called *scientific common sense*. The received view seems obvious to many scientists, to be without need of justification. The underlying question, however, is whether this interpretation is correct or not. Although the received view may be correct about the essential role of observation in scientific method, serious conceptual problems remain concerning its nature and status. In order to indicate more clearly what these conceptual questions are, we must briefly review the rationale underlying the received view.

The Baconian Tradition

According to most individuals, modern science began in the 17th Century. It was during this period that the scientific method was first clearly practiced and formulated; and it was in this period that individuals began to deliberate about the precise nature of this scientific method. During this period, therefore, there was "a search for method"--a search for an adequate account of the nature of the scientific method. One of the key individuals here was **Sir Francis Bacon**, whose work *Novum Organum* (Bacon, 1620/1985) many take to be the first systematic account of the scientific method.

Bacon had a negative view concerning most of the history of thought prior to the 17th Century, for in this period, Bacon believed, nothing could really be counted as knowledge, primarily because individuals were following the wrong method of knowing. The logic they were using was deductive logic, in particular, the deductive logic of the

Aristotelian. The problem with this method was that individuals would begin with very general principles and then deduce other principles, thus, proceeding in a purely deductive way. As we have seen, deductive logic guarantees the validity of an argument, but it does not address the issue of whether the original premises are true (from what source their truth comes); this is precisely the issue about which Bacon was concerned.

Bacon is probably most famous in the history of thought for his four **idols of the human mind.** It is because the mind is prone to worship such idols that it is incapable of arriving at true general principles. Bacon's program, therefore, was to point out the influence of such idols and to suggest a different method. This new method was that of inductive logic.

Because the human mind is constituted in a particular and fallible way, there are several ways in which it tends to be misled or deceived. These idols of the mind--idols of the *tribe*, idols of the *cave*, idols of the *market place,* and idols of the *theater*--distract us from true knowledge. In order to purge the human mind of these idols, Bacon advanced a kind of epistemological therapy. Instead of allowing the human mind to give into its natural bent--of premature speculation, unchecked theorizing, etc.--he advocated a different approach: begin by putting aside these natural tendencies toward theorizing and instead *observe nature in a systematic way.* Once purged of these idols, the human mind could then see nature as it truly was. Such observation would provide one with the necessary data, facts, and evidence, which would constitute the bedrock of all knowledge.

One begins, Bacon claims, by making observations of *particular* things. After careful and controlled observations, recorded in a series of tables--of presence, absence, degrees, etc.,[1]--one then proceeds to induce what he called a middle axiom or generalization. This act of induction was to be carried out in a slow and cautious way, with due regard being given to the constraints and evidence set out in the tables. Later, from such middle axioms one induced more general principles, but again in a controlled and systematic way. Thus, the picture that we get of Bacon's scientific method is the following (Figure 1). It is for this reason that Bacon is often taken (wrongly) to be the chief exponent of **inductivism**, the view that one should only use inductive logic in science.[2]

General Principles
↑
Middle Axioms
↑
Particular Observations

Figure 7.1. Baconian Induction

There are several kinds of inductivism. According to narrow inductivism, scientific method consists of the following steps: first, collect all the brute facts (free of theoretical interpretation); second, from these particular facts induce a generalization; finally, from these generalizations a theory can be induced. According to this kind of inductivism, scientific method consists solely of inductive logic, inferences proceeding from particular facts to more general laws and theories characterized by induction by simple enumeration. As many people have argued, including Bacon himself, this narrow version of inductivism is inadequate and must be supplemented with additional principles of (inductive) inference, e.g., eliminative induction. Hence, narrow inductivism has had few champions among scientists and philosophers of science.

The Empiricist Tradition

Sir Francis Bacon's espousal of inductivism provides one pillar of the received view—its logic. The other pillar is that of **empiricism** (see Chapter 2). Empiricism is a theory of knowledge advocated by several individuals, most notably John Locke (1632-1704), Bishop George Berkeley (1685-1753) and David Hume (1711-1776). "Empiricism" is a term coming from the Greek *empeiria*, which means "experience.". Empiricism is the view that knowledge comes from (sense) experience.

One influential version of empiricism that has been very influential is **radical empiricism.** According to it, since everything one knows comes from sense experience, all knowledge is reducible to sense-experience. "If you can't observe it, then you can't know it; if you can't know it, then it doesn't exist" might be a capsule summary of radical empiricism. Hence, empiricism is the epistemological basis of the received view.

The epistemological motivation behind radical empiricism is clear--it is designed to avoid **skepticism**. The epistemological skeptic is one who believes that *nothing can be known.* According to the skeptic, to have knowledge, there can be no doubt about its truth. But everything, in principle, can be doubted. But if so, then no knowledge is possible.

Empiricism is an attempt to answer skepticism and thus to show that knowledge is possible. Radical empiricism is one particular interpretation of empiricism. Agreeing with skepticism that knowledge must be absolutely certain, the radical empiricist believes she has found something that is indubitable—one's immediate sense-impressions. As one modern advocate of this position puts it:

> When I see a tomato there is much that I can doubt. I can doubt whether it is a tomato that I am seeing, and not a cleverly painted piece of wax. I can doubt whether there is any material thing there at all. Perhaps what I took for a tomato was really a reflection; perhaps I am even the victim of some hallucination. One thing however I cannot doubt: that there exists a red patch of a round and somewhat bulgy shape, standing out from a background of other color-patches, and having a certain visual depth, and that this whole field of color is directly present to my consciousness. What the red patch is, whether a substance, or a state of a substance, or an event, whether it is physical or psychical or neither, are questions that we may doubt about. Whether the something persists even for a moment before and after it is present to my consciousness, whether other minds can be conscious of it as well as I, may be doubted. But that it now *exists*, and that *I* am conscious of it--by me at least who is conscious of it this cannot possibly by doubted. And when I say that it is "directly" present to my consciousness, I mean that my consciousness of it is not reached by inference, nor by any other intellectual process. (Price, 1933, p. 3)

Since, according to radical empiricism, our sense-impressions are absolutely certain, knowledge is possible, and this knowledge will consist of our sensations.

For the radical empiricist, it is scientific observation that is thus certain. Anything over and above this, in particular one's interpretation of an observation, is uncertain. It is uncertain because one can always doubt one's interpretation. Whenever one interprets something, one is making a *judgment* about it and judgments are all liable to error. By contrast, a sensation is not an interpretation since it reports what is immediately there. Hence, knowledge consists of sensations, and all judgments should be avoided.

It is easy to see that narrow inductivism goes along with radical empiricism, for the facts which one initially collects are simple and certain observations. The ground floor for knowledge consists of particular facts. The only way to proceed beyond these data is by means of inductive inference. Inductivists have argued that induction can be safe and certain; for since the particular steps involved in induction are (or should be) gradual and cautious, there will be no jumping to conclusions but rather a steady progression from fact to fact to fact.

The role of *theory* in this entire account has an ambiguous role. On the one hand, theory comes *at the very end* and is not presupposed anywhere along the way. Hence, a theory will be something relatively certain. On the other hand, if we ask what particular *function* a theory plays, the answer can only be: "not very much," for according to narrow inductivism, *a theory is really just a convenient summary of a set of laws* (and contains no new terms beyond those found in the laws). Since a law is itself just a convenient summary of a set of facts, a theory ultimately will be just a convenient summary of observational data; as one individual (Pearson, 1911) put it, theories are just "conceptual shorthand" for taking down in brief and succinct form the longhand writing of experience.

Evaluation of the Received View

What can be said about the adequacy of the received view? Sometimes, scientists seem to be following procedures of data collection followed by inductive inference. For example, it seems to be characteristic of some of the kinds of research that scientists do in their laboratories and observatories, examples of which can be found in the works of Galileo, Boyle, Kepler, Harvey, Delong and Petit, Bode, and others. But as a description of how scientists *always* proceed and as an account of how they *ought* to proceed, this account is inadequate. Much of science does not seem to fit this model.

Furthermore, scientific method does not consist of passive, undirected observation "without rhyme or reason," coupled with simple induction. This is because there are very few brute facts; there are very few, if any, observations free of all theory and interpretation. It is true, of course, that some facts are relatively brute and free of interpretation; some scientific observations are more theory-free than others. But the notion that all observation is *completely* free of all theory (hypothesis, conceptualization, etc.) is a remnant of the past that no longer seems defensible. This does not mean that scientists should not be impartial and objective in their observation; clearly, they should resist jumping to conclusions not warranted by the data. But this does not mean that scientists should be passive blank tablets either.

Observations are Theory-Laden

The received view concerning the nature of observation is correct in giving an essential role to observation; however, it has an inadequate account of the nature of scientific observation.

According to several critics of the received view, observation is not free of theory, but rather *observation is theory-laden.* In fact, scientific observation involves theoretical and conceptual components from the very beginning, and such conceptual scaffolding cannot be eliminated.

By a *theory* defenders of this alternative view do not (necessarily) mean a full-fledged, explicit, formalized theory. What they do mean is not always clear, but they seem to use ‘theory’ to mean a variety of things including any general idea or hunch about what features of experience go together, what factors are explanatory of our experience, etc. These types of theories are clearly very generic and primitive, ones of which one may not even be aware. Perhaps a better term to use here would be *hypothesis, hunch*, or *expectation*; however, because such prior conceptions have much in common with the more developed entities called 'theories,' it may not be too misleading to refer to these expectations as theories.

Direct versus Indirect Observation

In most discussions of the nature of observation, it is customary to distinguish direct observation from indirect observation. In *direct* observation, one is observing something merely with one's senses, without the aid of instruments. Here, interpretation is at a minimum and no obvious theory seems to be present, e.g., a subject in a psychological experiment may be placed in a dark room and told to observe a point of light in the distance, or a scientist may observe that a pointer moves from '2' to '3', that a chicken pecks at a kernel of corn, that a rat presses a bar, that a piece of litmus paper turns red, that the bird before you is albino, etc.

In these cases, one is using one's senses to a maximum extent and the contribution of theory is minimal. This is not to say it is absolutely absent but only that it is minimally present.

By contrast, we usually call *indirect* observation those cases in which instruments are involved. All of these cases involve a great deal of interpretation. Thus, a scientist may report that she is observing an electron in a cloud chamber (as the movement of a dust particle is recorded), or that her subject is having a dream (because the subject is manifesting Rapid Eye Movements). These are cases of indirect observation because the observation is mediated by a third factor besides the object and the observer--the instrument or theory. In order to observe certain kinds of entities or events, indirect observation is thus necessary.

Instruments

We can distinguish at least four kinds of instruments: (l) *observing* instruments (e.g., microscope, telescope), (2) *measuring* instruments (e.g., balance beam, clock, carbon-14 dating, Geiger counter, photometer, galvanometer), (3) *recording* instruments (e.g., cathode ray tube, oscilloscope, photographs, liquid scintillation counter, Wilson cloud chambers, EEG recordings), and (4) *calculating* machines and computers.

Passing over the second and fourth cases, we can say that when one is employing an observing instrument such as a microscope, there is a direct extension of one's observational powers. These give one access to hitherto unseen realms--the microcosm and the macrocosm. Just as eyeglasses allow one to see things one could not see before, so observing instruments act like magnifying glasses. In this case, we see objects directly, since we are seeing these very objects and not their effects. In the case of recording instruments, however, we are not seeing the objects themselves but rather their effects. In these cases, we often say we are indirectly seeing the object--their cause. Thus we are indirectly observing an electron if we observe its tracks in a cloud chamber.

Some individuals have argued that whenever we use an instrument to observe something, we are presupposing a theory, e.g., a theoretical account of the nature of the instrument and how it works. Such an account is ordinarily a causal account, since the entities in question act on the instrument, causing us to have a certain percept by means of a long chain of events. These theories of instruments explain the process of observation and give an account of when observations are veridical and when they are not. They indicate what the nature of the object is (including its causal powers), what intervening causal mechanisms result in an observation, what the nature of this observation is, etc. Such theories of instruments may include a theory of matter, optics, physiology, electricity, psychology, etc.

Observations are "Theory-Laden"

A theory has several functions, but in this context one of its main functions is to *direct* the observer's attention to what is to be observed and to *select* certain features. A theory often allows one to interpret (to make sense out of) what one sees by first forming a meaningful whole into which sensations are placed. Theories are thus essential for observation. In fact, according to some individuals, all observation is "theory-laden" in the sense that a theory (partially) constitutes the very object of awareness. It is, therefore, misleading (according to them) to separate the underlying sensation from the theoretical interpretation. To illustrate what these individuals have in mind, consider the following figure (Figure 7.2).

If one were asked what one saw in this figure, one would not answer: a patch of ink. Some individuals might see this, but that is because they have not yet seen what is there to be seen--the bearded figure. Once one sees the bearded "Jesus," he can no longer see the drawing as a series of meaningless marks on paper; instead, the figure appears immediately and involuntarily it leaps out at you. Likewise, if one looks at Figure 7.3, it is easy to see two figures here: a "young woman" and an "old woman." As one's attention shifts from one part of the drawing to another, one at first sees one kind of woman and then another kind.

The point of these Gestalt figures is to illustrate the claim that observations are always laden with theory. To make an observation, it is not enough to be aware of a sensation (the set of ink-marks); one must be aware of these *as* a certain kind of thing. It is the underlying concept that is the theoretical ingredient contributed by the observer. What one literally can observe will depend upon one's prior expectation, training, mental set, cognitive equipment, in short, one's entire conceptual scheme. Lacking this cognitive structure, one will not be able to see what others can see. Hence, what one sees depends upon the conceptual contribution of the observer.

Similarly, in the case of scientific observation, what a scientist observes depends upon what conceptual resources, categories, interpretations, expectations, she brings with her. These theoretical elements determine if and what she sees. Hence, two scientific observers can look at the "same" data and see different things just as two individuals can look at the lines in Figure 7.3 and see different women.

Figure 7.2. The "Jesus" Figure

Figure 7.3 The "Old Woman-Young Woman" Figure

To illustrate how scientific background and training are necessary to scientific observation, consider Figure 7.4. What does one see? Many observers will see nothing meaningful here at all but rather just a jumble of lines. Other individuals, who have the suitable training and

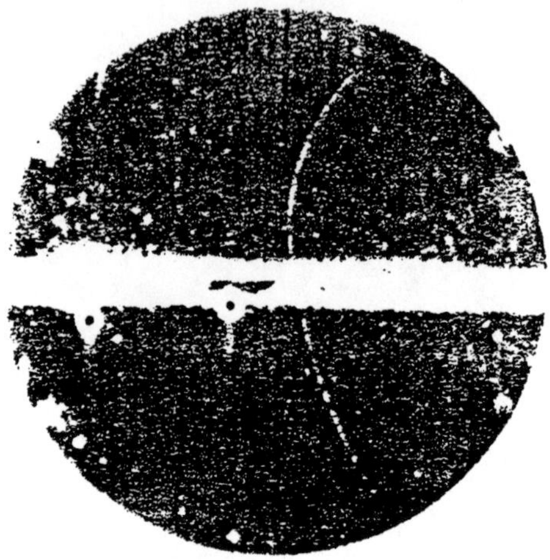

Figure 7.4. Cloud chamber track of a positron moving in a magnetic field directed toward the reader. The positron originated in the lower right-hand part of the cloud chamber, passed through a lead plate of thickness 6mm, and then continued on with decreased speed. C. D. Anderson took this photograph, marking the discovery of the positron, in 1932.[3]

theoretical knowledge, will see the track of a particle in a Bubble cloud chamber. What you can see, therefore, will depend upon one's prior theoretical knowledge and training. But if so, then scientific observation cannot merely be opening one's eyes and recording sense data. Instead, it requires a great deal of prior knowledge, theory, and training.

On this account, therefore, scientific data can no longer be seen as an infallible, absolutely certain basis for all knowledge. If observations are theory-laden and if some theories are false, then what we observe and what we record may not really exist. As theories change over time, reported data may be rejected since the objects reported do not exist (from a subsequent point of view). Later theories may result in reinterpretation and correction (even overthrowing) of data. But if so, then data cannot be held to be epistemologically sacred and inviolable.

Exercise 7.1

1. According to Bunge (1967), the only directly measurable magnitudes in astronomy are star brightness and the angle subtended at the telescope by two celestial objects. All other magnitudes, in particular distances and masses, are inferred with the help of theories and the above-mentioned angle data. Can astronomy accordingly be regarded as an *observational science*?

2. Examine the following items. Try to evaluate (1) the accuracy, (2) the reliability, and (3) the importance of the observed facts. To what extent does theory enter into these observations?

a. On September 11, 1942, at 23h, a new comet was discovered. Its position at the time of discovery was R. A. (right ascension) 23h 51m, Dec. (declination) + $6^{o}20'$, and its magnitude was 13. The same comet was observed again on November 6, 1942 at 20h 16m. Its position at this time was R. A. 4h 11m, Dec. $-0^{o}47'$. Its magnitude had not altered. The comet was moving slowly northward. When a third position is observed, the orbit of the comet will be computed.

b. The larva of the silver-spotted skipper butterfly, in building a house, fastens several leaves together with strands of silk. It generally remains in its house during the day, and comes out to feed during the night. As the caterpillar molts and grows bigger it leaves its old home and constructs a new and larger one. This is shown by the fact that many of these peculiar dwellings, among locust leaves where the caterpillar usually lives, are empty when examined. Whether or not the house is occupied, it is always as scrupulously clean as that of a Dutch housewife.

--Science News Letter, 1946

c. There, in the shadow of the leaves, was the praying mantis. Even to me it was a rather awesome-looking insect, with those powerful, spine-armed forelegs held in an attitude of seeming prayer; to its insect victims it must have been a veritable death-ogre. Its operation, as I observed it that autumn afternoon, was direct and simple--and a bit spine chilling. The mantis waited motionless while her unsuspecting prey--a grasshopper--wandered within range. Then those tongs of doom snapped on it like a trap, and then despite all struggles of the helpless victim the mantis proceeded to gnaw deliberately into its vitals. I began to feel very sorry for the grasshopper, even though I knew that the 'hopper was my enemy and the mantis my ally.

d. I still had the stale taste in my mouth. I felt a severe pain in the back of my neck, and a strange, throbbing noise in my ears. My right leg seemed paralyzed and, in the darkness, I felt that both of my hands were securely tied. It was no enviable position to be in.

3. When you look through an electron microscope, you are engaging in "indirect observation." When you are wearing glasses, is this indirect observation? What about looking through a windowpane?

Facts, Evidence, Data and Observation

As we have seen, the received view concerning the nature of scientific observation is correct about the importance of observation but wrong about their epistemic certainty. As the theory-laden view maintains, observations are fallible, and this is partly because they involve a conceptual framework (theory) for making sense out of them.

But does this mean, as some followers of the theory-laden interpretation claim, that "everything is a matter of theory," that everything provided by the observer in the way of conceptual integration and scaffolding is a subjective bias, that there is no objectivity in science? Certainly not. In looking at the Gestalt figures, for example, one can see the lines *as* a young woman or *as* an old woman, but one can't see the lines in any way one wants. What you see may be variable to a certain degree, but there is not infinite plasticity here; on the contrary, sensory observation places firm constraints on one's perception. It is thus a mistake to claim that all theories are biases; for if they were biases, we would not be able to determine that they are biases. Hence, scientific objectivity need not be thrown out. On the contrary, throughout the course of the history of science we have learned that certain theories, methods, assumptions are inadequate; and we have learned this because the empirical facts we have collected have led us to these conclusions. But, if this is so, then what account of these notions--observation, fact, evidence and data--should be given?

Data, we will suggest, are the end-results or outcomes of observation. Typically, they will be couched in the form of propositions claiming that something has a certain property (e.g., "the litmus paper turned red," "subject S scored 120 on the IQ test," "the averaged score of the experimental group was 120," etc.) Thus, data can consist of things like reported scores on a test, injections of a chemical into a subject, a report of the swimming behavior of fish, etc. Scientific observation is a process and data are the products of the activity. Facts, by contrast, are those things (states of affairs) that exist independently of the observer, those features of reality whose properties are formulated in a proposition. We can, thus, express the general relation between these the following way:

Facts → *Observation* → *Data*

Each of these concepts need to be discussed. We will begin with facts, which are the most basic, and then turn to observation and data.

Facts

A fact is an objective feature of reality. When a proposition is true, what the proposition states is a fact. What makes the proposition true is a certain state of affairs, a particular feature of the world. Sometimes 'fact' is used in a slightly different way, as "any proposition that is true" as in "it's true *that*. . ." A proposition typically has a that-clause attached to it. When we say, for example, "it's a fact that 3 is a prime number," we are expressing a proposition in the form of words. A proposition, as we have seen, is a logical entity, which is the content of linguistic statements. Although it is not the same as a linguistic expression, it seems to contain something like a linguistic or conceptual medium. Since propositions and reality should be distinguished (propositions are true because of the nature of reality). Facts in the sense of propositions, however, are derivative, since they depend upon facts in the primary sense, namely, facts as features of an independent reality.

This use of 'fact' is based upon a widespread assumption about the nature of truth, the ***correspondence theory of truth*** (see Chapter 2). According to this theory, a proposition is true if and only if it corresponds to reality. If a proposition adequately describes a feature of reality, then the proposition is true. There are, therefore, two important ways we speak about facts. Facts are features of reality and facts are propositions describing this state of affairs.

Particular Facts versus General Facts. Facts are something contrasted with laws (and generalizations), in the sense that facts are *particular*, whereas laws are *general*. For example, a particular fact might be an observation made by an individual at a particular time and place (e.g., at 3:10 p.m. on November 12, 1943 astronomer S observed planet P).

When individuals speak of "collecting facts", they seem to have this sense of fact in mind. Indeed, this is precisely what the inductivist tradition recommends: go out and collect particular facts about individuals. Later, one may come up with a generalization covering these particular cases such as: "all items of gold have a melting point of $x^{o}C$," or "all examples of homo faber became extinct 1 million years ago." This is an additional sense of fact, since here we are using "fact" not in the sense of a particular item of knowledge but in the sense of a generalization. Indeed, this seems to be the meaning of statements such as: "it's a fact that water boils at $100^{o}C$," "it's a fact that whales are mammals."

Are Facts Discovered or Created? If facts are (primarily) things that are independently real, it makes sense to say we discover them, that we come to know things we didn't know before. This does not necessarily mean that the inductivist and the empiricist are correct, or that facts are a-theoretical. Indeed, in order to discover a fact, a great deal of effort is ordinarily required, an activity that involves language, theory, social conventions, interactions with others, etc. In this process there may be a great deal of activity on our part, but this is compatible with the existence of real facts.

Recently, there has arisen a school of thought that has challenged this view. It is sometimes called *(social) constructivism.* A central thesis of this school is the claim that "scientific facts are constructed." Rejecting the received view concerning the nature of scientific method and the nature of facts, these individuals have been impressed by the key role scientific theories, beliefs, and class interests play in the scientific process. They have also stressed the inescapable social dimensions of science. Together, these two elements have led to the notion that in the process of scientific research, individuals "construct" scientific facts (and reality). By this they seem to mean not just that propositional facts and epistemic facts are constructed but that ontological facts are constructed, that in some sense "we make up reality."

What they mean by all of this is not clear. If they mean that our knowledge of scientific facts is the end product of a constructive activity on our part, this seems to be correct. Knowledge is a creative and constructive activity. Furthermore, our description of a fact is a social creation since, e.g., there are a variety of ways of describing it. It also seems true that the object of knowledge is also (partly) our construction since what is known typically involves ideas, concepts, theories, etc., which are our construction.

But having admitted this, it does not follow that an ontological fact is a construction on our part, that we make up facts in some kind of collaborative act of fantasy. Although we are free to construct any description or representation we wish, it does not follow that these will all be true. What we know is determined by those features of reality that require us to abandon and modify our theories. It is this same feature of reality that is the core sense of "fact." If reality had no role to play in the determination of facts, we would be free to make up reality any way we wish. This would be a kind of metaphysical **idealism**, the view that nothing real exists except for ideas in the mind of the individual.

Observation

In one clear sense, facts are features of an independent reality, those things against which we test our theories. The process of discovering these facts occurs through the process of observation, turning our sense organs onto the world, receiving information from the world, and recording the results. This process of observation is a complex and highly technical skill, not the simple kind of thing empiricists and inductivists thought it was.

Naive vs. Critical Observation. According to the empiricist interpretation, observation can be characterized as being passive, nondirected, and involving no interpretation. Hence, observation is passive. In order to observe something, all you need to do is open your eyes and let the sensations flow in. Crudely put, this is a kind of sponge model of perception, where the mind simply soaks up perceptions.

If the mind is passively filled with information, clearly it is not actively involved in the process of knowing. On this account, when the mind perceives something, it does not select or attend to incoming information; instead, whatever flows into it is recorded.

To say that perception involves no interpretation is to say that perception involves the recording of incoming sensations as they appear.

But scientific observation typically is not of things like "red squares" and "patches of yellow" (naive observation). Perhaps at the very beginning stages of scientific observation, when a scientist does not know *what* he is observing, this is true. But it is not the way we ordinarily observe things, nor is it the way scientists typically observe things. Scientific observation is ordinarily a much more complicated affair, not a matter of naive observation, but of *critical observation.* Unlike naive observation, it is active, directed, and critical.

When a scientist sets about to observe some phenomenon (or to conduct an experiment), she does not ordinarily passively wait for the phenomenon to appear.[4] Rarely (if ever) is scientific observation a passive affair, one in which the observer sits by and lets something happen to her, recording everything that she experiences.

Furthermore, in the very act of observing something, observation is a highly trained skill, mediated by numerous cognitive processes. To be able to make a scientific observation, one must first have the necessary background, training and experience, and part of this training consists in mastering a set of techniques and a complex body of background theory.

If scientific observation is an active process, it is also one that is directed. In observing something, one has to focus on something and single it out from its background. If a scientist is looking into an electron microscope, he will neglect or ignore everything else--the ticking of the clock, the telephone--and select only those percepts assumed to be relevant. As Mario Bunge (1962) puts it:

> The amateur naturalist will record heaps of minute facts which the field biologist will neglect; the latter will on the other hand record a few essential

> observations, many of which will have escaped the untrained watcher for want of a definite aim marked off by a conceptual system. (p. 163)

From a welter of sensations one must *select* which ones to observe. In this process of perception, something becomes the *figure* and something remains in the *background*. Suppose, for example, someone gives you the following command: "Observe!" The obvious answer is: "Observe *what?*" There are an indefinite number of things one can observe. Which of these should I observe? Clearly I cannot observe all of them; I must select which one to observe. In short, observation is a directed activity.

What determines this process of selective attention? Since there is an indefinite number of things one can observe, how does one decide which one to observe? The best answer to this question, I think, is that training, experience, practice, expertise and knowledge direct this process. These processes, in turn, constitute what we can consider to be a*cquired background knowledge*.

By virtue of one's past training, general experience in the field, and acquired competence, one will be able to observe certain kinds of things in a natural way. Indeed, what one actually sees will in some way be dependent on one's past cognitive experiences and knowledge. These experiences, in turn, will be impregnated with a vast amount of theoretical knowledge, knowledge that directs the entire process of observation and allows one to make sense out of it.

In short, unlike amateur observation, scientific observation is typically critical, not naive. In scientific observation, one does not naively accept appearances at their face value but subjects them to a rational analysis to see if what appears to be true really is.

Illusions. The difference between naive and critical observation can also be illustrated in the case of illusions. An illusion is an experience in which certain properties appear to be real but are not. Ordinarily, we are not taken in by these appearances because on the basis of our background knowledge we discount them. What initially appears to be real is, upon critical reflection, judged to be non-real. Similarly, in the case of scientific observation, what appears to be real may, upon critical reflection, turn out to be false. This is why, for example, the heliocentric theory of the heavens is not refuted by our ordinary experiences of the sun moving.

If experience needs to be evaluated and interpreted in order for naive observation to be transformed into critical observation, what determines this evaluation and interpretation? Typically, it is one's theoretical orientation and conceptual categories. These theoretical components allow one to make sense out of the original experience by indicating which aspects are real and which are not. This identification of the experience involves categorization and classification (see Chapter 8) and this, in turn, involves *bringing an experience under a more general set of interpretive concepts.*

When these more general concepts are connected together in a larger theoretical network, the beginnings of a theory emerges. This is why the layperson may see only a dark patch, but the biologist will see mitochondria. What the dark patch *is* will be determined by the set of concepts into which one can incorporate the appearance of a dark patch. By bringing it under these concepts, the dark patch becomes an instance of the more general concept: mitochondria.

Scientific Observation and "Constants." In scientific observation, appearances are often discounted or at least not taken at face value. But how does science distinguish what is real from what is mere appearance?

I suggest that in scientific observation, the *real* is whatever experiences coherently cluster together (such clusters being termed *generalizations* or *laws*). In science and in ordinary life, reality is distinguished from illusion by the fact that illusions lack the necessary *coherence* reality must have. If a visual experience (e.g., seeing the stick as bent in water) does not cohere with a tactual experience (e.g., feeling the stick as straight), the initial experience is typically declared to be an illusion. When a set of sensations do cluster together in certain kinds of invariant patterns, this is typically taken to be a mark of reality.

These relatively permanent patterns of experiences constitute what can be called *constants* among one's experiences. These invariant features are presented not only within one's own observations, but also between one's own observations and the observations of others. Scientific observation is largely concerned with discovering these constants within an individual's experiences (*intra-individual consistency* and coherence) and between the experiences of several individuals (*inter-individual consistency* and coherence). Furthermore, such intra-individual and inter-individual constancy is sought not only with regard to current observations but also between present observations and future ones.

Public Verifiability. Because scientific observation involves observational constants—constant features connected together in varying degrees of uniformity—several other characteristics of scientific observation follow. First of all, the observation of an individual scientist cannot, under most circumstances, be unique to that individual but must be *general.* It must be possible for other individuals to observe the same thing or to have the same observational experiences. In other words, scientific observation must be *public* and not private. This is often put in this way: what a scientist observes must be capable of being verified or confirmed by other individuals who are competent and qualified to make such observations. If a single individual makes an observation and other individuals cannot make this same (or relevantly similar) observation, then scientific observation would be something private to the individual. But if it is private to the individual, there would be no way to check to see if what the individual claims to have observed really did occur; there would be no way to check the truth status of the individual's observational report.

In requiring that other individuals be able to make the same (or relevantly similar) observations, it is not required that they actually do so in all cases. Sometimes, it may not be technologically possible to do so, or it may not be necessary to do so, e.g., if the observation is not a novel one. But if the observation is a novel one (or a particularly important one), we ordinarily require that others verify or confirm it; we require, in short, *replication* of the findings by others. Without replication, there is no guarantee that the initial observation is a repeatable one and hence objectively real.

The Objectivity of Scientific Observation. The preceding discussion has laid the basis for showing how scientific observation is objective, for it is the crucial feature of public verifiability that makes scientific observation repeatable and hence objective.

In the present context, what does "objectivity" mean and how does it differ from "subjectivity"? "Subjective" means "belonging to the subject" (the human person), whereas "objective" means "belonging to the object." This is a neutral sense of "subjective." In a narrower sense, "subjective" means "allowing something inside the realm of the subject to sway one from what is objectively real." A personal desire or wish may be the decisive factor in forming a judgment about something in the face of actual or potentially objective evidence. When an individual allows her personal beliefs or wishes to become the reason for believing something, she is being subjective (biased) in a pejorative sense. But clearly not everything subjective (in the first sense) is subjective (in the second sense), namely where one's own belief is determined by objective factors. Hence the common notion that "all beliefs or interpretations are biased" is false.

By contrast, "objective" refers to the realm outside of the subject--the interpersonal world, the physical world, the world of common experience. Being objective means letting these extra-subjective factors be the factors determining one's belief and not giving in to one's personal wishes; it means discounting (or being willing to discount) personal desires, wishes, and beliefs and being willing to change them because of reasons residing in this external realm; it means changing one's subjective judgment—accommodating them—to match the objective realm.

This is not to say, of course, that one should always bend one's beliefs to those of others or that one should become a social conformist. Indeed, we have numerous cases in the history of thought where it was the solitary individual who was right and the larger group wrong. We are not talking about conformity per se. We are talking about the difference between distorting reality for personal reasons (having no rational basis for what one claims), versus allowing the realm outside of oneself to be the final court of appeals.

Exercise 7.2

1. Suppose an individual scientist produces a certain result that is strikingly novel and theoretically challenging. Before the scientific community accepts it, should it be actually replicated? How often does this really happen? Suppose they try to replicate it and they fail? Are the original results invalid? Suppose the first individual replies that they lack suitable training and skill? (For an actual case study of the latter point, see Arthur Koestler, (19).)

2. Suppose a lone astronomer is awake one night and observes a new comet passing through the sky. She tells others about it but she didn't record it or photograph it. The unique comet cannot be seen at any other time. Should they reject this as an invalid datum or not? What about the criterion of replicability?

3. In psychology, subjects sometimes introspect and observe private mental states (e.g., I have a pain in my chest, I am seeing an afterimage). Since no one else can observe this pain or afterimage, should such data be allowed in science?

4. Consult a standard account on social psychology (e.g., Moscovici, 1985) and review Asch's studies of how false group opinions may lead an individual to change his correct judgment to an erroneous one. Is there a parallel case with scientific conformity? How can this be avoided?

Evidence. *Evidence* is fundamentally a relational term: something x is evidence *for* something else y. This in turn means: the presence of x increases the probability that y exists.

As we have seen, there are conflicting views about the meaning of 'probability,' in particular, whether it refers to propositions or things (events). Many individuals have argued that, strictly speaking, only propositions can be evidence for other propositions (since only propositions can increase the probability of something else). On this view, therefore,

> I. Proposition x is evidence for proposition y $=_{df}$ the truth of x increases the probability that y is true.

In this sense, a set of premises in an argument can provide evidence for the conclusion, since the premises constitute the grounds of the conclusion and the truth of the premises increases the likelihood that the conclusion is true.

On the other hand, *evidence* also seems to be used in science to refer to events in the natural world. Hence, on this account

> II. An event x is evidence for another event y $=_{df}$ the existence of x increases the likelihood of the existence of y.

Earlier, we distinguished direct observation from indirect observation. In cases of indirect observation, we observe something (indirectly) by means of something else, which functions as a sign of it. Here one thing is an indicator of something else in the sense that the presence of it increases the likelihood of something. We can say, for example, that we indirectly observe that it's going to rain because we look at a barometer. Here, the reading on the barometer is a sign (indicator) of rain; it is evidence for rain. In medicine and biology, fever, red spots on skin, and swollen glands are a sign of a certain illness.

In these cases, certain things are taken as indicators of something else, and the former constitute evidence for the latter. How strong an indicator it is and how strong the evidence is will depend upon the degree of lawful connection between them: the greater the degree of connection, the stronger the evidential relation between them. Furthermore, the more the lawful regularity approaches a causal relation, the stronger the evidential relationship is.

Evidence and Causality. Evidence and causality seem to be closely related concepts, in fact, they seem to be asymmetrically related: (ordinarily) *if y is a cause of x, then x is evidence for y*:

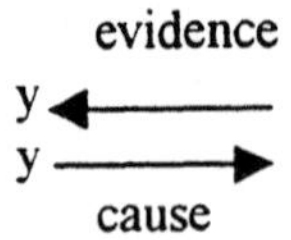

Thus, if a particular virus causes symptoms x, y and z, then the presence of x, y or z constitute evidence for the virus. The reverse relationship does not always hold, however, since something x can be evidence for something else y without the two being causally related. For example, a score on one's SAT test is evidence for how one will do in college, but the two are not causally related. Instead, they both seem to be effects of a third thing, say, cognitive ability, which is the cause of both. So, correlations between two things can sometimes function in an evidential relationship but this may be due to a covert cause at work (see Figure 7.5).

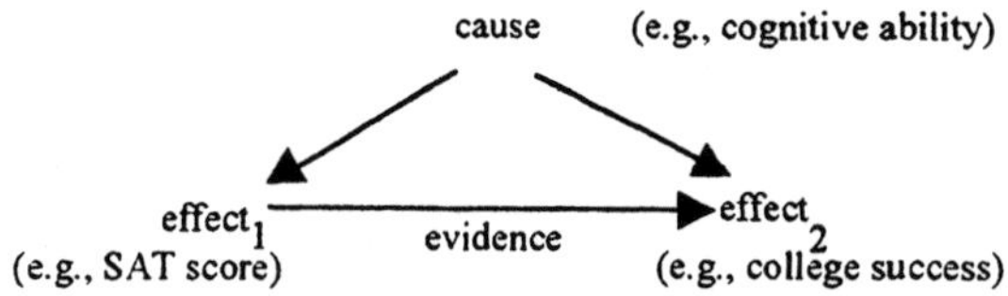

Figure 7.5. Common Cause

We can summarize this in the form of the following diagram (Figure 7.6)

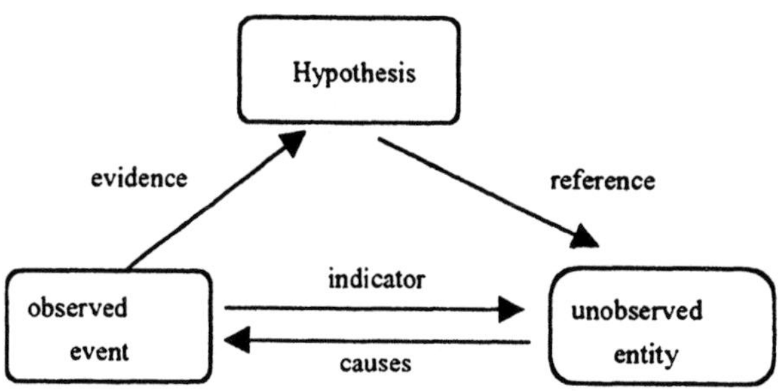

Figure 7.6 Relation between Evidence and Causality

A scientific hypothesis often *refers* to an (unobserved) entity (e.g., a virus). This unobserved entity might cause the observed event, which is the indicator of the unobserved entity. The observed entity is also evidence for the hypothesis concerning the unobserved entity. Thus, evidence and causality are closely related concepts.

Scientific Data

By means of the process of observation, scientists discover facts about entities, propositional facts that refer to features of reality. These objective facts ordinarily cause us to have certain observations. These in turn constitute evidence for these facts. Observation is thus a principal means of knowing. But although we have discussed the concepts of facts, observation, and evidence, we have said little about the concept of *data.*

A *datum*, we suggest, is the *outcome of observation, measurement, or experimentation.* Some data are the result of simple observations (e.g., observing a pigeon in the park, the sexual behavior of a particular species of fish, or the appearance of a new star in the heavens). It may or may not involve the use of instruments. But not all data are qualitative items of experience: many kinds of data are the result of experimental arrangement and complex measurement processes (e.g., scores on a pencil and paper test, pointer readings, the track of an electron in a bubble chamber, the frequency with which a junior member of the faculty addresses the senior member of the faculty, etc.). Nevertheless, scientific data seem to be the outcome of certain processes of observation.

Raw Data versus Analyzed Data. Two categories of data are often distinguished in science: **raw data** and data that has been organized, analyzed and interpreted (**analyzed data**). From one's observations, one selects features to attend to. When these features are observed, they become raw data. Raw data consists of certain immediate results of observation, which are typically recorded (e.g., the score of a subject on an IQ test, the weight of a chemical element in a test tube, photographs of a planet). In descriptions of an experiment, however, raw data are rarely given because a listing of the raw data is long and cumbersome and of little value. What scientists are really interested in is the question: what do the data *mean*? To answer this question, one must *describe*, *organize*, *analyze*, and *interpret* the data, turning raw data into analyzed data. Although we will have much to say about the processes of transforming raw data into analyzed data in the next chapters—processes involving description, qualitative classification, measurement, quantitative representation, etc.—we will at this point merely illustrate the difference between raw and analyzed data by reference to two famous examples from the history of science, one involving an experiment and one involving an observation.

First, consider the following table, which shows raw data obtained by Sir Robert Boyle on the compression of air:

Table 7.1 Robert Boyle's Data[5]

Volume of Air (Number of Spaces on short leg of J-tube)	*Observed Pressure* (inches of mercury)
48	29 2/16
46	30 9/16
44	31 15/16
42	33 8/16
40	35 5/16
38	37
36	39 5/16
34	41 10/16
32	44 3/16
30	47 1/16
28	50 5/16
26	54 5/16
24	58 13/16
23	61 516
22	64 1/16
21	67 1/16
20	70 11/16
19	74 2/16
18	77 14/16
17	82 12/16
16	87 14/16
15	93 1/16
14	100 7/16
13	107 13/16
12	117 9/16

These figures represent Boyle's' raw data. In themselves they mean very little until one begins to analyze and interpret them. Then their significance becomes apparent, since these data can be displayed graphically (as shown in Figure 7.7). In this form, the data show something they did not previously show, namely, that there is a lawlike relationship between pressure and volume, in particular, *pressure is inversely related to volume*. Here we see the advantage of analyzed data over raw data: analysis has the effect of pointing to relationships and patterns that may not have been obvious at first sight.

As another example, consider the following data (See Table 7.2), which illustrate how Johannes Kepler could have induced one of his famous laws of planetary motion, that the average distance (of a planet from the sun) cubed divided by the time (required to make a revolution around the sun) squared is a constant, i.e.,

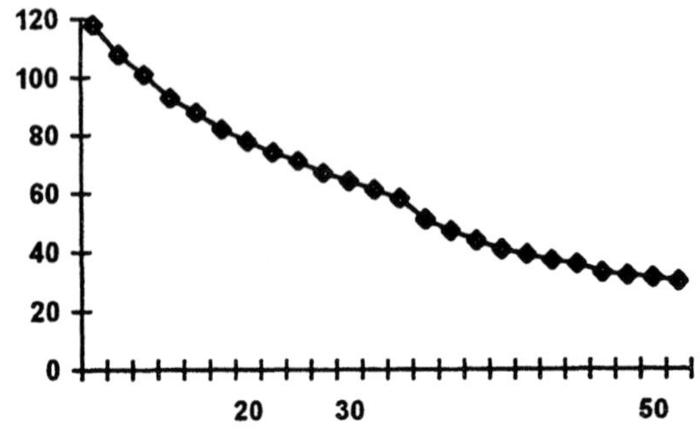

Figure 7.7. A Graphical Representation of Boyle's Data

$$\frac{\text{distance}^3}{\text{time}^2} = \text{a constant}$$

As Table 7.2 indicates, if one plots the ratio of distance and time, there are various possibilities: R/T, R^2/T, and so forth. But none of them represent a constant except for the ratio of R^3/T^2, in which case we find that there is a constant, namely (roughly) 6. This is one of the three laws of planetary motion Kepler discovered. As the above discussion illustrates, a scientist may sometimes discover a scientific law by arranging her observational data in certain ways, namely, in the form of mathematical relations represented by charts, etc. The data in this case were not raw data but data that had been subjected to certain kinds of arrangements, modifications, etc.

There are several steps involved in this process of organizing the data.[6] (1) First, the raw data (e.g., the actual numerals obtained) must be *refined*. Here, useless information is eliminated and the remaining data are put into conventional form, e.g., a standard nomenclature and scale. It is a custom among many behavioral scientists, for example, to eliminate certain kind of data if they happen to fall more than three standard deviations away from the mean. Likewise, if one is testing the effect of a drug on the IQ scores of healthy subjects and one subject happens to get sick that day, his score would be eliminated. (2) Next, refined data are often *reduced*. For example, instead of reporting the scores of all individuals or groups of individuals, *averages* of the experimental and control group are given. Likewise, it is customary among many different kinds of scientists to calculate not only averages

Table 7.2 Planetary Data and Kepler's Third Law (R: radius of planet's orbit; T: time of revolution)

Planet	R	T	R^2	R^3
Mercury	3.6×10^7	88	1×10^{15}	47×10^{21}
Venus	6.7×10^7	225	5×10^{15}	304×10^{21}
Earth	9.3×10^7	365	9×10^{15}	803×10^{21}
Mars	14.2×10^7	876	20×10^{15}	2839×10^{21}
Jupiter	48×10^7	4332	234×10^{15}	112900×10^{21}
Saturn	89×10^7	10760	785×10^{15}	695700×10^{21}

Planet	T^2	T^3	R/T	R^2/T
Mercury	7739	680775	408776	1.5×10^{13}
Venus	50490	11345123	299065	2×10^{13}
Earth	133400	47950837	254312	2.4×10^{13}
Mars	472100	324242703	206114	2.9×10^{13}
Jupiter	18770000	8×10^{10}	111565	5.4×10^{13}
Saturn	115800000	1×10^{12}	82351	7.3×10^{13}

Planet	R^2/T^2	R/T^2	R^2/T^3	R^3/T^2
Mercury	1.7	48660	1909588337	6.009×10^{18}
Venus	8.9	1331	396646207	6.011×10^{18}
Earth	6.5	696	179349989	6.010×10^{18}
Mars	4.2	300	61682189	6.015×10^{18}
Jupiter	1.2	26	2882716	6.015×10^{18}
Saturn	---*	8	603846	6.008×10^{18}

but also measures such as the *median*, the *mode*, the *variance* and the *standard deviation* (see chapter 9). This is sometimes called *mathematical or statistical reduction*. (3) Third, this refined data is *systematized* by means of several devices: graphs, histograms, tables, etc. (see chapter 9). (4) Finally, once one has reduced the data and/or systematized it, it is often necessary to analyze it statistically or mathematically. For example, after having calculated the means and standard deviations, one can proceed to determine if the two groups' means are statistically significant by using certain statistical tests (discussed in more detailed in chapter 14). Similarly, once having plotted the numerical values of a variable on a graph, scientists will engage in a mathematical analysis, e.g., attempting to fit a curve through the points on a graph. Once we have constructed a curve for the data, one can then determine what mathematical equation best represents this curve, e.g., a straight line equation ($y = a + bx$), a logarithmic relationship ($\log y = a + bx$), a power function ($y = mx^n$), periodic relationships (sine, cosine functions, etc.) and so forth.

It should be apparent, therefore, that ordinary scientific data, unlike raw data, are anything but given. Indeed, they are the product of a complex and trained activity--*data analysis*. It is true, of course, that the analysis of data is always somewhat risky and that if any question about the legitimacy of the data analysis arises, one attempts to settle it by looking at the original raw data. But as scientists will attest, raw data are rarely reported in experiments precisely because *particular* data are not as important as their general implications--the patterns and meanings to be found in the available information--analyzed data. Data (like facts) never speak for themselves; they cannot stand alone in their pristine form but must be clothed in theoretical dress.

For Further Reading

Standard textbooks in epistemology and philosophy of science deal with many of the issues discussed in this chapter, e.g., Hospers (1988, ch. 1-2) and Musgrave (1993).

For Bacon's views, see Bacon (1620/1985). On Aristotle's theory of scientific method, see Losee (1993) and then Aristotle's difficult *Posterior Analytics* (Aristotle, 1996).

For a historical survey of theories of scientific method, see the excellent article by Laudan (1969). Losee (1988), which is recommended to the beginner, and Oldroyd (1986) are full historical surveys.

On the "theory-laden" interpretation, the classics are Hanson (1958), Kuhn (1970) and Feyerabend (1993).

The two standard classics of social constructivism are Knorr-Cetina (1981) and Latour & Woolgar (1986).

Notes

[1]These tables anticipate the methods of experimental inquiry of John Stuart Mill (see Chapter 10).

[2] As a result of recent work on Bacon, it is now clear that he did not really believe in this extreme form. Instead, he appears to have advocated an inductive-deductive method: after coming up with middle axioms and general principles, one must then test them by deducing observable consequences from them.

[3] Figure and description taken from Holton & Roller (1958, p. 700).

[4] This is true even of the astronomer who is waiting for a certain time of the year to occur when a star can then be observed: in this case the astronomer is actively engaged in making numerous preparations in his observatory so that it will be possible to observe the star when it does finally appear. This is a case of what Feibleman (1972) calls *controlled observation.*

[5]From Hurd & Kipling (1964, p. 261).

[6]Several of these ideas are taken from Bunge (1967, pp. 183-184).

8

Qualitative Description:

Classification and Definition

Introduction

Observation plays a key role in the gathering of evidence and the collection of data since it is the only means through which information about the empirical world can be obtained. As we have seen, the narrow inductivist account of scientific method and the radical empiricist account of observation are oversimplifications and unduly restrictive in their interpretations (although they are correct in their claim about the fundamental roles observation and induction play in science).

We may agree, therefore, that the collection *of data* (factual evidence) is often an initial step in the scientific method. But the data that are collected are not just a random collection of brute data; meaning must be attached to them. Data have meaning when one can make sense out of them, and to make sense of them is to interpret them, to place them into appropriate relations with other categories of thought, to subsume them under appropriate cognitive structures, in short, to make them a part of a larger conceptual system. Raw data must, therefore, be transformed into what we have called *analyzed data.* This process involves description and recording of observations.

Scientific Description

The description of data requires a symbolic system that encodes and summarizes the relevant information. Such a description involves either a *qualitative* form or a *quantitative* one. (This chapter discusses the first and the next chapter the second.)

Qualitative description involves a symbolic representation of the data that is linguistic (or quasi-linguistic) in nature and contains a logical structure. In employing language, words are used and words are by their very nature general—applicable to many different instances. Hence, any qualitative description of data ("the paper turned red," "the infant approached its mother," "the pea plant was wrinkled and small") will of necessity involve *abstraction* and *generalization.*

When data are described by means of a set of words, certain properties of the data are selected for attention and others will be ignored. This is the very nature of the process of abstraction. Furthermore, in using words ("red," "mother," "wrinkled"), one will be employing a general symbol and not a uniquely referring one. In labeling some datum "red," for example, one characterizes the property that the particular datum has in common with other red things, and this common property is redness. Hence, linguistic description involves the employment of universal terms refers to the common attributes of individuals, thus ignoring the unique and peculiar features of the data. Abstraction is a feature present in any symbolic system of representation.

Furthermore, any qualitative description, being universal, will embody an underlying system of classification. For in applying a general word (e.g., "red") to one's observations, one is subsuming the particular instance under a general category (*redness*) and this involves categorizing it as an instance of the class of red things. To classify/categorize is thus essential to the process of description. Indeed, whenever we think at all, we are categorizing and labeling. Categorization thus lies at the heart of the creation of meaning and making sense of data. Those who would eschew the tendency towards generalizing and advocate the unique description of particular data can not employ linguistic or symbolic media; if individuals can be described at all, it must be by virtue of abstractive commonalties.

In addition to qualitative description (categorization), there is also quantitative description. Here, one is not describing the properties as qualities (types, kinds) but as quantities; one is representing the *magnitude* of the property, its *extent,* how *numerous* it is, etc. In quantitative description, the underlying symbolic labels are numerals and numeral-like entities, and the resulting descriptions will be things like numerical summaries (frequencies, extent, and functional relations). Such quantitative classification and measurement are the subjects of the next chapter.

Classification

To classify is to categorize: to sort things out into groups and systematic arrangements, to indicate what *kind* of thing something is; it is to bring an item under a more general concept. Generalization, which is the forming of generalizations from particular data, is thus an essential feature of the scientific method.

When we generalize from a set of data to a lawlike relation (e.g., Boyle's Law, Kepler's Laws), we are engaging in inductive inference. A scientific generalization covers our data but it does so in a more convenient and economical way (e.g., Kepler's laws of planetary motion summarize a great deal of knowledge about the position and distance of the various planets); and in so doing, data are reduced to a simpler form. The means by which this occurs is through a *grouping* of the data; this orders the data and allows us to see an underlying pattern to them (as our example of Boyle's Law and Kepler's Law shows).

When we classify, we form a class; we group individuals together into larger classes and hence make an inductive inference from particular to general. In doing this, we create *unity out of diversity*; we reduce the multiplicity of heterogeneous data by finding an underlying unity to it; we discover the "one" in the "many."

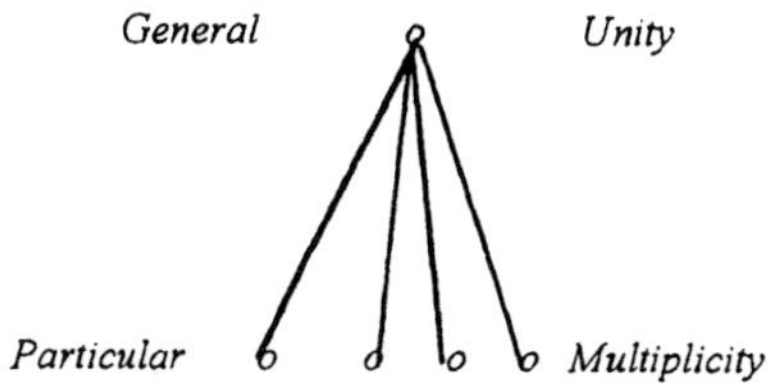

Classification, thus, proceeds upwards from particular individuals to the general; it is the bringing together of things that are similar and the separation out of those things that are different. Classification allows the human mind to conceive and to retain in memory the diverse set of properties found in heterogeneous data.[1]

The Role of Abstraction

If the purpose of classification is to discover "unity in diversity," how does this occur? Although there is no mechanical way of discovering such a unity,[2] the process of classifying can be described as a process of abstraction.

In abstraction, one lifts out what is common to members of a certain group and ignores those features unique to each individual. Thus, to use Whitehead's (1967, p. 19) example, the property "five" is common to five fishes, five children, five apples, and five days. When we abstract the number five, we seize upon that which is common to all these groups and ignore the properties peculiar to each fish, each child, each apple, each day.

We can also perform a *second-order abstraction*: one can abstract a property from an abstraction, that is to say, after abstracting the property five, and after abstracting the number three from, say, the Stooges, the Godhead, the angles of a triangle and the blind mice, one can then perform an abstraction on five and three: focusing on the common property these two share--being a number--we can abstract number from them.

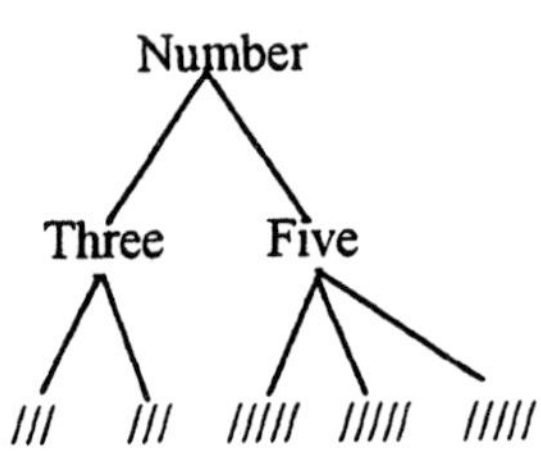

Thus, in the case of abstraction, we can form a hierarchy of classes and subclasses in which each subsequently higher level is an abstraction of what is common at the lower level:

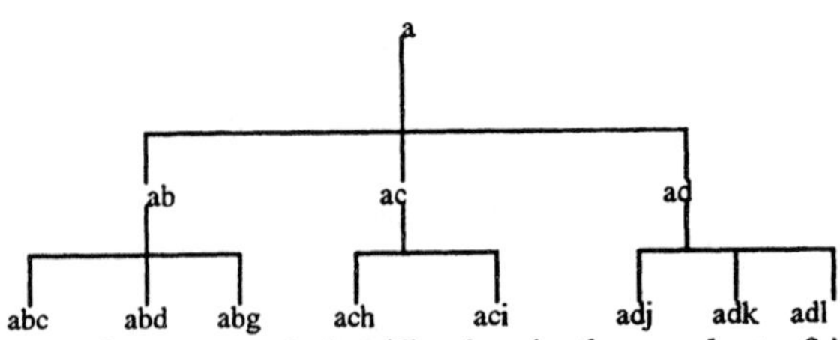

To illustrate the power of classification in the conduct of inquiry, we will briefly discuss an example of its use from the history of chemistry.

A Case Study in Classification: The Periodic Table of the Elements

Throughout the history of modern chemistry, a great deal of knowledge had accumulated concerning the chemical and physical properties of chemical elements. As more and more of these elements were discovered, the need arose to find some way to order them into some kind of system. Although the elements seemed to be quite different and to have very heterogeneous properties, this diversity (it was hoped) would be merely superficial and some underlying pattern or unity among this diversity would be found. After unsuccessful attempts by individuals such as William Prout, Jöns Berzelius and others, **Dmitri Mendeleef** came up with a first approximation to the modern periodic table.

How could all the chemical elements be grouped together into some kind of overall pattern? Mendeleef began his discussion of this problem (in his *Principles of Chemistry,* published in 1868) by pointing out that certain *principles of classification* would not work (e.g., those involving optical, electrical and magnetic properties). Mendeleef suggested that "the attempt to construct a system appears to be premature as long as there exists no hypothetical basis which could serve as the foundation of a strict system" (Mendeleev, 1869/1962, p. 87). Mendeleef then proposed this hypothetical basis: *the size of the atomic weight determines the nature of the elements.* If one grouped the chemical elements together according to the atomic weight—Mendeleef was wrong to choose atomic weight since the correct answer is atomic number, but he was very close since atomic weight and atomic number are closely related—there appeared to be a periodicity: "one element follows the other in the order of an arithmetical sequence" (Mendeleef, 1964, p. 90). For example:

Li= 7	Be= 9.4	B=11	C=12	N=14	O=16	F=19
Na=23	Mg=24	Al=27.4	Si=28	P=31	S=32	Cl=35.3
K=39	Ca=40	. . .	Ti=50	V=51	. . .	. . .

The modern version of the Periodic Table is in Table 8.2. Note that the modern principle of classification is not atomic weight (mass) as Mendeleef proposed, but rather atomic number. This was largely due to the work of Henry Moseley, who suggested (in 1913) that the properties of chemical elements are determined by their atomic numbers and hence that certain problems of chemical classification

Table 8.1 Mendeleff's (1872) Periodic Classification of the Elements

Group	I	II	III	IV	V	VI	VII	VIII
Higher oxides & hydrides	R_20 ----	RO ----	R_2O_3 ----	RO_2 H_4R	R_2O_5 H_3R	RO_3 H_2R	R_2O_7 HR	RO_4 ----
Series 1	H(1)							
Series 2	Li(7)	Be(9.4)	B(11)	C(12)	N(14)	O(16)	F(19)	
Series 3	Na(23)	Mg(24)	Al(27.3)	Si(28)	P(31)	S(32)	Cl(35.5)	
Series 4	K(39)	Ca(40)	---(44)	Ti(48)	V(51)	Cr(52)	Mn(55)	Fe(56), Co(59) Ni(59), Cu(63)
Series 5	[Cu(63)]	Zn(65)	---(68)	---(72)	As(75)	Se(78)	Br(80)	
Series 6	Rb(85)	Sr(87)	?Yt(88)	Zr(90)	Nb(94)	Mo(96)	--(100)	Ru(104), Rh(104) Pd(106), Ag(108)
Series 7	[Ag(108)]	Cd(112)	In(113)	Sn(118)	Sb(122)	Te(125)	I(127)	
Series 8	Cs(133)	Ba(137)	?Di(138)	?Ce(140)	----	---	---	
Series 9	----	----	----	----	----	-- -	----	
Series 10	----	----	?Er(178)	?La(180)	Ta(182)	W(184)	----	Os(195), Ir(197) Pt(198), Au(199)
Series 11	[Au(199)]	Hg(200)	Tl(204)	Pb(207)	Bi(208)	----	----	
Series 12	----	----	----	Th(2310	----	U(240)		

Table 8.2 Modern (1984) Periodic Classification of the Elements

Group		I	II	III	IV	V	VI	VII	VIII	O
Period	Series									
1	1	1 H 1.008								2 He 4.003
2	2	3 Li 6.940	4 Be 9.013	5 B 10.82	6 C 12.011	7 N 14.008	8 O 16	9 F 19		10 Ne 20.183
3	3	11 Na 22.991	12 Mg 24.32	13 Al 26.98	14 Si 28.09	15 P 30.975	16 S 32.056	17 Cl 35.457		18 A 39.944
4	4	19 K 39.1	20 Ca 40.08	21 Sc 44.96	22 Ti 47.9	23 V 50.95	24 Cr 52.01	25 Mn 54.94	26 Fe 27 Co 28 Ni 55.85 58.94 58.71	
4	5	29 Cu 63.54	39 Zn 65.38	31 Ga 69.72	32 Ge 72.6	33 As 74.91	34 Se 78.96	35 Br 79.916		36 Kr 83.8
5	6	37 Rb 85.48	38 Sr 87.63	39 Y 88.92	40 Zr 91.22	41 Nb 92.91	42 Mo 95.95	43 Tc [99]	44 Ru 45 Rh 46pd 101.1 102.91 106.4	
5	7	47 Ag 107.88	48 Cd 112.41	49 In 114.82	50 Sn 118.7	51 Sb 121.76	52 Te 127.61	53 I 126..91		54 Xe 131.3
6	8	55 Cs 132.91	56 Ba 137.36	57-71 Lanthanide series	72 Hf 178.5	73 Ta 180.95	74 W 183.86	75 Re 186.22	76 Os 77 Ir 78 Pt 190.2 192.2 195.09	
6	9	79 Au 197	80 Hg 200.61	81 Tl 204.39	82 Pb 207.21	83 Bi 209	84 Po 210	85At [210]		86 Rn [222]
7	10	87Fr [223]	88 Ra [226]	89— Actinide Series	104 Unq [257]	105 Unp [260]	106 Unh [263]			

The top number is the atomic number, the bottom number is the atomic weight. Brackets indicate approximate values.

could be resolved if the basis of classification were switched from atomic weight to atomic number.

In Mendeleef's table, there were several gaps indicating missing elements, e.g., between Ca and Ti there ought to exist a chemical element whose atomic weight was between 40 and 50. Here, then, was a definite *prediction* that followed from Mendeleef's system of classification. More importantly, not only did Mendeleef predict the *existence* of a hitherto unknown element, he also predicted its chemical properties. As he put it (Mendeleef, 1962)

> With the periodic and atomic relations now shown to exist between all the atoms and the properties of their elements, we see the possibility not only of noting the absence of some of them but even of determining, and with great assurance and certainty, the properties of these as yet unknown elements; it is possible to predict their atomic weight, density in the free state or in the form of oxides, acidity or basicity, degree of oxidation, and ability to be reduced and to form double salts and to describe the properties of the metalloorganic compounds and chlorides of the given element; it is even possible also to describe the properties of some compounds of these unknown elements in still greater detail. (p. 744)

He predicted, for example, the existence of what we now call germanium (which he called ekasilicon--the gap missing in his table: Group IV, row 5).[3] (This was discovered by Wickler in 1887.) He made equally definite predictions of two other elements which were subsequently discovered--gallium (III, 5) and scandium (III, 4).

Mendeleef could not explain why there was such a pattern, but he thought there would eventually be a physical explanation for this periodicity. Such an explanation was to come later, in the twentieth century, when quantum chemistry explained the periodic table in terms of electron configurations.

Other Examples of Scientific Classifications

There are several other important examples of classification in science. For example, **taxonomy** is that branch of biology concerned with the question of how plants and animals should be classified and arranged. The modern system dates back to the work of **Carolus Linnaeus**, who (in his *Systema Natura* [1737]), advanced a "system of nature" based on Aristotle's philosophy. It had five tree-like branches: class, order, genus, species, and variety. Unlike the underlying assumption of Linnaeus--that species were fixed--modern biological taxonomy takes its point of departure from evolutionary theory, with its stress on the plasticity of species. As one leader in this area writes: "Classification consists mainly of three operations: (1) recognizing and describing related smaller and larger groups of organisms according to the principles of populations and of phylogeny; (2) fitting these groups into a formal hierarchy; and (3) providing names for the various groups" (Simpson & Beck, 1965, p. 502.) The systematic hierarchy Simpson and Beck refer to is the following (in descending order of generality):

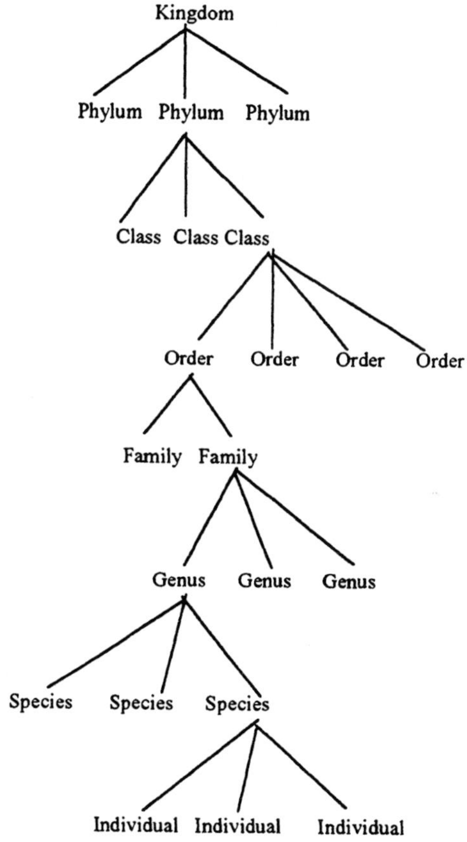

In addition to astronomy (with its classification of stars), we can find an important example of classification in psychiatry, where mental disorders are classified into nineteen hierarchical categories, each of which have, in turn, sub-classes similar to the tree diagram found in biological taxonomy.

Classification and Division

The topic of classification is inseparable from that of another concept—*division.* Indeed, they are mirror images of each other: classification is grouping individuals into larger and larger classes (bottom-up), whereas division is dividing groups into smaller and smaller ones (top-down). Classification is an inductive process in which one proceeds from particular to the more general, whereas division is often said to be a deductive process in which one proceeds from the more general to the less general. Both may be said to have the typical structure of a genealogical tree, in which there are subordinate, superordinate and coordinate relations between the members.

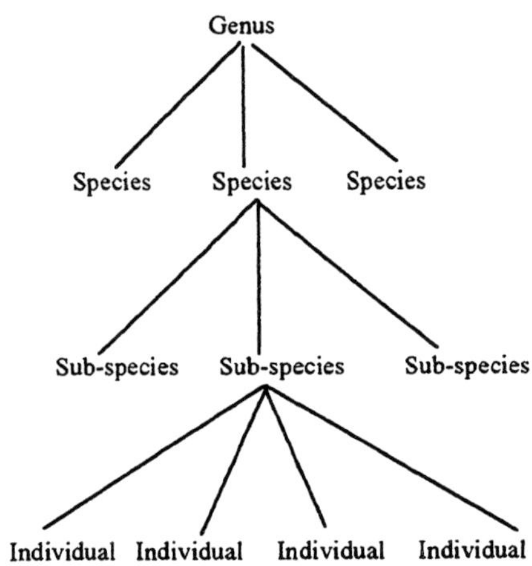

Figure 8.2 Genealogical Tree

When one classifies an object in terms of genus and species, one is grouping individuals into a sub-species, then subspecies are grouped into species, and finally species are grouped into genera until one reaches the highest genus. In division, one proceeds in the opposite direction. Beginning with the highest genus, one divides this into smaller genera, which are divided into species Next, each of these species is further sub-divided into more particular species, the entire affair constituting a hierarchical structure. Thus, division is the inverse of classification.

Aristotle's Theory of Classification and Division

Historically, the system of Aristotle (called Aristotle's theory of Predicables) has been extremely influential as a theory of classification and division. It involves the core concepts: essence, genus, property, accident, and differentia.

Aristotle's system is based upon the notion that a subject (A) and predicate (B) may be convertible and commensurate. If our original proposition is "A is B," it is *convertible* into "B is A." A and B are said to be *commensurate* if the conversion is true. Thus 'equilateral triangle' (A) is commensurate with 'equiangular triangle' (B) because every equilateral triangle is an equiangular triangle (A is B) and every equiangular triangle is an equilateral one (B is A).

Some predicates are commensurate and some are not. Of those that are commensurate, some express the *essence* of something—their *definition*—and some do not—
merely a *property*. Of those predicates that are not commensurate, some express part of the essence—the *genus* or *differentia* of the individual--and some do not--their *accidents*. See (Figure 8.3)

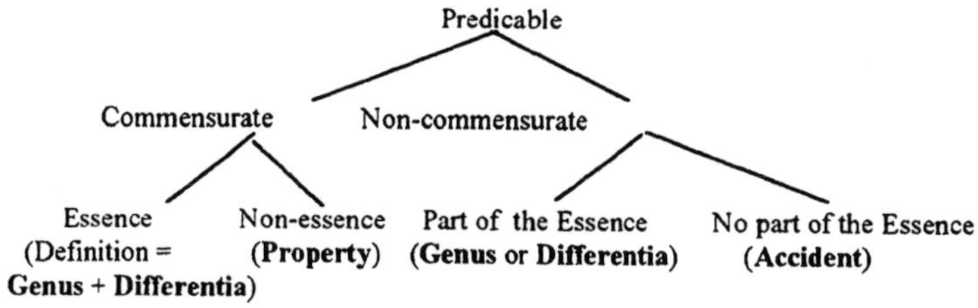

Figure 8.3 Aristotle's Predicables

Aristotle believed that individuals belonged to natural kinds, each natural kind being a species. Each natural kind had a set of properties--their *essence*--defining membership in that class. The essence of something, according Aristotle, is that set of properties that something must have in order to be a member of that class. Thus, the essence of being a human, we might say, is that of being a rational animal, since if one lacked these properties one could not be a human being and if one possessed these properties, this would automatically justify one's being included in this class. The essence of something is, thus, a real definition of that thing and is expressed as: genus + differentia.

In specifying the definition of something (e.g., man = rational animal, momentum = quantity of motion), one must mention the *genus* and the *differentia*. The *genus* of something is the general class under which different kinds of things fall: animal, material body, motion. Genus is part of a thing's definition or essence but it does not serve to differentiate the various kinds of things falling under it, since other things besides man is an animal, other things besides momentum have motion. *Material body* would be the genus under which several species fall. These species differ between themselves because each possesses a distinctive differentia, something that differentiates it from other species falling under the same genus.

Some attributes are commensurate but are not essential to the thing. These Aristotle called *properties*. Properties of things follow from their essence; e.g., as a consequence of man's essence, he has the property: the ability to learn grammar. Similarly, because of the essence of organisms, it follows that they are contractile, irritable, assimilate food, reproduce, etc. A property is not a part of something's essence but it is an invariable and universal part of it.

In addition to something's essence, there are accidental features (*accidens*) something can have. The possession of these accidents (e.g., man's baldness) does not make any difference to a thing's essence. Accidents are not-commensurate and not part of the definition.

The Principles of Classification and Division

In constructing a classification or logical division, certain rules should be followed: classification must be *uniform*, *exhaustive,* and *disjoint.*

First of all, the division/classification must be *uniform.* This means that there must be a single basic principle underlying the classification, one applied consistently to all the members. In the case of Mendeleef, for example, this uniform principle was that chemical elements were classified in terms of atomic weight.

Second, the classification must be *exhaustive:* it must apply to every member of the classification. No member can fall outside the system of classification, nor can there be a cross-division, which would result, for example, if we were to classify cars into public cars, private cars, and taxis, in which case some cars would fall into more than one classification. In Mendeleef's table, every single known chemical element has only one place.

Finally, the system of classification must be disjoint or *mutually exclusive*: no individual member will belong to more than one subclass. In Mendeleef's table, every chemical element appears in only one location.

Kinds of Divisions

Basically, there are three kinds of divisions: logical division, dichotomous division, and mixed division. A division involving a genealogical tree is a *logical division*. It is formed by following the logical rules for constituting divisions just discussed. It results in a hierarchy or genealogical tree in which there may be several branches leading from a common trunk.

In contrast to a logical division, a *dichotomous* (or *bifurcate) division* has an exhaustive and exclusive division: each genus is divided into two members, one of which possesses the property in question whereas the other does not. Thus, the class of animals is divided into two mutually exclusive classes—vertebrates versus non-vertebrates (invertebrates) as in the following:.

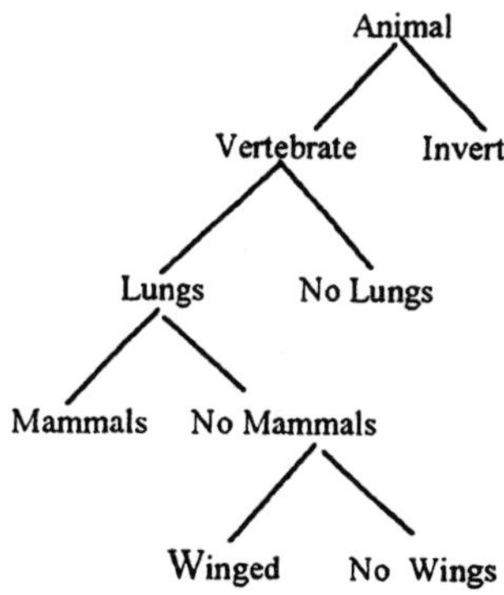

Finally, in a *mixed division,* one part of a division may be dichotomous, whereas another part may be "branching" as in a tree structure. For example:

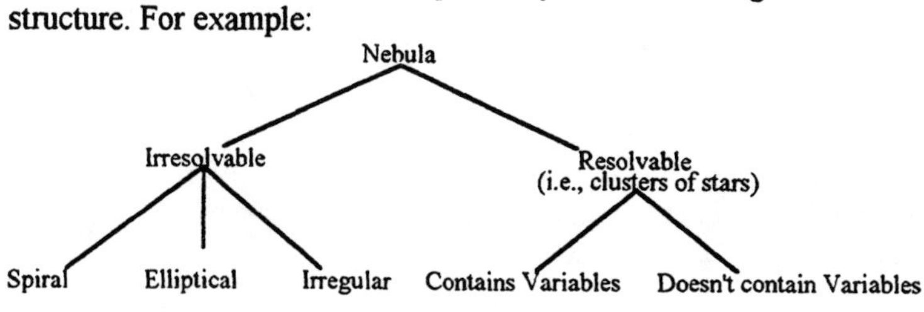

The Scientific Value of Classification

The case of Mendeleef's formulation of the periodic table of the chemical elements provides a useful example of the advantages of classification. Three such advantages are apparent.

1. Unity in Diversity. First, a classification groups data into more manageable proportions. Bringing an array of data into an overall pattern, united together by an overall plan provides a structural organization for the data. This is what happened in the case of Mendeleef when the diverse and heterogeneous assortment of chemical elements were shown to be closely-knit members of a single family with distinct relations between them. Just as an anthropologist might discover the complex pattern of kinship systems in an exotic tribe, so Mendeleef discovered the complex organizational pattern underlying the many chemical elements. This is a typical pattern in science--the search for "unity in diversity." Classifying scientific facts, observations, and data according to an underlying principle often shows this unity.

One should not think, however, that simply inspecting the data will automatically result in the classificatory pattern emerging. Mendeleef's scheme of classification was a *hypothesis* he advanced, one that was close but not quite correct. Since there is an indefinite number of ways in which one can classify data, there are several different hypothetical classifications that could have been applied to the chemical elements. The narrow inductivist's dream, that a set of data would automatically generate the classification itself, is another will o' the wisp.

2. New Predictions. Secondly, classification systems lead to new predictions. Mendeleef was able to predict not only the mere existence of new elements but also several of their properties.

3. Correlation of Properties. The reason classification can lead to the making of new predictions is the same reason classification leads to the discovery of new laws. Although classification may be a useful mnemonic device, it is much more than that. When one places an object or individual in a class, it is typically being claimed not merely that this individual belongs to this single class but that this membership in the class indicates, first, the existence of a set of properties making up this class and, second, that this set of properties is related to another set of properties. When, for example, potassium, sodium, caesium, rubidium, and lithium are grouped together into the class of alkaline metals, we do so because they share certain common properties: (l) they combine in a certain way with oxygen, (2) they decompose in water at all temperatures, (3) they form strongly basic oxides soluble in water, (4) their carbonates are soluble in water, (5) each metal forms a chloride, etc. (Here, analogy is crucial.) These properties are fundamental properties of the alkaline metals. Suppose now that we discover another metal which we classify as alkaline because it has properties (1), (3), (4) mentioned above. Because it has these properties, we infer that it should also have other properties, e.g., (5). Here, we are supposing that sets of properties tend to cluster together.

Similarly, in comparative biology, anthropology, and archaeology, we suppose bone and skeletal structure to be well correlated with certain biological and psychological capacities. Suppose we find a certain skeletal remains. From this we can infer what the likely biological and psychological capacities of this creature were, e.g., it was a meat-eater and tool-user, had a language, and so forth. The correlation of properties allows us to make this inference, with the certainty of inference depending on the certainty of the correlation.

Mendeleef saw that the chemical element Titanium had chemical properties similar to those of Carbon and Silicon and hence placed Titanium together with these latter elements into the *fourth* vertical column and not, as other individuals had suggested, into the third vertical column. It was the underlying assumption that *classes are related to each other* that led to the making of the classification and thus to the generation of new predictions. Hence a third advantage of classification is that it leads to the discovery of correlations between properties.

From Correlation to Causation

The typical course of events in science is for correlations to give way to causation. Classificatory schemes can thus lead to the generation of new hypotheses concerning *why* certain groupings occur the way they do. The question facing Mendeleef, for example, was this: why do the chemical elements fall into the classes they do?, i.e., What is the principle or plan underlying the arrangement? A first step in answering this question was to arrange the elements in terms of their atomic masses. Later, this mistaken view was replaced by one in which the elements were arranged in terms of atomic number. But this in turn led to the quest for an explanation about their atomic number--the number of protons in the nucleus of each atom of an element. If we knew the atomic number of an element, we could predict its electron configuration.[4] But physicists and chemists had to wait for the work of Edwin Schrödinger (1887-1961) before they could explain exactly how atomic number is related to electron configuration.

This is a typical pattern in the history of science. An initial stage of description leads to an explanation of these empirical correlations. This has occurred most notably in the physical and biological sciences and to a lesser degree in the behavioral and social sciences, where we have well-developed systems of classification but no known (or widely accepted) explanation of this classification (for example, of the mental disorders catalogued in the DSM-IV).

Diagnosis as Classification

Classification is a useful device to aid in the scientific process of diagnosis. In medical diagnosis, for example, one attempts to identify the underlying disease from its external symptoms. This typically involves the determination of what cause underlies the disease. The way in which this works is by means of classification. When one is diagnosing something, one is attempting to discover the class to which a certain specimen belongs. Given a particular system of classification, for example, a hierarchical one involving logical division, the question facing the diagnostician is where to place the case under consideration. It is an attempt to identify the present example by subsuming it under a certain class. If one can place it in the appropriate class, one will not only have identified it, one will be able to tell a great deal more about it, including (perhaps) its etiology, prognosis, and possible cure. This is due to the fact that properties tend to cluster together into meaningful patterns.

For example, a psychiatrist or psychotherapist may have to diagnosis an individual's mental state, behavior disorder, or mental disease, by reference to a set of unusual behaviors and mental states: depression, inappropriate affect, delusions of grandeur, hyperactivity, loss of sleep, etc. To what class of mental disorder does the person belong?

Classification is an invaluable tool in this process of diagnosis because of the correlation of properties. If an individual is diagnosed as belonging to a certain class (e.g., affective disorder) on the basis of *some* of the characteristic symptoms accompanying this set, it is reasonable to expect the individual to show other characteristics (e.g., distractibility, behaviors likely to lead to painful consequences, etc.), and to be treatable by techniques which work for that category of individuals (e.g., a certain kind of psychotherapy).

In order to diagnose and, hence, to assign a specimen case to a particular niche in a classification scheme, one needs characteristic marks or criteria. These features function as *differentia,* enabling us to assign an individual to a particular slot.

Two kinds of diagnostic systems may be distinguished. In the first kind, a dichotomous classification scheme is employed. Here, we have diagnosis which is certain, because at each choice point we have a bifurcation and every individual must fit into one of the options. For example:

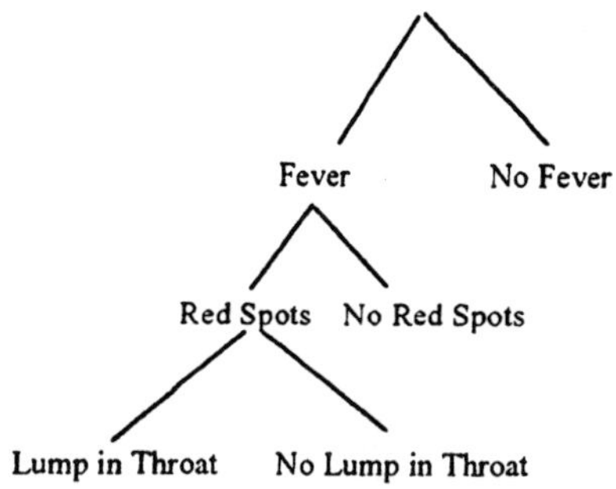

Here, the differentia or characteristic marks are either possessed by something or they are not. This type of diagnostic system is certain and gives unequivocal results. But such a system is rare.

The second kind of diagnostic system is one in which several symptoms cluster together, each cluster indicating a somewhat different disorder:

Ordinarily, such clusters will be such that a patient may not manifest all of these constituents and one will have to diagnose the disorder (A or B) on the basis of *some* symptoms being present (x, y versus u, v). In the latter case, we have a *probabilistic indicator.*

The picture is even more complicated when we consider the fact that some of these symptoms are usually weighted more heavily than others, e.g., fever may not be weighted very heavily, but bacteria in urine is. When one considers the complexities of diagnostic reasoning, a form of reasoning in which the doctor must be able to see patterns in the symptoms, it appears remarkable that individuals are as successful as they are.

Is Classification Natural or Artificial?

For centuries, logicians, philosophers and scientists have debated the question of whether classifications and divisions are natural or artificial. Does our system of classification express the order of things in nature, or is it artificial, a matter of convenience, in which there are an indefinite number of ways of classifying?.

Aristotle advocated the position that classes and divisions were natural, that nature came already sorted out into classes and into a hierarchy of classes--his scale of nature. Each individual substance (according to him) possessed an *essence* and this essence made up a natural class. Aristotle's view is thus a variety of **realism,** the view that *classes are real and exist in nature.* Historically, the view is opposed to **nominalism**, the thesis that *only individuals are real.* According to the nominalist, general properties are just *names*—hence the term *nominalism*—that we apply to certain sets of individuals.

As a consequence of his realism, Aristotle was lead to the view that biological species were real and fixed, a view shared by many pre-19th century biologists, including Linnaeus. Such a view was challenged by modern evolutionary theory, according to which species are not fixed but change over time. Hence, there are intermediate forms. Since biological species are in the process of continual evolution, many biologists have advanced the view that biological species (and classes in general) are created by human decision and there are various ways of classifying things, no one of which is true. This view is sometimes called **conventionalism,** the view that every system of classification is just a convention. According to some conventionalists, classification is arbitrary and subjective.

Which view is correct? Are scientific classes merely a matter of personal preference or is there a single correct way to classify them? These two alternatives are extreme positions, marking the opposite ends of a continuum. Like many extreme positions, neither one seems to be correct. There are several ways of classifying individuals but not all of these ways are equally good ones. In the case of Mendeleef, e.g., there were several ways of classifying the chemical elements, but some ways proved to be better than others. It is of course possible that some day, chemists will change this system of classification. But if they do, it will be because there are good reasons for doing so; it will not be for arbitrary reasons.

Exercise 8.1

For each of the following items, discuss various ways in which they might be classified. Indicate the principles used for division into subclasses.

1. Vehicles
2. Dogs
3. Horses
4. Airplanes
5. Vegetable
6. Educational institutions
7. Books
8. Works of Art
9. Religions
10. Colleges
11. Sciences

Definition

The first step in the scientific method is often that of observation. Observation is the means by which we obtain data and hence discover new facts. These facts, in turn, function as empirical evidence for various things, as the basis for generating a scientific law, as the basis for generating a hypothesis, as the empirical test between two competing theories, and so forth. Observation thus gives us facts. We have already seen that raw data are ordinarily transformed into refined (analyzed) data and that such data sometimes constitutes a scientific generalization (e.g., as points on a line having a particular mathematical form).

When we generalize from a set of data to a law-like relation, we are clearly engaging in inductive inference. This is possible because we are able to come up with a scientific generalization that covers our data but does so in a more convenient way, just as Kepler's laws of planetary motion summarizes a great number of particular facts about the position, distance, and velocity of the planets. In going from particular data to a general law, our data are reduced to a simpler form and the means by which this occurs is a grouping or arrangement of the data. This is the key element in an inductive inference: a grouping of the data simplifies and orders the data and allow us to see an underlying pattern, just as Kepler's law allowed us to see an underlying pattern to our planetary data. A key ingredient in the scientific method, therefore, is the step from the collection of data to the forming of a generalization.

Sometimes one can immediately see a pattern to the data whereas at other times one needs to go through additional, intermediate steps. As we have seen, one such step is the kind of grouping of data that occurs in classification. Classification divides the data into classes or groups and hopefully facilitates the generation of a law or hypothesis. Similarly, a concept closely related to classification is that of definition.

Definition is a more formal process classification, and explicitly involves the use of language. But, like classification, it is concerned with those properties defining membership in a class. Because the principles underlying classification and definition are so closely related, we will treat them together.

The process of forming a definition is a subject of considerable importance in science. Science is concerned with precision and accuracy in its recording of the data, but the precision and accuracy of terms and concepts are just as important. An essential part of the scientific method is the *introduction of new concepts* in the form of laws, hypotheses, theories. But here there is no less a need for clarity and rigor than in the collection and measurement of data. But to be concerned with the precise meaning of one's terms is to be concerned with the correct definition of them.

The Logical Structure of Definition

To give the definition of a term presupposes the term or concept is unclear; to clarify is to give its meaning; and to give its meaning, one ordinarily gives another word or phrase that is equivalent to it. Thus the logical form of a definition is the following:

Word to be defined $=_{df}$ words defining it.

Where "$=_{df}$" means "equals by definition." This symbol is sometimes replaced by "≡"—the biconditional connective—expressing the relation: if and only if. Thus

Word to be defined ≡ words defining it.

Because the above phrases are rather cumbersome, it has become customary to introduce two new terms to stand for these concepts: the **definiendum** is the word or concept to be defined, and the **definiens** is the word or words that define it. Thus in the definition

Bachelor $=_{df}$ unmarried male

"bachelor" is the definiendum and "unmarried male" is the definiens. It will be convenient to use this notation in the remainder of the discussion of definition.

Real versus Nominal Definition

Another preliminary question that must be dealt with is the question: Are definitions definitions of words or things? According to Plato and Aristotle, the most important kind of definition was one involving a statement about the essential properties of a thing—a **real definition.** In nature, Aristotle thought, individual things had certain properties making up their essence—a set of properties making it the kind of thing is was, such that if a thing lacked any of these properties, it would not be that kind of thing. Thus human beings had an essential nature (e.g., rational animal) that was unchanging and existing independently of our thoughts. One of the fundamental tasks of science was to discover this essence and hence to provide real definitions. Individuals who believe in real definitions believe that (at least some) definitions are not arbitrary or conventional, since statements expressing a definition are either correct or not (true or false), depending on whether the proposed definition correctly characterizes the underlying nature of an entity or not.

Beginning in the 17th century (but really going back to the Middle Ages), another group of individuals argued that definitions are always definitions of words—**nominal definitions**—and not of things (since they did not believe in essences). Most individuals who believe the only kind of definition in science is or should be a nominal definition also believe that definitions are matters of linguistic convention, that there can be various definitions of the same word, and that there is no such thing as a correct or true definition.

The issue of realism versus nominalism is an issue that arises in the context not only of definition, but, also, as we have seen, in classification. Indeed, it underlies a great number of issues in scientific methodology and philosophy of science. Although a vast amount of material has been written about this subject, the position taken in this book is that neither an extreme realism nor an extreme nominalism is a tenable position. Sometimes definitions are merely matters of names to which are attached conventional meanings, for example, when a scientist introduces a new concept. But on the other hand, there are examples of real definition in science, as when scientists discover that there are objects that have the properties indicated in the definition. For example, chemists give definitions of the chemical elements. These definitions, however, are not arbitrary verbal conventions, but real definitions. How a chemist defines a chemical element depends upon what properties he discovers this element to have; these properties make up the essence of the chemical element. Oxygen, for example, is an element that has atomic number 8, atomic weight 15.9, solubility in water of 31 ml per liter of water at 20°C, density of 1.4 g., boiling point of −183°C, and melting point of −118°C. This would seem to be an example of what Aristotle had in mind by a real definition. Such a real definition might change if subsequent physical evidence required it, but such evidence would constitute a discovery about the underlying nature of oxygen, and not a mere change in words.

Ambiguity and Vagueness

Individuals usually ask for the definition of a concept when there is some question about what it means. But there are several ways in which a word or concept can be unclear. We will mention two such ways—ambiguity and vagueness.

> A word is **ambiguous** if it has at least two meanings and it is unclear which is meant

Thus the word bank is ambiguous since it has several meanings: there is the bank of a river; there is the place where you put your money; there is the activity of "banking one's car," etc. Often, of course, the context will indicate which of these meanings is intended; but sometimes this will not be true.

One important type of ambiguity concerns what is called the *process-product ambiguity*, the distinction between the *activity* of doing something, and the *result* of that activity. When a word can refer to either the process or the product, and it is not clear which is meant, we have an important kind of ambiguity.

The word science is itself an example of this process-product ambiguity, since "science" can refer to the activity of doing science (e.g., engaging in scientific research, creating a theory), but it can also refer to the results of this activity (e.g., laws, theories, etc.). In discussions of the nature of "science," it may be unclear which of these is meant.

On the other hand, a word may be vague.

> A word is **vague** if it is unclear which object is to be characterized by this term since the presence versus the absence of the term is fuzzy.

Consider, for example, the term bald. This word is vague, because it is unclear which individuals are bald and which ones are not. Yul Brenner was bald and Farrah Fawcett is not. But what about cases in between, e.g., George Costanza?

Other vague terms include: "dark," "warm," "steep," "old," etc. A useful model for understanding this is the continuum model. Some concepts are discrete or dichotomous: they are either completely present or completely absent, e.g., pregnancy. But other terms are "more or less"; they fall on a continuum where one end point gradually fades off into a middle region and this in turn gradually fades off into the opposite end point. Thus black is at one end point, and white at the other and in the middle is a gray area where something is neither black nor white. Vague words also fall on a continuum: having no hair at all is at one end point and having a full head of hair falls at the other extreme. But both of these fade off into a middle, fuzzy area where it is not clear where to draw the line between being bald, and not being bald.

What can be done when there is no natural dividing line between the presence versus the absence of a property? One must draw the line somewhere, but is this completely arbitrary? In some cases yes and in some cases no. Where to draw the line will depend upon one's purposes and the goals of inquiry, the accumulated store of background knowledge one can draw upon, etc. One technique for removing vagueness is to stipulate where one will draw the line: "by "bald" I will mean anyone with less than 25 hairs on his head." Sometimes there will be empirical evidence relevant to this stipulation and sometimes there will not be. Common sense would suggest one should make use of all the available scientific evidence in coming up with a stipulation.

Exercise 8.2

How might a scientist remove the vagueness from the following words?

a. Physical health.
b. Mental health.
c. Life.
d. Death.
e. Homosexuality.
f. Happily married.
g. Progress.

h. Young.
i. Old.
j. Alcoholic

Extension-Intension; Denotation-Connotation

Definitions can be advanced in order to remove ambiguity and/ or vagueness. They can also be advanced when one is introducing a new, technical term, or when it is necessary to make a concept more precise. But what are the rules and principles for defining a word? What are the different kinds of definitions? These questions are part of what a theory of definition should be. We will begin with the latter question and then proceed to answer the former question.

To begin with, we must first introduce several preliminary distinctions regarding the various types of definitions.

> The **denotation** of a word is the class of *objects* to which the word can be correctly ascribed

Thus the denotation of the word scientist is the class of individuals of whom it is correct to say they were or are scientists: Copernicus, Kepler, Galileo, Newton, Darwin, Mendel . . . If we use set-theoretical notation, we could say:

> Scientist = {Copernicus, Kepler, Galileo, Newton, Darwin, Mendel . . .}

with the set being open-ended. A word denotes an object or set of objects in the sense of naming them or referring to them.

A concept closely related to that of denotation is extension. Although logic textbooks sometimes write as if denotation and extension were the same notion, this seems to be wrong.

> The **extension** of a word is the set of sub-classes of things that the word includes.

Thus the extension of the term scientist would be the set: physicist, biologist, sociologist, etc. Here the extension of "scientist" is not a set of individuals, but a set of sub-classes over which the term scientist extends.

Extension and denotation are contrasted with Intension and connotation. But whereas extension and denotation can be distinguished, it is less clear that intension and connotation can.

> The **connotation** of a word consists of those (essential) properties such that these properties determine the objects to which the word can be correctly applied.

Thus the connotation of the term scientist might be *follower of the scientific method.* It is the set of properties that determine membership in the denotation of this term.

Just as a term denotes one or more individuals, so a term is said to connote or imply a set of properties. Literally "connote" means to lead one to think of a set of properties (to "note along"). This has led some individuals to define connotation as the set of properties one thinks of when one thinks of the term. Thus "communism" might connote—what we will all *subjectively connote*—Russian, totalitarian state, repressor of freedom, enemy of America, etc., even those these concepts are not part of the strict or literal meaning of the word—what we can call the *objective connotation.*

The intension of a word is closely related to a term's connotation; in fact they are often identified. But since connotation and denotation are distinguished, so intension should be contrasted with extension.

> The **intension** of a word is those properties all and only members of the word's extension/denotation share.

Thus the intension of the word scientist would be all and only those properties possessed by physicists, biologists, sociologists, etc. (or possessed by Copernicus, Kepler, etc.).

Inverse Relation between Intension and Extension

What is the (extensional) relation between extension and intension? They are inversely related: as the intension of a term increases, the extension decreases, and as the extension of a term increases, its extension increases. To say the intension of a term increases is to say that more and more properties are added to it, that it is qualified in more and more ways. As a concept is qualified in more and more ways, it becomes progressively narrower in extent and fewer objects and subclasses can satisfy the description. An increase in intension thus proceeds from the most general to the less general (by adding subsequent properties). If *scientist* is the property under consideration, then as we proceed to qualify this term by adding more and more properties, we restrict its extension. The extension or denotation of "scientist" includes a great number of very different kinds of individuals—Galileo, Einstein, Darwin, Mendel, Freud, Tinbergen, Merton, Marx, etc. So this term has a large extension. Suppose now we increase its intension by adding one property to it, say "physical," so that we have "physical scientist." Now, the extension is narrower, since it "extends" over (includes) fewer individuals, namely, only those who are physical scientists, and not those who are biological, psychological and social scientists. Now suppose we add still another property to *physical scientist*, say, "twentieth century physical scientist." The result of increasing the intension is that the extension is further reduced, since Galileo and Maxwell are now excluded from the list.

Scientist	{Galileo, Maxwell, Einstein, Bohr, Darwin, Mendel, Freud, Marx, . . .}
Physical Scientist	{Galileo, Maxwell, Einstein, Bohr, . . }
20th C. Physical Scientist	{Einstein, Bohr, . . .}

Now, look at it the other way. Suppose we start with an arbitrary class—a single individual, say, Snow White. This is the extension of a certain concept, which has a corresponding intension. The intension is the set of concepts Snow White possesses: female, human being, imaginary character in a fairy tale, character in a Walt Disney film, beautiful, a friend of dwarfs, biter of a poisoned applied, being kissed by a Prince, etc. In short, we have a small extension with a large intension. Now, suppose we add to the extension by including Goldilocks. The extension—Snow White, Goldilock—is increased, but at the cost of decreasing the intension, since it will exclude the properties of Snow White that Goldilocks does not have. The resulting intension is: female, human being, imaginary character in a fairy tale, character in a Walt Disney film, beautiful, etc. As we further increase the extension, we progressively decrease the intension.

Exercise 8.3

1. Arrange each of the following groups of terms in order of increasing intension.

A. animal, feline, lynx, mammal, vertebrate, wildcat.

B. alcoholic beverage, beverage, champagne, fine white wine, white wine, win.

C. Integer, number, positive integer, prime, rational number, real number.

2. Indicate, for each property or thing, whether it is the denotation, extension, or connotation-intension of the given term..

A. Man.
 Animal
 Abraham Lincoln
 Erect stance
 Mohammed Ali
 Rationality
 Neanderthal
 Makes symbols

B. Scientist
 Engaged in research
 Einstein
 Collecting of data
 Experimentation
 Physicist

C. Schizophrenia
 Mental disorder
 Split personality
 Dr. Jekyll and Mr. Hyde
 Bizarre behavior

D. Medical Doctor
 Practices medicine
 Healing
 Dentist
 Gives medical treatment
 Osteopath

Kinds of Definition

A theory of definition should contain a discussion of the various kinds of definition. Historically, logicians and methodologists of science have used the distinction between denotation—connotation and extension—intension to construct a typology of definitions. Unfortunately, this typology has not been very consistent, nor logically rigorous. In fact, the classification system is one in which there is overlap between various types of definition. As we have seen, a system of classification should be not only consistent, but also mutually exclusive. Unfortunately, this has not been the case with traditional classifications of definitions. So, the reader cannot expect to be able to separate the various kinds of definitions into neat, water-tight compartments; in fact, a single example can often fit into several categories. Although this is unfortunate, it is not as bad as it may seem, since it indicates the complexity of our subject matter.

In our particular classification scheme, we will begin with those definitions that fall under the general category of *connotation-intension,* and the proceed to discuss definitions falling under the category of *denotation-extension.*

Connotative Definition. We have already characterized the notion of a term's connotation. We can then characterize a connotative definition in the following way:

> A *connotative definition* of a term is a definition by reference to those essential properties of the term, such that these properties determine the objects to which the term can be correctly applied.

For example, a connotative definition of "triangle" would be:

> Triangle = plane figure bounded by three straight lines.

In an objective connotative definition, we define a word or concept by mentioning a set of essential properties that are the word's meaning. As such, connotative definitions have been thought by many to be the only real kind of definition. Part of the reason underlying this belief stems from the fact that a definition is typically thought to involve the essence of the term and, hence, to presuppose real definitions.

Synonymous Definition. When we ask for the definition of a word, we often are asking for another term that means the same as the original word. The relation—means the same as—is the relation of *synonymy* and it is a relation between two different words. If someone does not know what "bachelor" means, one could give him a synonymous definition:

> "Bachelor" means the same as "unmarried male."

When one does not know the meaning of a foreign phrase, e.g., "Wissenschaft," one will ordinarily give the English equivalent:

> "Wissenschaft" = "science."

In a synonymous definition, one word is defined by reference to other words, typically by reference to a single word, whereas in a connotative definition, a word is defined by reference to a set of properties it possesses.

Ordinarily, the way in which one gives a synonymous definition is to look the word up in a dictionary to see what it means—a case of lexical (reportive) definition (see below). One will need such a synonymous definition only when the meaning of the word is unclear or unfamiliar, and one wishes to know what this word means. This will often occur with technical words in science involving theoretical definitions (see below).

Exercise 8.4

Give synonymous definitions for each of the following terms: abdominous, epistemology, quid pro quo, exacerbates.

Definition by Genus and Differentia. Synonymous definitions are not available for all words. When this occurs, another kind of definition is often used, one closely related to connotative definitions. This is Aristotle's technique of definition by genus and differentia.

As we have seen, Aristotle believed that a definition should be a report of a thing's essence. This, in turn, should involve, first, a mention of the genus to which the thing belongs and then its differentia—that property or set of properties that distinguishes it as a particular kind of species from other, related species falling under the same genus. For example,

Man $=_{df}$ Rational Animal

is one example. Man belong to the genus *animal* (along with other species of animals), but, unlike these other species, man is *rational.* This property is the difference separating humans from those animals that are not rational, e.g., ducks.

Similarly, a university is defined as an institution of higher learning (genus) qualified to award the Ph.D. degree (differentia). There are other institutions of higher learning that do not award the Ph.D. degrees, but instead just the master's degree or the bachelor's degree; so awarding the Ph.D. degree differentiates Universities from Colleges.

Exercise 8.5

Construct examples for the following terms by matching the definiendum with an appropriate genus and difference.

Definiendum	*Definiens*	
	Genus	*Differentia*

Bachelor
Banquet
Boy
Brother
Child
Foal
Daughter
Ewe
Father
Giant
Girl
Husband
Lamb
Mare
Midget
Mother
Pony
Ram
Sister
Snack
Son
Spinster
Stallion
Wife

Offspring
Horse
Man
Meal
Parent
Sheep
Sibling
Woman

Female
Male
Married
Married
Very Large
Very small
Young

Denotative Definition. If a term has a denotation, then we can speak of the denotative definition of that term:

> A *denotative definition* of a word is a definition of the word by reference to its denotation

Thus, a denotative definition of "scientist" would be a definition involving the particular individual men and women who are scientists. Sometimes individuals prefer to speak of extensional definitions instead of denotative definitions. But according to our usage, an extensional definition and a denotative definition are different. In the case of "scientist," for example, an extensional definition would be a listing of the various classes of scientists: physicist, biologist, chemist, etc. (This kind of definition is not encountered very often.)

There are several example of denotative definitions one can give. We will mention three important types: enumerative, example, and ostensive.

Enumerative Definition. An enumerative definition specifies the denotation of a term by listing *all* the individuals to which the term applies, i.e., its complete denotation. The obvious problem with such a definition is: in order to be useful, the denotation must be finite, and in fact rather small. It would be virtually useless, for example, to define "philosopher" by attempting to enumerate all the philosophers that have ever lived. It's quite different, however, to list a few and then to add "and so forth." This leads us to a second kind of denotative definition, one where we enumerate just one case.

Denotative Definition by Example. Suppose one is faced with the task of defining the phrase: "happily married couple," or "true love." Faced with the obvious difficulty of providing a connotative definition, one might respond: "I can't define this concept in the way indicated. But I can give you a good example of it: Ronald and Nancy" Here one is defining the phrase by giving an example of it. Similarly, if one had to define "rigorous science," one might offer another example by definition and reply "mathematics."

Ostensive Definition. A definition very easy to confuse with the above is an ostensive example. In an ostensive definition, one does not define a term by mentioning other words; here one defines a word by physically pointing to (gesturing to) a thing that is a clear-cut example of it. For example, if someone did not know the meaning of "taupe," one might show him an example of it by pointing to the sweater one was wearing..

In an ostensive definition, one shows the denotation by pointing to an instance of it; one does not use other words to describe the example. Suppose one pointed to several examples of a term, e.g., in defining "dog," I point to Spot, Lassie, etc. Here one would be giving a kind of ostensive definition by pointing to several examples of the concept.

According to many philosophers, some words can only be defined ostensively. Suppose, for example, one does not know the meaning of the term "crimson" or "pain." According to them, no set of words could convey to the person what these terms meant; one would have to have an experience of crimson or pain in order to know what these words meant. These experiences would constitute the ostensive definition of the term.

Contextual Definition. Another kind of definition we will mention is a contextual definition. As the name suggests, in a contextual definition, a word is defined, not in general (i.e., for all cases) but only in one particular set of circumstances or context. For example, one might give a contextual definition of "average" by claiming that "for present purposes, I will define 'average' income to be . . . ." or "the average American has 2 ½ children." Here, an individual is specifying the meaning of the term ("'average") in a limited context; it does not have this meaning in other contexts. Similarly, one might read: "'more scientific than x' will mean anything employing mathematical formulas," or "'force in Newtonian physics' means . . . ." In this latter case, one is giving the contextual meaning of "force" where it is the entire theory of Newton that is the context.

Operational Definition. One of the most important kinds of contextual definitions is called an operational definition. In an operational definition, the meaning of a concept is given by specifying: (1) a set of operations the scientist perform in order to measure or test something, and (2) the ensuing result observed. This is true of properties such as: length, hardness, solubility, etc.

Operational definitions were introduced into science explicitly in the early part of the 20th century by individuals who maintained that many concepts in science were unclear or imprecise (e.g., "simultaneous," "intelligent"). According to them, one would make the concepts precise by giving an operational definition of it. Thus, the "length' of an object is operationally defined by specifying the measuring operations used (e.g., applying a ruler to the object), and seeing how long the object is in relation to the ruler. This would provide a contextual definition involving the length of a particular object. Similarly, in psychology, the term "intelligence" was to be operationally defined by first mentioning the measuring operations (e.g., giving an individual a paper and pencil test), and then calculating the resulting score (e.g., "120"). This wold provide an operational definition of the phrase: an individual has an intelligence of 120".

Exercise 8.6

1. Formulate operational definitions for each of the following.

a. The term "weighs 120 pounds" in "Joan weighs 120 pounds."

b. The expression "believes he will win a scholarship next year" in "Frank believes he will win a scholarship next year."

c. The term "acid" in the expression "this liquid is an acid."

d. The term "alive" in the expression "the patient is still alive."

e. The term "magnetic" in the expression "this object is magnetic."

f. The term "hungry" in the expression "the baby is hungry again."

g. The term :"prime number" in the expression "7 is a prime number."

2. Tell whether each of the following definitions are connotative, synonymous, genus and differentia, enumerative, example, ostensive, contextual or operational.

a. An object is fluorescent, if, when you put it under an ultraviolet lamp, it gives off light whereas it did not before.

b. The provinces of Canada are British Columbia, Alberta, Saskatchewan, Manitoba, Ontario. Quebec, Newfoundland, New Brunswick, Nova Scotia, Prince Edward Island, the Northwest Territories and the Yukon Territories.

c. The "Big Dipper" is that object up there.

d. The constellation *Ursa Major* is the Big Dipper.

e. "Myopia" means "nearsightedness."

f. Marsupials are animals such as kangaroos, wombats, bandicoots, and opossums.

g. Marsupials are an order of animals in which the females have an abdominal pouch for carrying their young.

h. A series of numbers is a geometric progression if the same answer results from dividing the second number by the first, the third by the second, the fourth by the third, and so on to the end of the series.

i. A body is to be called electric if it is "capable of attracting small bodies when rubbed."

j. "Lexicography" means "the act, art., or work of writing a dictionary."

k. Acoelomate is an animal without a coelom.

Use of Definition

One way of classifying definitions is by their type. Another way concerns the use to which they may be put. Of course, what use a definition has will depend on the type of definition it is. In fact, there is a considerable overlap with respect to the various ways in which definitions are classified.

Reportive (Lexical) Definitions. One use to which a definition may be put is to report what the standard usage of some term is. If a student encounters an unfamiliar word, e.g., "exacerbates," she will typically look up its meaning in a dictionary. Such a definition is a reportive definition, also called lexical definition, because the dictionary entries are statements about how a certain language community uses the ward—they are reports of the standard meaning of a term. A reportive definition can be a synonymous definition, but they can also include other types of definitions.

Because this type of definition is used to report the standard usage of a term, they can be true or false. It is true, for example, that "bachelor" means "unmarried male" and false that it means "unmarried female." Reportive definitions can evolve over time and change their meaning but this latter fact does not detract from the point that, unlike other definitions, they are true or false (at a particular point of time).

Stipulative Definition. Suppose I look up the meaning of a word in a lexicon and don't like the meaning of the word. I could redefine it; I could stipulate what I will mean by that term. In so dong, I would be creating the meaning of the word. I can create stipulative definitions for words already present in my language (e.g., "democracy" will mean "American capitalist system" or "true" is going to mean "what the majority believe"), I can also introduce entirely new words into my language. The latter technique is often employed in science when new (and often technical) terms are introduced. Many of the technical terms of science have been introduced this way, e.g., "by the term I.Q. I will mean mental age divided by chronological age"; "mitochondria will mean a particular part of the biological cell that functions in a certain way during respiration," etc. In short, stipulative definitions often function as theoretical definitions (see below).

Unlike reportive definitions, which are true or false, a stipulative definition is neither true nor false. When one indicates how he proposes to use a term, one cannot claim that this new convention is false; one can only require that the person be consistent in their use of the term.

Precising Definition. Because many words, both in common speech and in scientific discourse, are unclear, the need often arises not just to indicate the (new) meaning of a word, but to make the meaning of a term more precise. In these cases, an individual may introduce a definition whose function it is not to create an entirely new word, but rather to take an old concept and to "nail down it meaning" a little more.

This is typically accomplished by making the term less vague (ordinarily if a definition eliminates ambiguity, it will do so by a stipulative definition). To do this, a precising definition must indicate what the cutoff point will be between the presence of the concept and its absence: it must create the boundaries of the concept. It will then be possible to decide how to classify borderline cases. For example, one might decide to define "bald" to be "anyone who has no more than 25 hairs on his or her head." This is stipulating what "bald" will mean (in the present context). The point of dong this is to permit one to separate all the intermediate cases in the gray area. Likewise, a scientist might believe that the term "alive" is vague and try to remove it by advancing a precising definition:: "alive" means "any organism whose heart is beating without artificial support."

Like examples of stipulative definition, precising definitions are not true or false. One can choose to make a term more precise in a variety of ways, the only question being which way is more fruitful. This latter question can often only be decided as more empirical evidence accumulates. It may then turn out that a particular way of dividing cases according to a certain cutoff point is scientifically fruitful (or fruitless), because one discovers the cutoff point is highly correlated with other variables in which one is interested, and which are important for understanding the issue at hand. For example, one may stipulate that "juvenile" will mean those individuals above 13 years of age" because 13 is the average age of initial ovulation and this is an important scientific variable. Likewise, one may make the term "economic security" more precise by specifying that it shall mean "anyone who makes over $50,000 a year because the latter figure is the amount of income one needs in order to have a comfortable standard living in 1998. In short, although it is partly a matter of convention how one makes a term more precise, this decision will have important bearings on future scientific research, and theorizing and the latter, in turn, will often indicate that a decision was well-founded or not.

Theoretical Definition. There is one kind of definition that occurs in science more often than in any other field of human endeavor. This is a theoretical definition. A theoretical definition, we have said, is a stipulative definition, since one is stipulating what a new word will mean. But, it is narrower than an ordinary stipulative definition, because this kind of definition essentially involves a theory in order to understand the meaning of the term. For this reason it is often called a contextual definition. Hence, in order to understand the meaning of a theoretical definition, whose meaning occurs in the entire context of a theory, one must understand the background theory. For example, one might offer the following as theoretical definitions.

> Enzymes are special kinds of protein that control the different steps in the synthesis or breakdown of an organic molecule.

> “Nucleus” means the central part of an tom, the fundamental parts of which are the proton and the neutron. It carries a positive charge and constitutes all the mass of the atom. It is what remains after the orbital electrons have been stripped away.

It should not be thought that there is a very sharp line dividing the various uses of a definition. On the contrary, an individual may advance a definition whose purpose is to stipulate a precising definition, to report a theoretical definition, to make a precising definition, and possibly even to stipulate a theoretical definition that is precising: e.g., "force' will hereafter mean "anything causing acceleration. "The fundamental difference, however, is that a precising definition typically will use an old or accepted term from ordinary language and attempt to make it more precise, whereas a theoretical definition will (typically) introduce a new term into the language (e.g.., "cathexis") although this is not always the case as the examples of "force," "work," and "energy" show.

Are theoretical definitions true or false, or neither? Clearly, theoretical definitions are supposed to be real definitions (on one popular interpretation). Consequently, they can be said to be true or false: they are true if they describe the underlying nature of the entity correctly; they are false, if they don't.

Exercise 8.7

For each of the following, indicate how the definition is being used: reportive, precising, stipulative, or theoretical. If there is more than one use, indicate which of them is the primary one.

1. In chemistry, the term "sublimation" refers to the process whereby a solid changes directly into the gaseous state without liquefaction."

2. A judge to a lawyer defending a thirteen-year-old charged with burglary: " in this district all charged persons under sixteen years of age are juveniles, and hence are not subject to the jurisdiction of this court, but rather to the jurisdiction of the juvenile court."

3. A teacher to her class of sixth-graders: "The word 'juvenile,' which you just came across in this books, means 'a young person.'"

4. The electric volume-density at a given point in space is the limiting ratio of the quantity of electricity within a sphere whose center is the given point to the volume of the sphere, when its radius is diminished without limit."

5. From now on, any substance that is reddish, has the melting point of 1083°C, and the atomic number 29 shall be called copper.

6. "Blind" means "vision corrected only to 20/400 or less."

7. "Superego" is "the portion of the psyche responsible for moral conduct."

8. By "telegnosis" I will mean "information acquired by watching television."

9. A feminine characteristic is one that involves nurturing or aesthetic characteristics.

10. Heat is the kinetic energy of molecules.

Rules for Defining Terms

A theory of definition should contain a set of rules or criteria for evaluating the adequacy of a proposed definition. Such rules would allow an individual to determine if a proffered definition was a good one or not, and these same rules would allow him to construct more adequate definitions. Although these rules were originally advanced as criteria for evaluating definition by genus and differentia, they have, subsequently, been applied to all cases of definition, especially lexical definitions.

1. **A definition should state the essential attributes of the term.** Although this first rule makes sense only from within the realist framework of essences, logicians have typically thought it should apply not only to definitions involving genus and differentia, but to all definitions, especially connotative and theoretical definitions. Of course, when one is providing a denotative, operational, or stipulative definition, this rule seems not to have much point: the latter definitions seem to be conventional definition for which there are not essences in nature.

This point is well taken, but unless one is careful, it can be formulated in a misleading or exaggerated way. Sometimes, a definition in science can be merely a convenient device for summarizing a set of cumbersome sentences. This is true of terms such as "I.Q," "centripetal acceleration," "S.E.S.," and so forth. These are simply abbreviations for longer combinations of concepts and are obviously conventional. This same applies (to a lesser extent) to operational definitions. How one operationally defines a term is a matter of experimental convention, since there is an indefinite number of ways in which measurement procedures and experimental conditions can operationalize a concept. But what is important, as we have already pointed out, is that not every way of operationally defining a concept is as good as any other way. Which particular operational definition is a good one will depend upon subsequent empirical evidence; an operational definition will prove to be useful in science if it correlates well with other empirical evidence involving the same, or related, concepts. This is the case, for example, with respect to the traditional operational definition of "hunger" as involving hours of food deprivation and subsequent eating behavior. This operational definition correlates well to other definitions of hunger involving, say, blood glucose level or activity in the ventro-medial hypothalamus.

Hence, subsequent scientific evidence may indicate that one particular operational definition is not as good as another one because the empirical operations do not cluster together into regularities. This, in turn, indicates there is some empirical feature of reality that accounts for the presence or absence of such lawful regularities of data. Ultimately, therefore, operational definitions are not a matter of arbitrary convention, but are rooted in the structural regularities to be found in nature. These structural features, in turn, are known only by means of our theories, which involve theoretical definitions.

Theoretical definitions purport to be real definitions and not merely conventions. That is to say, when one offers a theoretical definition of a concept, say, *gold* or *electron*, this definition claims that the properties listed in the definiens are the real properties of this entity. Hence, if science is going to discover the real properties of independently existing entities, these entities must have something like a real nature, or, as Aristotle would have put it, an essence. Ultimately, therefore, our definitions must deal with the essential properties of things.

2. **A definition should not be circular.** Few individuals would quarrel with this criterion. If one advances a definition of a term, presumably this term was one that was unclear. Now, if one proceeds to use this very term in the definiens, how will the individual know what is being talked about? If one defines "human" as "anything with human-like characteristics," this is a circular definition of "human," and is of little value. Likewise, with "science is an activity engaged in by scientists." In short, one should not use the definiendum—the term being defined—in the definiens—that which is the definition.

3. **A definition should be neither too broad, nor too narrow.** According to this criterion, a definition should express an equation (and if and only if relation) involving necessary and sufficient conditions. Hence, the two must be logically equivalent.

A definition should, first of all, not be *too broad.* That is to say, a definition should not include cases to which the term does not apply.

Thus, if one defined "science" as "any academic subject," this would be too broad since the definition includes examples, such as art, not ordinarily thought to be a science. If one wishes to criticize a definition, therefore, one way to do so is to show that it is too broad, and this is equivalent to showing that the biconditional is false, since you have produced an example where one side of the biconditional is true and the other side false; e.g., art is not a science, but art is an academic subject.

A definition, on the other hand, should not be *too narrow*, that is to say, it should not exclude cases to which the term does apply. Suppose, for example, I define "science" as "those academic disciplines that employ the experimental method." This would be too narrow, because it would rule out certain examples of science, e.g., astronomy, geology, that do not employ the experimental method. Since it is accepted that these examples are cases of science, the definition is too narrow. This would be equivalent to showing that the above definition, expressed as biconditional, is false; something is a science but is not experimental.

4. **A definition should not be expressed in ambiguous, obscure, or figurative language.** Obviously, if the original word is unclear, one will want a definition to remove this lack of clarity. But, if a definition is advanced that is itself ambiguous or obscure, this will hardly be of help. To be told, for example, that "life is neither the wick nor the was of a candle, but the burning" may be metaphorically enlightening—see Chapter 11—but it will not be of much help understanding the nature of life from a scientific perspective. The same is true of: "man is an ape with possibilities."

It should not be thought, however, that a proposed definition is unclear just because it is very technical. A technical definition may not be very clear to the ordinary layperson, but quite clear to those acquainted with the technical language. In fact, this is typically the case with very precise but technical scientific disciplines.

5. **A definition should not be negative where it can be affirmative.** Since a definition is supposed to explain what something is (or means) instead of what it is not (or what it does not mean), this criterion stipulates that definitions, except for unusual circumstances, must always be expressed in a positive or affirmative way. If I want to know what a gene is, and I am told that it is not a neuron, this will not be of much help to me.

Sometimes, however, it is impossible to define terms except in a negative way; e.g., "bald" = "lacking hair on one's head." Fortunately, these cases are rare.

Exercise 8.8

Criticize the following definitions according to the rules for defining a term. Note: some definitions are not defective.

1. "Square" means a plane figure consisting of two isosceles right triangles having a common hypotenuse.
2. "Cause" means something that produces an effect.
3. "Antidote" means a remedy to counteract the effects of arsenic.
4. "Economics is the science which treats of the phenomena arising out of the economic activities of men in society."

—J. M. Keynes, *Scope and Method of Political Science*

5. A depressant is not a stimulant.
6. a. "Velocity" means the same as "speed."
 b. In physics, velocity is how fast something travels.
 c. In physics, velocity is the rate at which a body's position changes with time.
7. a. Science is an organized body of knowledge.
 b. Science is where you test things empirically
8. Scientific method is any systematic approach to research.
9. Intelligence is what intelligence tests measure.
10. Man is a featherless biped.

Exercise 8.9

Definitions are often used and misused in arguments. For each of the following, (1) point out the premises and conclusions, (2) point out any definitions, (3) indicate their type, and (4) criticize each of the definitions using the rules discussed above.

1. A team of researchers from the University of Colorado has found that religious people who are drawn to the mystical qualities of life make the safest drivers. According to their report, the subjects of their study were 264 airmen at Lowry Air Force Base in Denver, whose accident records were known. According to Vernon Allport's definition, the religious man is one who is concerned with the supernatural meaning of all existence and particularly his place in it. Subjects who belonged to this type according to the test had collision accident scores lower than those of the non-religious type. The study concluded that this non-religious type (The Theoretic and Aesthetic) had higher scores because of a general tendency to be less conventional,

psychologically more complex and conflicted, and more ready to resort to complex defense operations as a protection against anxiety.
2. The development of leisure in a civilization permits time for the development of the arts, philosophy, history and the sciences. When these areas of human endeavor become developed, then the civilization can truly be said to be a "high" civilization. Thus, the test of the quality of a civilization is the quality of its leisure. What then about leisure in America? In America we have a great deal of time but we do not truly have leisure. Because leisure is indeed an affair of mood and atmosphere rather than simply of the clock. Leisure is not a chronological occurrence but a spiritual state. It is unharried pleasurable living among one's native enthusiasms. Leisure is those pauses in our lives when experience is a fusion of stimulation and response. Genuine leisure yields at once a feeling of vividness and a sense of peace. In America we have plenty of time to kill, but we do not have leisure.

For Further Reading

A useful selection from the work of Mendeleef can be found in Runes (1962, pp. 739-746) and in Hurd and Kipling (1964, pp. 83-100). I have found no complete philosophical discussions of Mendeleef and the periodic table.

On biological taxonomy, of which there is a considerable literature, see Simpson (1961). In Psychiatry, see American Psychiatric Association (1980).

The subject of definition is covered in most standard textbooks on logic. In addition, Robinson's (1950) is highly recommended.

Notes

[1]As cognitive psychologists put it, heterogeneous and discrete data are difficult to remember. But if we codify this knowledge in the form of bits or chunks, the human mind can remember this more easily. It is thus a function of the human mind to transform discrete information into more organized forms, one example of which is classification.

[2] This is contrary to what Jevons (1905, p. 690) suggests. In answering the question, How should we group together individual objects?, Jevons replies: "Group them in every way in which they can be grouped and then observe in which method of grouping the correlation of properties is most conspicuously manifested." The problem is that there is an indefinite number of ways of

grouping objects; to suggest that one do it "in every way in which they can be grouped", is to suggest something impossible.

[3] He predicted, for example, that it had an atomic weight of 72 with a boiling point of 160° C, with oxide-forming properties, etc. Most of these predictions were (basically) correct.

[4] For example, knowing that the atomic number of sulfur is 16, we know it has 16 electrons and hence that its electron configuration is: $1s^2 2^2 2p^6 3s^2 3p^4$.

9

Quantitative Description

Descriptive Statistics and Measurement

Introduction

One can classify an individual qualitatively or quantitatively. In qualitative classification, an individual object, event or process is put into a category; the fundamental questions here are whether an individual belongs to the category or not and how this category relates to other categories. Ordinarily, with respect to many categories, one assumes that membership in the category is an all-or-none affair: either one belongs to the category or one does not. Furthermore, in such discrete classification, which we call categorization, the class in question is typically a qualitative one and not a quantitative one; the underlying assumption here is that there are no degrees of the categorical property. For example, the properties—red , table, male, religious scientist, flower, Republican—are such that each is either completely or totally present in an individual or it is not.

In *quantitative classification,* by contrast, an individual is placed in a numerical or quantitative class, e.g., hot, wealthy. Here, although a member either completely belongs to the class in question or does not, the property in question *can vary in amount or degree,* e.g., one object can be hotter than (wealthier than) another, so that there is a *comparative degree* of assessment. Furthermore, it seems possible not only to employ comparative quantitative assessment; one can also describe the amount of the property in question (e.g., the actual degree of heat, the quantity of annual income). Here, one is not merely employing comparative degrees; rather, one is measuring the quantity or magnitude in question. These properties are not discrete properties but *continuous* ones.

Determining how much (how many instances) of the property is present involves the process of measurement, of determining how hot something is, how wealthy someone is, etc. Measurement includes both the notions of counting the number of instances of a property and also determining the amount of something (measurement proper).

Descriptive Statistics

Counting

A *numeral* (e.g., "12," "XII") is a physical symbol representing a *number* (e.g., *Twelve*), which is an abstract property; in addition, there is also *the number of* something (e.g., the apostles). In counting and measuring, one assigns a numeral to the number of something. In counting the number of apostles, for example, one assigns a numeral ("12") to this group, but the number of apostles is not normally thought to be the same as the number Twelve. In addition to counting the number of apostles, one can also determine how many apostles were right-handed, how many were fishermen, etc. Here, we are determining frequencies of membership in a class.

When one counts the number of objects in a group, one is determining *how many* there are (e.g., twelve). This is different from the question of *how much* of a quantity is present (e.g., 4 lbs.). In counting, one first sets up a dichotomous set of classes (e.g., male vs. female), assigning individuals to these two classes, and then counting the number of discrete individuals present in each.

Counting involves an operation called a *mapping* of individual objects (members of the class) onto the set of natural numbers (really natural numerals): 1, 2, 3, ... such that the two sets are in a particular kind of (one-one) relationship: every object gets assigned one and only one natural numeral and no natural numeral gets assigned to more than one individual. In this way, one can determine the *total number* of individuals present. By counting, one can determine the number of individuals present in one class (e.g., triangles) and in another class (e.g., squares). The science most directly concerned with the process of

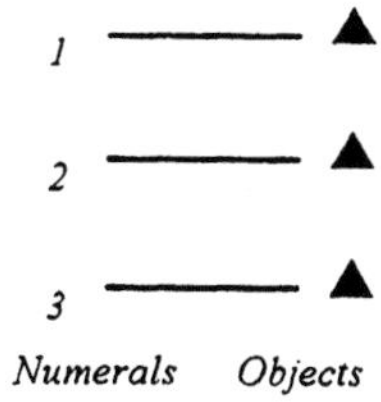

counting and the legitimate conclusions one may draw from these operations is statistics.

Descriptive versus Inferential Statistics

Statistics may be defined as the methodology of collecting, describing, and analyzing data and drawing conclusions from them. There are two branches of statistics--descriptive statistics and inferential statistics. **Descriptive statistics** is concerned with the logic behind the collection, classification, description, and analysis of numerical data obtained from observations and experiment. As such, it is concerned with the appropriate methods of organizing and summarizing the information contained in this data. This involves the construction of tables, charts, graphs, and the calculation of descriptive, summarizing indices such as *average*, *percentile*, *standard deviation*, etc.

By contrast, **inferential statistics** is concerned with the question of what conclusion may legitimately be inferred from such descriptive statistics, in particular, what conclusion may be drawn about a population from information contained in a sample of that population. For example, from a sample of 250 individual subjects contacted over the telephone about their political views, what may be inferred about the political views of the entire population?

Inferential statistics is a relatively new field, developed largely in the 20th century, and especially in conjunction with experimentation and experimental design. Statisticians have developed several arguments concerning the logic of statistical inference, especially inferences about statistical hypotheses and their appropriate tests. Because of this, we will devote a separate chapter to it (Chapter 14) but only after we have discussed the hypothetico-deductive method and the logic of the experimental method. For now, we will simply summarize the basic principles of descriptive statistics and how they relate to quantitative classification and the numerical summary of data.

Sample vs. Population

The fundamental entities in statistics concern a **sample** and a **population**, with descriptive statistics being concerned with numerically describing characteristics of the sample and the population. The logic behind the use of a sample is obvious. In some cases, one obtain information about the entire population by observing each individual in the population and determining whether these individuals have a characteristic feature or not. This would be a **census**, a version of complete (perfect) induction. As we saw earlier, such inductive inferences are of limited value and this is also true of a census. Wallis & Roberts (1962) give several reasons for using a sample instead of a census: (1) the measuring process can destroy that which one is investigating, (2) the cost could be prohibitive , (3) a sample may be a more accurate than a census, (4) a census may be logically impossible, and (5) the population may be inaccessible. For these and other reasons, it is typically necessary to employ a sample.

A sample is always a sample of a population. Although originally a population meant a class of individual people (e.g., the class of all college sophomores), it was later extended to cover any class of objects that could be counted and still later to include a group of numerical observations. Hence, in the broadest sense, a **population** is the collection of all actual and possible individuals, items, numbers, or data under consideration (e.g., all possible scores on an IQ test), and a **sample** is the part or subset of this population actually selected or observed (see Figure 9.1). The characteristics of a population are known as **parameters**, whereas those of a sample are termed **statistics.** The former are denoted by Greek letters (μ, σ) and the latter by ordinary letters ($\bar{x}$, s, etc.).

There are various ways of sampling and hence various kinds of samples, e.g., random sampling, cluster sampling, stratified sampling, etc. In *stratified sampling* the population is divided into subpopulations (strata), e.g., women and men, and sampling is done from each of these strata. *Cluster sampling* is the successive sampling of sets and subsets in a hierarchical fashion. In a *convenience sample*, one does not sample randomly; instead, one samples whatever is available, e.g., one accepts volunteers to serve as subjects in a psychological study. (For the risks associated with this kind of sampling see Chapters 13 and 14.) But the crucial issue in all kinds of sampling is this: Is the sample *representative* of the population? If it is not, clearly an inference to the population will be fraught with danger, whereas if it is, we may be reasonably confident in our conclusion. For our purposes, the most important kind of sampling is **simple random sampling**. Here, each

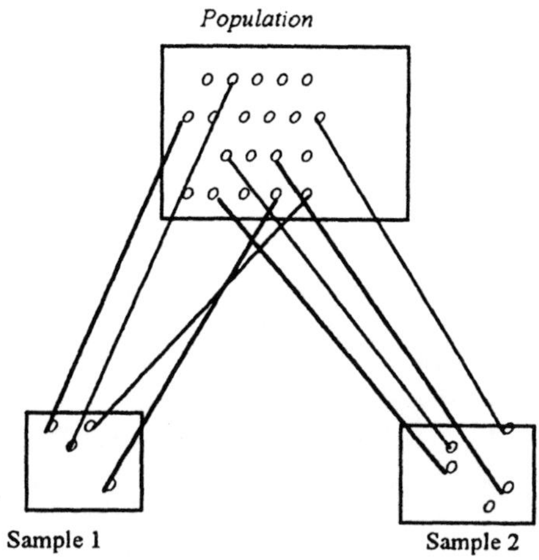

Figure 9.1 Population and Sample

possible sample has an equal chance of being selected, with the selection process being determined by some kind of randomizing device or a table of random numbers. Simple random sampling can either be done *with replacement* or *without replacement.*

If a sample is a random one, one of the essential epistemic features of a good sample will have been meet, namely, that it will be a **representative sample** of the population--it will be an accurate reflection of the relevant features of the population. It is possible, of course, for a random sample not to be representative. Merely by chance, for example, a random sample might turn out to be uncharacteristic of the population. But according to statistical theory, if a sufficiently large number of random samples are taken, these samples tend to be representative of the population at large, with the degree of representativeness depending on the size of the sample and the number of samples. Since the key properties of the population concern the numerical values of one or more quantitative properties (parameters), the key question here is *whether the numerical properties (statistics) of a (random) sample are representative of the numerical properties of a population (its parameters)?* As we will see, several important theorems address this question, providing the basis for the theory of statistical inference.

Organizing Data

Once having selected a sample, one can then collect raw numerical data from this sample by means of observations (e.g., a questionnaire) or an experiment (e.g., subjects' reaction time to visually presented stimuli). This collection of numbers needs then to be *organized* in some way, since it is cognitively difficult to deal with the entire collection of data and to determine if there are any significant relationships present. In short, one has to make sense out of raw numerical data just as one has to make sense out of raw qualitative data (Table 9.1) . This ordinarily occurs by *grouping*--by putting data into numerical groups--and determining the frequency of occurrence in each class by counting the members. In such a case, we would have a **frequency distribution**. (Table 9.2). However, we might also be interested in the percentage of each class present, in which case we would construct a **relative frequency distribution**, a listing of classes together with the ratio of the frequency of that class to the total number of data (i.e., the relative frequency). Most statisticians interpret probability in terms of relative frequency; hence, one can translate a (relative) frequency distribution into a **probability distribution.** One could also construct a cumulative frequency distribution**,** which results from summing the frequencies of all classes less than the class limit. A cumulative relative frequency distribution, obtained by dividing the cumulative frequency by the total number of data (the **cumulative relative frequency**). These types of groupings are illustrated below. Suppose, for example, a physician collected the following data on the ages of people with diabetes. Ordinarily, one groups numerical data by 10's, resulting in these classes: 0-9, 10-19, etc. Using these class intervals, one could determine the frequency, relative frequency, cumulative frequency, and cumulative relative frequency of each class, resulting in Table 9.2.

Table 9.1. Hypothetical Data Collected concerning the Ages of People with Diabetes

48415783415559
613848797577 7
54234756796861
64455382683870
10608376216543

Such a table is sometimes called a **grouped data table.**

Table. 9.2 Frequency, Relative Frequency, Cumulative Frequency, and Cumulative Relative Frequency of Ages of Diabetes

Age	Frequency	Relative Freq.	Cum. Freq.	Rel. Cum. Freq.
0-9	1	1/35 = .03	1	.03
10-19	1	1/35 = .03	2	.06
20-29	2	2/35 = .06	4	.12
30-39	2	2/35 = .06	6	.18
40-49	7	7/35 = .2	13	.38
50-59	6	6/35 = .17	19	.55
60-69	7	7/35 = .2	26	.75
70-79	6	6/35 = .17	32	.92
80-89	3	3/35 = .09	35	1.0
	35	1.00	35	1.0

If one graphs such a frequency table, one obtains a **histogram** in which the classes are represented on the horizontal axis (the abscissa) and the frequencies (or relative frequencies) on the vertical axis (the ordinate). The frequency (or relative frequency) is represented by the height of the vertical bar. A histogram of the frequency of diabetes appears in Figure 9.2.

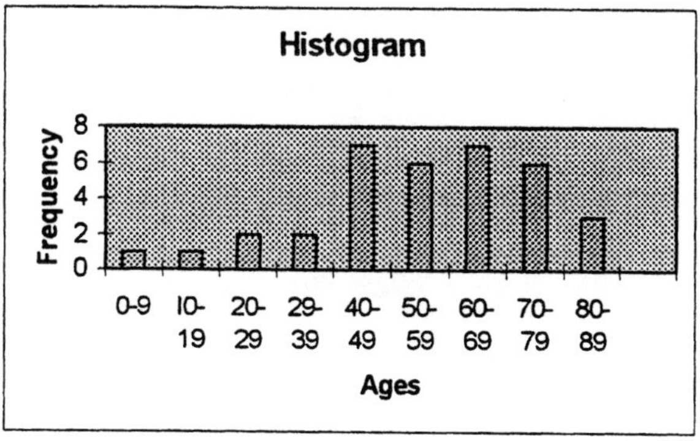

Figure 9.2 Frequency Histogram of Diabetes and Age

Although Figure 9.2 shows the frequency of the age of diabetes onset, one can also plot the relative frequency (instead of the frequency) of the age of diabetes onset.

Another way to graph a set of data is by means of a **frequency polygon.** Here, a point is plotted above each class mark at a height equal to the frequency of the class and connecting lines (Figure 9.3) joins the points.

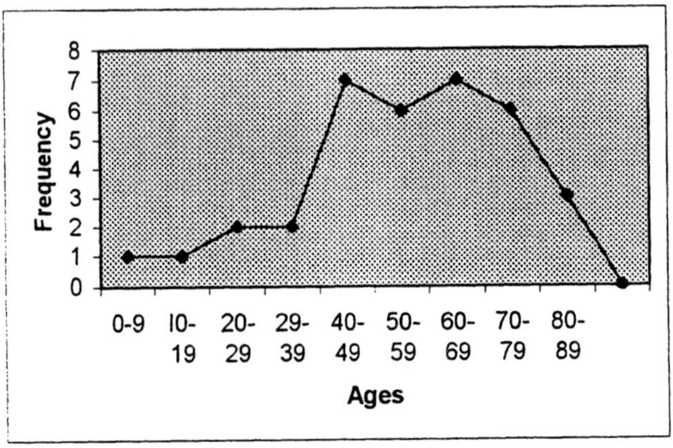

Figure 9.3 Frequency Polygon of Ages of Diabetes

One can also construct a cumulative frequency (and cumulative relative frequency) curve or polygon. In the latter case, the curve is often called an **ogive** (See Figure 9.4).

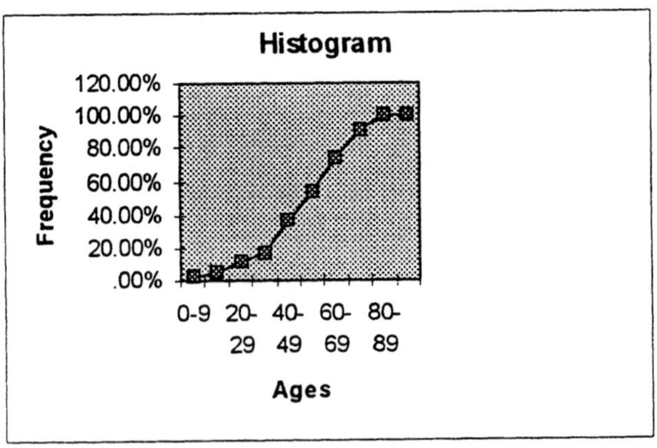

Figure 9.4. Cumulative Relative Frequency of Age of Diabetes

As the number of cases increases and the class intervals become smaller and smaller, the number of rectangles in a histogram increases;

as this process continues, the tops of the rectangles become more and more like a curve.

Distributions can have different shapes. In particular, the kind of curve representing the distribution can take on a variety of forms: bell-shaped, triangular, uniform, J-shaped, right skewed, bimodal, multimodal, etc. (see Figure 9.5). If one draws a smooth curve through

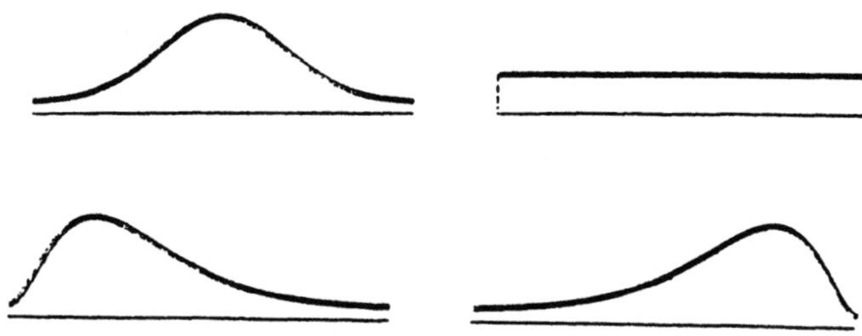

Figure 9.5 Common Distribution Shapes

the corresponding histogram representing the data, the histogram can be said to have a characteristic shape. For example, the distribution shape of our diabetes example has the form of being left-skewed.

Random Variable

In probability theory and statistics, the concept of a random variable plays an important role. A **random variable** is a numerical quantity whose value depends on chance. For example, the height of a person chosen at random is a random variable, since the particular value of this variable (e.g., 6 ft.) depends on chance. Coin tosses, drawing from a deck of cards, etc., are random variables, whereas a biased coin, a weighted roulette wheel, etc., would result in nonrandom variables since all of the values of the variable do not have an equal chance of being selected.

There are two kinds of random variables: discrete random variables and continuous random variables. A discrete random variable is a random variable whose possible values have a finite[1] number of values, e.g., gender, the outcomes of a coin toss, the number of cars owned, etc. A continuous random variable is a random variable whose possible values form an infinite[2] set of values, e.g., height, weight, and speed.

We can form a probability distribution and a probability histogram for a discrete random variable. Ordinarily, the probability of a discrete random variable is just the relative frequency (percentage) of its various values represented, for example, in a histogram. The probability distribution of a continuous random variable is curve, the most important kind of which is the normal distribution (the normal bell-shaped curve).

Exercise 9.1

In each of the following exercises, construct a grouped data table, with equal intervals (starting with the class interval specified). Include frequency, relative frequency, cumulative frequency, and cumulative relative frequency columns. Then construct a frequency histogram, polygon, and cumulative relative frequency curve. What is the distribution shape: roughly symmetrical, right-skewed, or left-skewed?

1. The United States National Science Foundation collects data on the ages of recipients of science and engineering doctoral degrees. A sample of one year's recipients yields the following ages. (Start with the interval: 20-24)

37	28	36	33
37	43	41	28
24	44	27	24

2. A biologist is studying the gestation period (duration of pregnancy) of domestic dogs. Fifteen dogs are observed during pregnancy and are found to have the following gestation periods (in days.)

62	61.4	59.8	62.2	60.3
60.4	59.4	60.2	60.4	60.8
61.8	59.2	61.1	60.4	60.9

(Start with the class 59-59.4).

3. A pediatrician who tested for cholesterol levels of several young patients was alarmed to find that many had levels over 100 g per 200mL. The readings of 20 patients with high cholesterol levels are presented below. (Use a class width of five, starting with 95-99).

210	209	212	208
217	207	210	203
208	210	210	199
215	221	213	218
202	218	200	214

4. As reported by the College Entrance Examination Board in *National College-Bound Seniors*, the average verbal score on the SAT in 1991 was 422 points out of a possible 800. A sample of 25 verbal scores for last year follows:

346	496	352	378	315
491	360	385	500	558
381	303	434	562	496
420	485	446	479	422
494	289	436	516	615

(Start with the class 250-299)

5. IQs measured on the Stanford Binet Revision of the Binet-Simon Intelligence Scale have a mean of 100 points and a standard deviation of 16 points. Suppose that 25 randomly selected people are given that IQ test and that the results are as follows. (Start with the interval: 80-85.)

91	96	106	116	97
102	96	124	115	121
95	111	105	101	86
88	129	112	82	98
104	118	127	66	102

Descriptive Measures

In addition to organizing and summarizing data by means of tables and graphs, one can also give numerical descriptive summaries of them. These descriptive measures are of two kinds: measures of central tendency and measures of dispersion.

Measures of Central Tendency

Measures of central tendency indicate the (approximate) center, average, or typical value of the data sets: the mean, the median, and the mode. The **median** of a set of data is that number that divides the bottom 50% of the data from the top 50%. To do this, arrange the data in increasing order and then determine the middle value. If the number of data is odd, the mean is the data value exactly in the middle; if it is even, then the median is the mean of the two middle data values. For example, if our data set consists of the observations: 1.8, 2.0, 1.8, 1.9, 2.0, the median is 1.9. If our data were 1.8, 1.7, 2.0, 1.9, 2.0, 1.8, then the median would be 1.85.

The **mode** is that numerical value of a data set that occurs most frequently, e.g., in the set 1.8, 2.0, 1.8,1.9, 2.1, the mode is 1.8. If no value in the set of data occurs more than once, the data set has no mode; if there is a tie for the most frequently occurring value, a data set has two modes. In 1.8, 2.0, 1.9, 1.7, 2.1, there is no mode. In 1.8, 2.0, 1.8,. 2.0, 1.9, 2.1, there are two modes: 1.8 and 2.0.

The **mean** of a set of data is the sum of the data divided by the number of pieces of data:

$$\text{Mean} = \frac{\text{Sum of data}}{\text{Number of pieces of data}}$$

For example, if one's data set consists of observations 1.8, 2.0, 1.8, 1.9, 2.0, the mean is 1.9:

$$\text{Mean} = \frac{1.8 + 2.0 + 2.0 + 1.9 + 2.0}{5} = 1.9$$

If x = our set of observations (consisting of $x_1, x_2, \ldots, x_n$), then the mean

$$\text{Mean} = \frac{x_1 + x_2 + \ldots + x_n}{n}$$

which we can symbolize as:

$$\text{Mean} = \frac{\Sigma y}{n}$$

where 'Σ' means sum all of the values of y.[3]

Samples have a mean and so do populations. The sample mean is symbolized by $\overline{x}$ and the population mean by μ

Measures of Dispersion

Whereas measures of central tendency describe the center or typical value of a set of data, measures of dispersion indicate the spread or variation of the data. The **range**, for example, is the difference between the maximum and the minimum values of the data. The interquartile range, typically used in conjunction with the median, is the difference between the first and the third quartiles, i.e., the middle 50% range of the data.

For our purposes, however, the variance and standard deviation are the most important measures of dispersion. The **variance** of a set of data is a measure of how much each particular datum varies from the mean. For example, given our data of 1.8, 2.0, 1.8, 1.9, 2.0, and our mean of 1.9, the sample variance of this set of data (abbreviated as s^2) is determined by taking each datum (x) and the mean ($\overline{x}$), calculating their difference ($x - \overline{x}$) squaring it $(x - \overline{x})^2$, summing them all up $\Sigma\,(x-\overline{x})^2$, and then dividing by the number of data (n):

$$s^2 = \frac{(1.8 - 1.9)^2 + (2.0 - 1.9)^2 + (1.8 - 1.9)^2 + (1.9 - 1.9)^2 + (2.0 - 1.9)^2}{5}$$

$$= .01$$

or symbolically:

$$s^2 = \frac{\Sigma(x - \bar{x})^2}{n}$$

This is the variance of the sample (s^2). Because the sample variance (and standard deviation) are often used as estimates of the population variance (σ^2) and standard deviation (σ), and because dividing by n tends to underestimate the population parameters, in calculating the sample variance one divides by n-1 instead of n, i.e.,

$$s^2 = \frac{\Sigma(x - \bar{x})^2}{n-1}$$

In our example, this would give us a value of .0125 instead of .01. At this point in our discussion, however, we will continue to use n instead of n-1. The population variance is:

$$\sigma^2 = \frac{\Sigma(x - \bar{x})^2}{N}.$$

The **standard deviation** is just the square root of the variance and it has two forms, one for the sample and one for the population. For the sample, the standard deviation (s) is:[4]

$$s = \sqrt{\frac{\Sigma(x - \bar{x})^2}{n - 1}}$$

In our example, this would be .1 ($\sqrt{.01} = .1$). For the population, the standard deviation is:

$$\sigma = \sqrt{\frac{\Sigma(x-\mu)^2}{N}}$$

Exercise 9.2

Using the problems in Exercise 9.1, determine the mean, median, and mode of each set of data. Then calculate the variance and standard deviations.

Z-Scores

The standard deviation measures the degree of dispersion or spread of a set of data. Given the standard deviation and the mean of a set of data, one can determine the standing of any particular individual datum in relation to the mean and standard deviation of that group of data. This numerical measure is called a standard score or **z-score**, the number of standard deviations the particular numerical datum is away from the mean. It is computed by subtracting a particular datum value from the mean of the data and dividing by the standard deviation of the data set:

$$z = \frac{x - \bar{x}}{s}$$

For example, suppose your final grade in a large class is a B-, which is roughly (say) 2.75 GPA on a 4-point scale. How well did you do in comparison with the rest of the class? Perhaps the class in general did very well (in which case you did not do as well as the class as a whole did) or the class may have done very poorly (in which case you did pretty well). Suppose the average grade in the class was a C+ (2.25) and the standard deviation was .2 GPA. Then your z-score would be:

$$z = \frac{2.75 - 2.25}{.2} = +2.5$$

What does this mean? It means that your grade is 2½ standard deviations higher than the class mean. This is obviously good, but how good is this? It can be shown (via a famous rule—Chebychev's rule)[5] that 75% of all the data lie within two standard deviations on either side of the mean ($\bar{x} \pm 2s$, i.e., $\bar{x} + 2s$ and $\bar{x} - 2s$), that 89% of the data lie within three standard deviations of the mean, and so forth (see Figure 9.6).

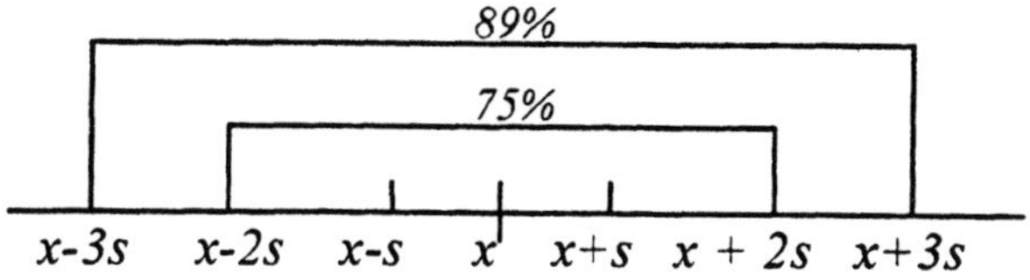

Figure 9.6. Chebychev's Rule

Suppose, for example, one has a z-score of +2.0. Then 75% of the area lies between -2 and +2 standard deviations of the mean and 25% of the scores fall outside this area. A z-score of +2.5 indicates that about 82% (more precisely, 84%) of the class grades are below your grade, which is very good (see Figure 9.7). A similar evaluation of other grades (e.g., 1.6, 2.0) can be made by reference to their z-scores, which would indicate their relative standings. For example, a z-score of -2 would indicate the score was two standard deviations below the class average. In this case, your GPA is higher than 84% of the rest of the scores of the students.

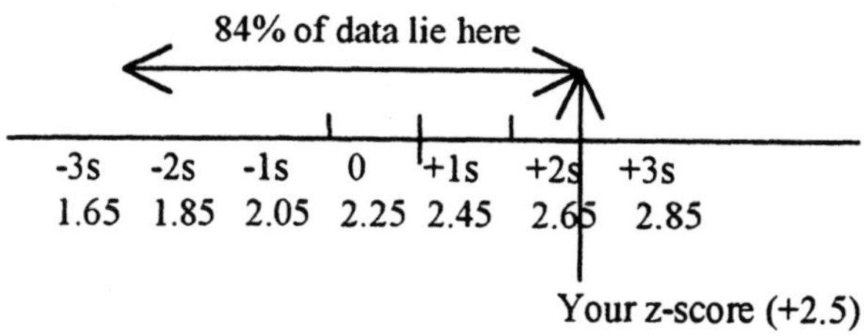

Figure 9.7. Relative Standing of a Z-score.

Normal Distribution

Every student is familiar with the concept of the normal curve (the bell-shaped curve), which was briefly mentioned above as the probability distribution of a continuous random variable. The *standard normal curve* (or z-curve) has the following properties (see Figure 9.8): (1) the total area under the curve is 1; (2) the curve extends indefinitely in both directions approaching but never touching the a mean of 0 and a standard deviation of 1.0. horizontal axis; (3) it is sym-

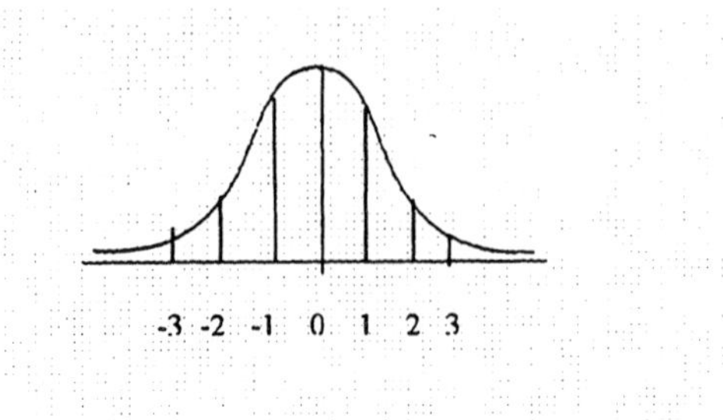

Figure 9.8. Standard Normal Curve

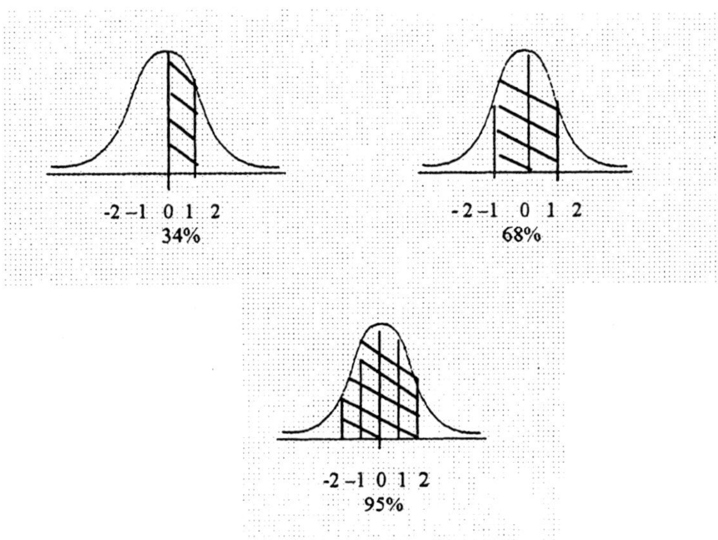

Figure 9.9. Percentages of areas of normal curve.

symmetric about the midpoint 0; and (4) it has a mean of 0 and a standard deviation of 1.0

How much of the area (the distribution of the data values) falls in the region between the mean and +1.0 standard deviations? Between -1.0 and +1.0? This is shown in Figure 9.9. This means that 95% of the area under the standard normal curve (the data values) lie between -2 and

+2 standard deviations from the mean. (This should not be confused with the values for Chebychev's Theorem.)

Scientists collect data and categorize them in certain classes, often forming a grouped data table with a corresponding histogram. Since one does not take a non-denumerably infinite set of values, one's histogram will be a discrete probability distribution. However, discrete probability distributions often approximate continuous probability distributions. In fact it is often convenient to superimpose a normal curve onto a discrete probability histogram.

Let us return to our earlier discussion of z-scores. Using the concept of a normal curve, a z-value of +1 indicates this score fell one standard deviation above the mean, a z-value of .25 indicates it fell 1/4 of one standard deviation above the mean, etc. In terms of the area under the normal curve, a z-value of +2 means that 95% of the area lies between -2 and +2 standard deviations and hence that 5% of z-values fall above your score (1-95%).

Exercise 9.3

In the examples for Exercise 9.1, determine the z-scores as indicated.

1 For the USNSF example, calculate 32, 36.
2. For the gestation example: 59, 63
3. For the pediatrician example 204, 210.
4. For the SAT example: 397, 460.

Correlation and Regression

Up to now, we have been discussing representations of a single variable, e.g., the age of having diabetes. But obviously a scientist will also be interested in the relation between two or more variables, e.g., the relation between the age of having cancer and smoking behavior, the relation between the temperature of a gas and its pressure, the relation between IQ and grades, etc.

Sometimes this bivariate relation will be causal but sometimes it will be correlational. Two 19th century thinkers—Francis Galton and Karl Pearson—laid the foundations of statistical correlation.

Sir Francis Galton (1822-1911) was a pioneer in the study of intelligence. He believed that genius was inherited and in attempting to study how it was inherited over generations, he investigated how height was inherited over successive generations. In one of his tables, he listed children's heights and parent's heights in terms of their frequency of occurrence. (This was the origin of the notion of a scatterplot.) He discovered that tall parents tended to have children who were taller than average but not as tall as their parents were; short parents tended to have children shorter than average but not as short as their parents. In summary, children's heights tended to drift back to the mean. A follower of Galton, Karl Pearson (1857-1936) called this drifting back or "regression to the mean" by the term r, which is also known as a *correlation coefficient.* (Pearson did a study similar to Galton and found that the correlation between fathers' heights and childrens' heights was .5). Although the original sense of r as a regression to the mean has been lost, r is still used as a label for the correlation between two variables.

Although one can list the relations between two variables in a table and frequency distribution, it is often more useful to represent them by a graph—a scatter diagram. A scatter diagram or plot shows the distribution of the particular values of two variables where each dot in the scatter diagram represents one pair of values--one from the horizontal (x) axis and one from the vertical (y) axis. For example, it is natural to expect that there is a connection between number of hours spent studying and one's grade point average. Suppose, for example, we had the following data:

	Hours of Study(x)	GPA(y)
Student #1	42	3.3
Student #2	23	2.9
Student #3	31	3.2
Student #4	35	3.2
Student #5	16	.9
Student #6	26	2.4
Student #7	39	3.7
Student #8	19	2.5
Mean	$\bar{x} = 28.9$	$\bar{y} = 2.9$
Standard Deviation	$s_x = 8.8$	$s_y = .55$

From this data, we can construct the scatter diagram shown in Figure 9.10. Clearly there is a relationship here between the amount of time spent studying and one's GPA, namely, the more time spent studying, the more the scatter per week, the greater one's GPA. In a scatter diagram, the more the scatter or spread, the weaker the relationship, and the narrower the scatter or spread, the stronger the relationship. But what is the precise degree of relationship between the two variables, the degree to which there is a co-relation? This is answered by the **correlation coefficient**[6]--a numerical value indicating the strength of the linear relationship between two variables. See Figure

9.11. The correlation coefficient (**r**) can vary from -1.0 to +1.0, with 1.0 and +1.0 indicating very strong linear relationships—one negative and one positive--and 0 indicating the absence of such a relationship. A correlation coefficient can be calculated in various ways, but one useful way is in terms of standard- or z-scores: first convert each value of x to a corresponding z-score, convert each value of y to a corresponding z-score; multiply each respective pair of z-scores, sum all of them up, and divide by n.

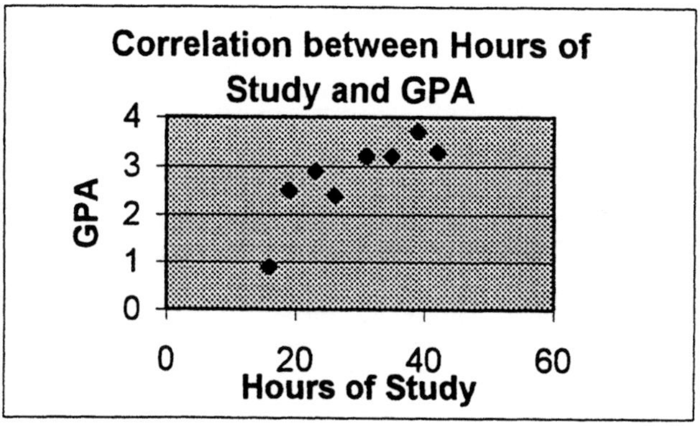

Figure 9.10. Scatter Diagram of Hours studying and GPA

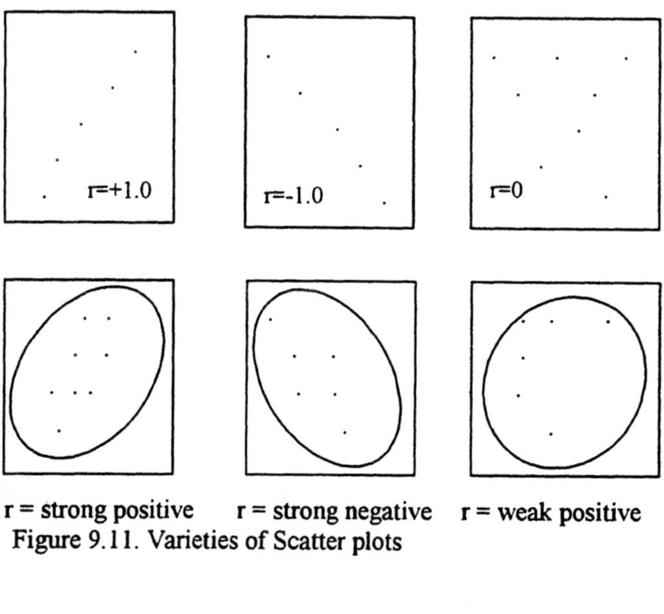

r = strong positive r = strong negative r = weak positive
Figure 9.11. Varieties of Scatter plots

For example, what is the correlation coefficient between hours of study and GPA? First, we calculate z-scores for each value of x and y. For the x variable (hours of study), the z-scores are:

$$\frac{42-28.9}{8.8} = 1.49 \quad \frac{23\text{-}28.9}{8.8} = -.62 \quad \frac{31\text{-}28.9}{8.8} = .24 \quad \frac{35\text{-}28.9}{8.8} = .69$$

$$\frac{16-28.9}{8.8} = -1.47 \quad \frac{26\text{-}28.9}{8.8} = -.33 \quad \frac{39\text{-}28.9}{8.8} = 1.15 \quad \frac{19\text{-}28.9}{8.8} = -1.13$$

For GPA, the z-scores are:

$$\frac{3.3 - 2.9}{.55} = .73 \quad \frac{2.9\text{-}2.9}{.55} = 0 \quad \frac{3.2\text{-}2.9}{.55} = .55 \quad \frac{3.2\text{-}2.9}{.55} = .55$$

$$\frac{1.9-2.9}{.55} = -1.18 \quad \frac{2.4\text{-}2.9}{.55} = -.9 \quad \frac{3.7\text{-}2.9}{.55} = 1.45 \quad \frac{2.5\text{-}2.9}{.55} = -.73$$

Next, we can place them in the following table:

x z-score	y z-score	xy
1.49	.73	1.09
-.62	0	0
.24	.55	.13
.69	.55	.38
-1.47	-1.18	2.65
-.33	-.9	.3
1.15	1.45	1.67
-1.13	-.73	.83

and then calculate r:

$$r = \frac{(1.09) + (0) + (.13) + (.38) + (2.65) + (.3) + (1.67) + (.83)}{8} = .88.$$

r is the sum of the z-scores of x multiplied by the z-scores of y divided by n or (in symbols):

$$r = \frac{\Sigma z_x z_y}{n}$$

When there is a perfect correlation between two variables, the correlation is 1.0 and all of the scores occur on a straight line. In Figure 9.12, the correlation is obviously very high and the data points do not lie on a straight line, but they do cluster around a straight line. Such a line is called a **regression line** and it is the straight line that best "fits" the data points. How is the regression line determined? The standard method is to use the **least squares criterion,** which is that line for which the sum of squared errors is the smallest. An error is a divergence between the actual score and the score as predicted. So, according to the least squares method, one should fit the regression line

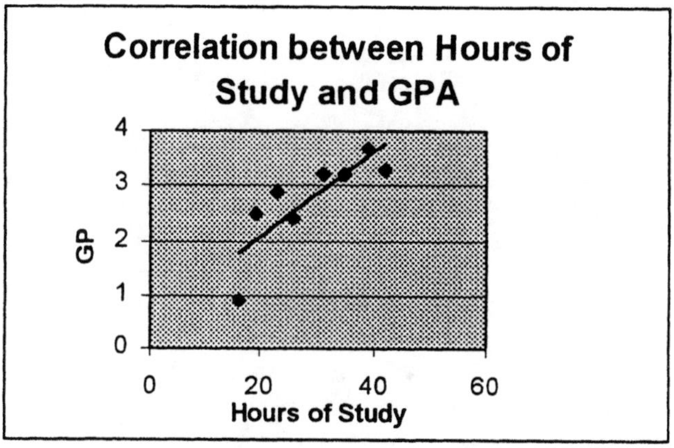

Figure 9.12 Scatter Plot with Regression Line

to the data points such that the sum of these squared errors (deviations between the actual score and the projected line) are the smallest. The regression line has a corresponding equation, called a **regression equation** and has the form of a linear equation (a straight line on a graph): Y = a + bX (where a is the y-intercept, b is the slope, X is the independent variable, and Y is the dependent variable).[7] In our example of GPA scores, the regression equation is: Y = 1.3 + .055X. This would mean that 1.3 is the intercept--the point where the regression line crosses the y axis (GPA)—and .055 is the slope—how steep the line is, i.e., how much the y-value on the straight line increases (or decreases) by one unit when the x-value increases (or decreases) by one unit. For example, in our regression equation, whenever the x-value (hours of study) increases, say, from 25 to 26 hours of study, the GPA scores increase by .055--from 2.68 to 2.73.

Regression equations are powerful tools in science. How can they be calculated? Here is one way. First, determine r (as above). Then, the slope (b) equals r times the standard deviation of y divided by the standard deviation of x:

$$b = \frac{rs_y}{s_x} = \frac{(.88)(.55)}{(8.8)} = .055$$

The y-intercept (a) is the mean of y ($\bar{y}$) minus the mean of x ($\bar{x}$) times the slope (b):

$$a = \bar{y} - (\bar{x}b) = 2.9 - (28.9)(.055) = 1.3$$

(There are other standard ways of calculating this, but this form of calculation has the advantage of intuitive appeal since respective z-scores rather naturally express the notion of correlation.)

What is so useful about regression equations? By means of them, one can make predictions about possible values of one's dependent variable (Y). For example, suppose one wants to know the likely GPA of someone who studied 40 hours per week. Given the above regression equation, we can predict that such an individual would get a GPA of 3.5 (Y = a + bX = 1.3 + .055X. = 1.3 + .055(40) = 3.5). Since X is the independent variable (hours of study) and Y is the dependent variable (GPA), Y is also called the **response variable** and X is called the **predictor (explanatory) variable** because the latter (hours of study) is used to predict (or explain) the value of the response variable (GPA). In keeping with this terminology, if one squares the correlation coefficient (r), thereby producing r^2, one has the **coefficient of determination**, which is often interpreted as the percentage of total variation in the observed Y variable explained by the regression line. For example, if there is a correlation (r) of .88 between hours of study and GPA, then the coefficient of determination (r^2) is .77. This is the amount of the variance of the observed y-values (GPA scores) explained by (predicted by) the regression line (regression equation between hours of study and GPA), in this case 77%. In short, the degree of linear association between the two variables explains 77% of the total variance of the GPA scores.

Given a scatter diagram together with a regression line (or regression equation) that fits the set of data points to a certain degree, there are two other important operations related to the data points: **interpolation** and **extrapolation.** In both of these cases, one is hypothesizing the values of missing data. Interpolation occurs when one hypothesizes or guesses the value of missing data between the lower and upper bounds of one's data, and extrapolation occurs when one makes a hypothesis outside the lower or upper bound.

After collecting a set of observations plotted as data points on a scatter diagram, a scientist will want to raise questions about the relation between the independent variable and the dependent variable. Assuming there is a relationship between them (i.e., a correlation of sufficiently high numerical value), what precisely is the quantitative relationship between them, and is the relationship strong enough to warrant the conclusion that it is an instance of a scientific law (or law-like generalization)? The search for such relationships is an example of **quantitative induction:** from a particular set of measurable data, can one induce a more general numerical relationship?

Given a set of data points on a scatter diagram (Figure 9.13), it is possible to fit a regression line to these data points according to the method of least squares. But it is also possible to fit several different

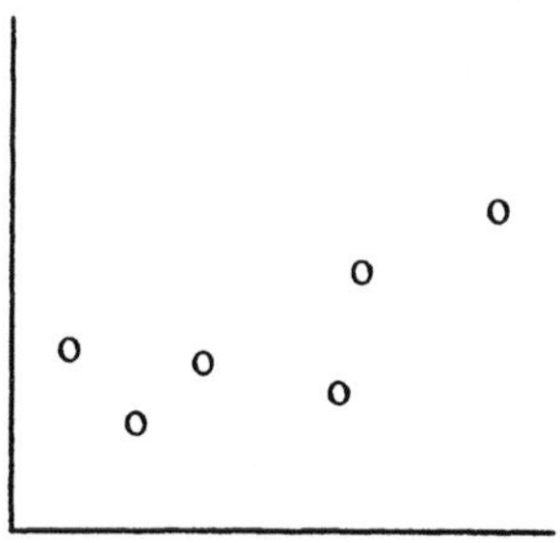

Figure 9.13. A set of data points

straight lines to the data points (using different regression equations). Hence, we have the problem of **curve fitting**: which curve (functional equation) fits the data best? Since there is an indefinite number of functions that will fit the data, which one should we choose?

In Figure 9.14, one equation (y=f(x)) is a straight line equation, which fits the data points. y=g(x) is a different functional equation resulting in a curvilinear line passing through the data points. Here, one is seeking a general (mathematical) formula to cover the particular set of data. What kind of mathematical curve will best fit the data? Various kinds of curves (and corresponding mathematical equations) are possible. In fact, an indefinite number of mathematical equations can be fitted to any set of data. Which one, therefore, should be chosen?

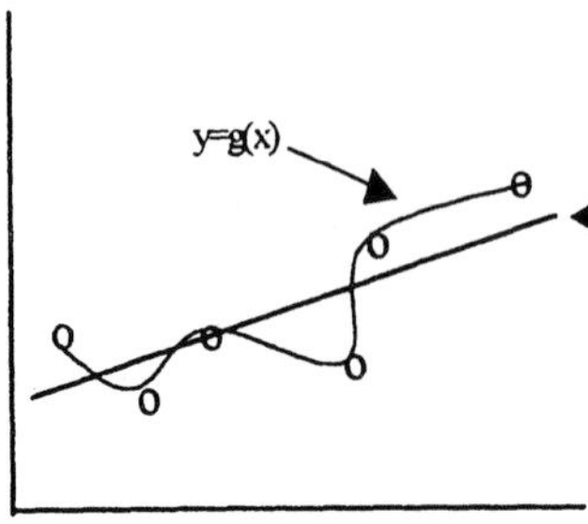

Figure 9.14. Data points to which two curves can be fitted

Ordinarily, it is assumed that the curve should be a smooth one (i.e., a continuous function) and not a discontinuous one. Secondly, it is assumed that one should use the *simplest* continuous equation, proceeding from a constant (y = a), to a linear equation (y = a + bx), then to an exponential equation ($y = ax^n$) or logarithmic equation (y=log x), then to a periodic equation, etc. (see Figure 9.15). These criteria are, at best, rough guidelines and do not constitute mechanical rules for discovering the best fitting mathematical representations.

Exercise 9.4

In the following examples, (1) plot a scatter diagram for the data and draw a regression line. (2) Use the least squares method to determine a regression equation. (3) Describe the relationship between the independent and the dependent variables. (4) What does the slope represent? (5) What are the response and predictor variables? (6) Compute r and the coefficient of determination (r^2). (7) What percentage of the total variance in the y-values is explained by the regression? (8) Calculate the predicted value (as indicated).

1. An article read by a physician indicated that the maximum heart rate an individual can reach during intensive exercise decreases with age. The physician decided to do his own study. Ten randomly selected people performed exercise tests and recorded their peak heart rates. The results are shown in the following table, where x denotes age, in years, and y denotes peak heart rate.

X	30	38	41	38	29	39	46	41	42	24
Y	186	183	171	177	191	177	175	176	171	196

(Use the regression equation to predict the heart rate of a 28 year old.)

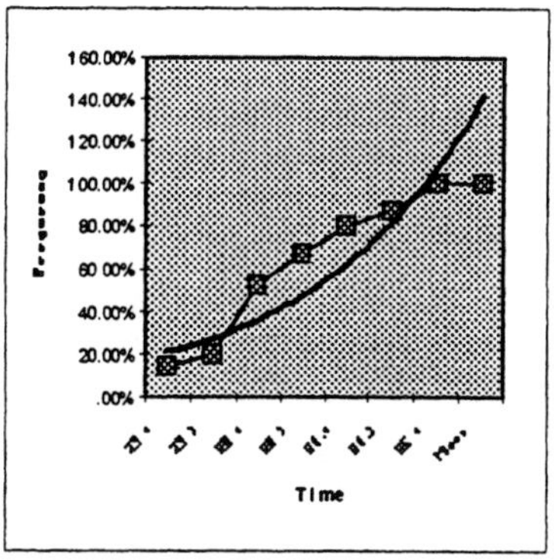

Logarithmic

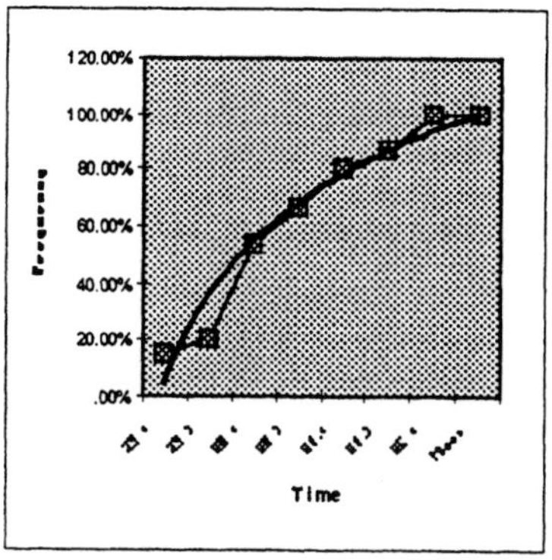

Exponential

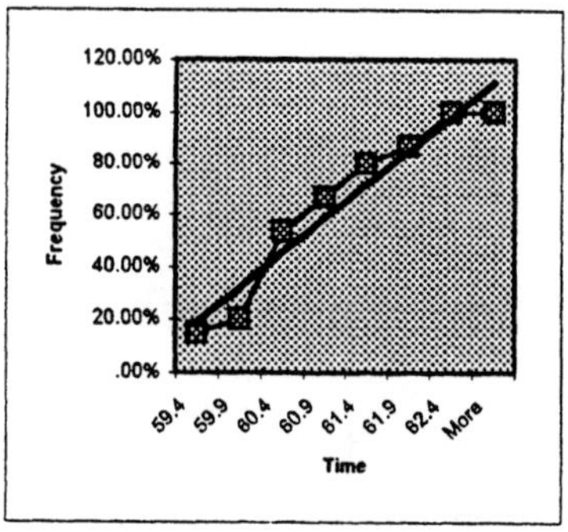

Linear

Figure 9.15. Graphical Illustration of Different Mathematical Equations

2. An economist is interested in the relationship between the disposable income of a family and the amount of money spent annually on food. For a preliminary study, the economist takes a random sample of middle-income families of the same size (father, mother, and two children). The results are as follows, where x denotes disposable income, in thousands of dollars, and y denotes food expenditure, in hundreds of dollars.

X	30	36	27	20	16	24	19	25

Y	55	60	42	40	37	26	39	43

(Use the regression equation to predict food expenditure of an individual whose disposable income is 33.)

3. The following table is adapted from a study by Doll and shows per capita [i.e., for each individual]consumption of cigarettes in various countries in 1930, and the death rates from lung cancer for men in 1950. (In 1930, hardly any women smoked; and a long period of time is needed for the effects of smoking to show up.)

Country	Cigarette Consumption	Deaths/million
Australia	480	180
Canada	500	150
Denmark	380	170
Finland	1100	350
G. B.	1100	460
Iceland	230	60
Netherlands	490	240
Norway	250	90
Sweden	300	110
Switzerland	510	250
USA	1300	200

Use the regression equation to predict the death rate of a country, whose cigarette consumption was 800.

Causation versus Correlation

A correlation is a mathematical relation between two variables. The IQs of parents are correlated with the IQs of their children, smoking is correlated with lung cancer, how well the stock market does is correlated with who wins the superbowl, etc. But as some of these examples suggest, *correlation is not the same as causation.* Just because two variables are correlated, this does not mean that one was the cause of the other. Shoe size, for example, is correlated with intelligence, but shoe size does not cause intelligence, or vice versa. There is a high correlation between smoking and lung cancer. But is smoking the cause of lung cancer or are they due to the presence of a third variable—some biological trait in the individual? Most individuals (e.g., the Surgeon General) would argue that smoking does cause lung cancer. But how would you show on scientific grounds that smoking causes lung cancer? Clearly, something more is needed than just showing they are correlated. Indeed, correlation is necessary for causality, but it is not sufficient.

There are several approaches to the question of how one distinguishes mere correlation from causation. First, there are various statistical techniques that have been developed to assist in the identification of a cause from a correlation. These statistical techniques involve the notion of *partial correlation,* in which correlation is partialled out among subcomponents that are correlated. There are two main types: (1) path analysis (also known as causal modeling or structural equations), and (2) cross-lagged panel correlation.

Second, one can increase one's confidence that two correlated variables are causally related by performing an experiment instead of relying on observational data. If one uses smoking (or ingestion of nicotine) as one's independent variable, one can then study lung cancer as the dependent variable. This can't be done in humans but it could be done in, say, mice. If one can change the independent variable—smoking—and this has an effect on the dependent variable—cancer—then one's confidence that smoking is the cause is increased. It is assumed that there is what is called a causal arrow, that the direction of causality is from cause to effect and not the other way around.

Third, there are theoretical grounds for arguing that certain correlations are not causes. There is no theoretical way in which having lung cancer could cause one to smoke; there is not a relevant or plausible causal chain linking the two together. But there is a theoretical linkage between smoking (inhaling nicotine) and lung cancer. Clearly, our theoretical knowledge can exclude certain things as causes and increase one's confidence that other things are causes. This is a form of *eliminative induction (s*ee Chapter 10): eliminate all correlated variables as possible causes except for one and this one is probably the cause.

There is no simple way to distinguish causation from correlation. But by using the above approaches, one can make a plausible case that not only is, say, smoking is correlated with , smoking is the cause of cancer. Needless to say, in making such a claim one should always do so cautiously and with an open mind.

Measurement

Most of the variables in the physical sciences, e.g., temperature, weight, density, hardness, mass, length, etc. are quantitative variables whose values are determined by various measurement techniques. To a lesser extent, this is true of the biological and social sciences whose variables include, e.g., genetic dominance, intelligence, aggressiveness, social conformity, political preference, etc. Some of these are quantities and some are qualities. The social sciences appear to be as committed to the importance of measurement as are the physical and biological sciences, but the subject matter of the physical sciences lends itself to measurement much more easily than does the subject matter of the social sciences.

Extensive vs. Intensive Measurement

Quantities vary in degree: one liquid is hotter than another one, one object is smaller than another; one person is smarter than another. Here, the relevant notion is that of something being *greater than*--the comparative concept of quantitative evaluation--in which one individual has more of a quantity than another individual has. 'Greater than' is a relation between individuals and along with this relation go two others: *less than* and *equal to.* Ordinarily, if A is greater than B, then B is less than A, and B is not equal to A. From these relations other ones can be introduced: if A is greater than B, then B is not greater than A (*asymmetry*). If A is greater than B, and B is greater than C, then A is greater than C (*transitivity*), etc. Hence, there are certain order relations between measurable variables.

There is a strong and a weak sense of order: a relationship is a partial order if it is asymmetric and transitive; a relationship is a complete order if it is asymmetric, transitive, and any class can be compared to any other class with respect to asymmetry and transitivity. Part of the logic of measurement is concerned with a formal examination of such ordering relations.

Some properties are intensive properties and some are extensive properties. Intensive properties are ordered (either partially or completely) in a series, e.g., a set of personal preferences for candidates for president. Here, the preferences are arranged in an order but nothing is said about how much a person prefers one candidate to another one. Intensive properties, therefore, involve just an order or ranking. Other examples of intensive properties include: hardness, temperature, density, intelligence, introversion, beauty, etc. (As we will see later, intensive properties allow for nominal and ordinal scales of measurement.)

By contrast, extensive properties possess a stronger quantitative feature, for not only are there degrees comparative, but one can also determine how much these comparative degrees are. For example, one can determine how much the length of an object is, how heavy an object is, etc.

To be an extensive property, a property must be *additive*, i.e., there must be a physical operation of addition of one object to another such that a certain outcome occurs. For example, in the case of weight, one can physically add the weight of one object to the weight of another object by placing one object on top of another object on a balance scale and determining how heavy it is. Suppose we determine that objects B_1 and B_2 are equally heavy (by a prior set of operations involving a balance). Then if we place B_1 and B_2 together in one pan and A in the other pan and if they balance each other, then A is as heavy as B_1+B_2; A twice as heavy as B_1 (or B_2), so that $A = 2B_1$, etc. This physical operation of addition corresponds to the mathematical operation of addition, allowing one to apply certain mathematical properties to the quantity in question, e.g., division, multiplication, etc. Other possible operations of physical addition include the operation of mixing liquids together, placing one object at the end of another object, adding one event to another, and so forth.

Although certain physical operations of addition are possible with some properties, they are absent in the case of others. For example, there is no physical process of adding the temperature of one object to the temperature of another body, or of adding the intelligence of one person to the intelligence of another person. In short, in order for a property to be an extensive property and not merely an intensive one, the process of physical addition must be possible. Hence, temperature and intelligence are intensive properties but not extensive ones.

Fundamental vs. Derived Measurement

Another important distinction concerns the difference between **fundamental** and **derived measurement**. Some kinds of measurement seem to presuppose other kinds of measurements. For example, the measurement of density, which is a ratio of weight/volume, cannot be done directly but only by measuring other properties, namely, weight and volume. The result of this is a constant called density (W/V = density). Other examples of derived measurement include: viscosity, temperature, solubility, velocity, acceleration, I.Q., social esteem, group cohesiveness, etc.

Fundamental measurement does not presuppose other kinds of measurement, e.g., length, time, electric resistance, mass, and weight. These measurements are not derived from the measurements of other properties. Properties such as volume seem intermediate since they assist in measuring density (and hence appear to be fundamental measurement) but are, in turn, measured by length. For some examples, therefore, the distinction between fundamental vs. derived measurement appears to be relative.

How do we distinguish fundamental from derived measurements? Two accounts have been given. According to one account, a measurement is derived if it presupposes other measurements either definitionally (as in density = weight/volume, force = mass x acceleration, or IQ = MA/CA) or empirically (since there is no known way to measure the property directly, e.g., solubility). According to a somewhat different account, a fundamental measurement has the property of physical additivity (an extensive property), whereas a derived measurement does not (an intensive measurement.) On this account, the measurement of weight is fundamental since there is a physical operation of addition applicable to it. However, there does not seem to be any such physical operation possible for density and temperature, which would be derived measurement.

Scales of Measurement

In the behavioral and social sciences (and to a lesser extent) in the physical and biological sciences), scales of measurement have been widely discussed (and debated). Earlier in this century, S. S. Stevens (1946) introduced a famous distinction between several types of scales of measurement--nominal, ordinal, interval and ratio scales--and his classification has been widely adopted by behavioral and social scientists. (In the physical sciences, it is more customary to distinguish intensive vs. extensive scales of measurement.)

A **nominal scale** is a scale in which numerals are assigned to discrete categories of objects, e.g., the system of assigning numerals to football jerseys or the assignments of telephone numbers to households. Here, the numerals are merely labels, tags, or names for the different categories, e.g., the class of females get assigned '1' and the class of males '2.' Hence, such a scale is a way of determining the identity and difference between classes of objects, but there is no ordering of numerals, nor can they be compared quantitatively other than by frequency. In fact, a nominal scale appears to be a technique for the categorization of individuals into two or more discrete classes and for the counting of their frequency. Although there is considerable room for doubt whether this is really an example of measurement, it has become a standard practice to include it in discussions of scales of measurement.

An **ordinal scale** is one in which objects are ranked in an order, e.g., lst, 2nd, 3rd, etc., as in a race. Although there is a comparative relation here, involving 'greater than' and 'less than', true quantities do not appear to be present. The hardness scale of Mohs is a classical example of an ordinal scale: objects are ranked in terms of 'harder than' by means of the operational test of whether one object can scratch the other object. The result is a rank order, with diamonds being the hardest (receiving a numerical value of 10), down to the last object talc (receiving a numerical value of 1). Other examples of ordinal scales include: density, temperature, intelligence, preference order, leadership ability, and SAT scores. In the behavioral and social sciences, ordinal scales are quite common and are present whenever one is asked to rank order an object (e.g., a candidate), according to, say, a five-point scale, with one being the least favored and 5 being the most favored.

Unlike a nominal scale, an ordinal scale involves intensive properties; however, it does not involve extensive properties since objects are merely rank-ordered and are not additive. Ordinal scales also seem to be examples of derived measurement.

If nominal scales are questionable examples of genuine measurement, ordinal scales are a much better example since they measure a comparative degree (more than). But because ordinal scales do not permit the determination of the quantity or amount of a variable (how much), it remains a weak form of measurement. By contrast, interval and ratio scales are much stronger forms of measurement.

Interval scales are scales of measurement in which there are equal intervals between the ranks, permitting the determination of degrees of the quantity. For example, the Fahrenheit scale of temperature, weight scales, paper and pencil tests, etc. all have equal intervals of degrees, pounds, correct scores. etc. In order to have equal intervals, one must have a standard unit of measurement (e.g., a gram, an inch), which determines the amount of the quantity in question. An object has, say, twice as much of a quantity as another object does if the difference between the two objects is twice the standard unit. Although there are several examples of interval scales in the physical sciences, it is somewhat less clear that there are any such examples in the behavioral and social sciences (although some social scientists have argued that IQ tests, personality tests, attitude tests, and ability tests can give rise to interval scales).

The key issue here appears to be the following: although there are equal intervals on one's physical scale so that 20-10 degrees equals 90-80 degrees, it may be less than clear that there is a corresponding difference in the quantitative property being measured. It seems questionable, for example, whether the difference between 20 and 10 degrees of temperature is the same as the difference between 90 and 80 degrees of temperature, just as it is doubtful whether the difference between IQ scores of 70 and 80 is the same amount as the difference between IQ scores of 130 and 140 (the difference seems non-linear to most). Hence, there is a lack of a correspondence between the measuring scale and the quantity being measured.

Ratio scales are the rarest scale of all (at least in the social sciences); they involve all of the above properties plus the existence of an absolute zero amount of the quantity in question, e.g., the absolute temperature scale, length, time, volume, weight. In the social sciences, for example, it has been conjectured that perhaps certain kinds of psychophysical scales are ratio scales, as are psychological tests in which there can be a zero amount of error, reaction time, etc.. Although some have claimed this is true, others have disagreed and have concluded that only in the physical sciences do you find ratio scales.

Exercise 9.5

1. The pressure of a gas is proportional to its temperature and inversely proportional to its volume. How are these three quantities measured? Do the same for income being proportional to the age of an individual.
2. How is the age of an individual measured? The age of a tree? The age of a bone? The age of the universe? Are these fundamental or derived measurements? How do clocks "measure" time?
3. Probability is a quantity that is measurable. How is it measured? Is it extensive or intensive? Fundamental or derived?
4. Which of the following are extensive and which intensive?

Number	Temperature	Intelligence	Mass	Humidity
Desirable	Pressure	Heat	Beauty	Poverty
Velocity	Pleasure	Acceleration	Energy	Color
Piety	Morality	Sense of humor		

5. What kind of scale is being used in the following statements?

The Rockies have done better than the Nuggets have.

John's GPA is .26 lower than Mary's is.

Susan is twice as smart as her sister is.

20 people favored candidate A and 10 people favored candidate B, so candidate A is twice as good as candidate B.

Machine A is charged to 40 volts and machine B is charged to 80 volts, so the voltage between them must be 40 volts.

For Further Reading

All standard statistics textbooks discuss the nature and types of samples as well as the other topics in statistics covered in this chapter. An excellent beginning textbook is Freedman, et al., (1991). Next, read one of the many standard textbooks, e.g., Weiss (1996), which is an excellent one.

For an excellent introduction to the topic of measurement, see Cohen & Nagel (1938, ch. xv), followed by Ellis (1968).

Notes

[1] Or (technically speaking) countably infinite (denumerable), i.e., they can be mapped onto the natural numbers.
[2] Or noncountably infinite (non-denumerable), i.e., they cannot be mapped onto the natural numbers as in the interval between 0 and 1.
[3] Summation notation (Σ) means: "the sum of." Instead of saying: $x_1 + x_2 + x_3$, etc., we can say Σx: "sum all of the values of x."
[4] A well-known shortcut formula for calculating the standard deviation is the following:

$$s=\sqrt{\frac{\Sigma(x^2)-\frac{(\Sigma x)^2}{n}}{n-1}}$$

[5] Named for the Russian mathematician Pafnuty Lvovich Chebychev (1821-1894).
[6] This is also called the Pearson product moment correlation coefficient or Pearson r--named after Karl Pearson, who used the term r for regression (after the work of Galton).
[7] The y-intercept is the point on the Y-axis where the regression line crosses, and the slope is the ratio between the increase of x-values and the increase of y-values.

10

Eliminative Induction

Mill's Methods of Experimental Inquiry

Laws of Nature and Causality

In the inductive step from particular observations to a general conclusion, we have seen that several processes may intervene: one may subject raw data to various kinds of transformations to produce various degrees and kinds of analyzed data. These transformations may involve description, classification, quantification, measurement, interpretation, etc. The result of analyzing raw data is analyzed (interpreted) data and these data, in turn, may be subjected to various types of transformations to produce generalizations, both qualitative and quantitative. For example, as we saw in the case of Boyle's law, we may graph the relation between two sets of data expressed as variables and discover there is a quantitative relationship between them. But even if we are not able to come up with a precise mathematical equation describing our data, hopefully we will have found some kind of relationship between two or more classes of data, e.g., a relation between atmospheric pressure and the boiling point of water, between velocity and distance, between income and education, etc. In the conduct of inquiry, therefore, one of the main goals is the discovery of lawful regularities (generalizations). Many of these are called scientific *laws*, e.g., Ohm's law, Boyle's law, Kepler laws, Mendel's laws, the Hardy-Weinberg law, the law of supply and demand, etc.

But whether scientific generalizations attain the status of *laws* or not, they either express uniform and invariable associations between two or more variables (*universal laws*) also called *deterministic laws*--or statistical relations between variables where the association concerns a frequency of cases (*statistical laws*), also called *probabilistic laws,* e.g., the rate of radioactive decay = kN. Scientific generalizations may be expressed in *qualitative* terms ("bad money drives out good money") or *quantitative* terms,

$$\oint B \bullet dl = \mu_o I$$

but in either case what is being expressed is a *regularity*. Such regularities may be precise and exact or they may express a vague tendency ("people tend to be selfish"). Furthermore, most individuals distinguish *true scientific laws* (Newton's law of gravity) from *accidental (mere) generalizations* ("all the cars in the parking lot are Porsches") on the basis that true scientific laws (unlike accidental generalizations) do not mention any particular individual objects, nor any region of space or time, etc.

One important kind of a scientific law is a *causal law* (e.g., smoking causes cancer, dominant genes cause certain kinds of phenotypic traits, anxiety causes the selling of stock, etc.). Indeed, much of the conduct of inquiry is concerned with the discovery of causes, e.g., the cause of cancer, the cause of AIDS, the cause of racial bias. An important issue here concerns how one discovers the cause of some phenomenon or effect, i.e., what methods to employ in this kind of inquiry. Sir Francis Bacon and John Stuart Mill both devoted a great deal of attention to this question and provided accounts of what the appropriate methods were, of how (beginning with certain kinds of data) one should proceed to draw a conclusion about the cause of some phenomenon. But before discussing these methods, we should say something (however brief) about the general nature of causality.

Necessary and Sufficient Conditions

In an earlier chapter, we discussed the difference between necessary and sufficient conditions. Something P is a *necessary* condition for something Q just in case: if P is false, then Q is false: ($\sim P \supset \sim Q$). Something P is a *sufficient* condition for something Q just in case: if P is true, then Q is true: ($P \supset Q$). These concepts can be used to express the notion that something P is the cause of Q. On the standard account,

> *P is the cause of Q if and only if (1) P is a sufficient condition for Q and (2) P is a necessary condition for Q.*

To say P causes Q is, first of all, to say that if P occurs, then Q will occur, i.e., that P is a sufficient condition for Q. This allows for causality as the production of the effect: in order to produce Q, just produce P (e.g., in order to produce pre-grade school children who do well, first produce an enriched, challenging and supporting environment prior to kindergarten).

If a germ (e.g., tubercle bacillus) is the cause of a disease (tuberculosis), then the germ is a necessary condition for the effect: if the germ is not present, then the disease will not be present. Hence, a cause is also a necessary condition for the effect. This gives rises to the notion of causation as *elimination*: if one wants to eliminate the effect, then eliminate the cause, e.g., if one wants to eliminate the existence of small pox, then just eliminate the cause of small pox.

If we combine both of these conditions, then a cause is a necessary *and* sufficient condition for the effect. Not everyone would claim that a cause must be a necessary and sufficient condition for the effect; some would only claim that some causes are sufficient conditions, and some causes are necessary conditions. Indeed, as we have seen, if P is a sufficient condition for Q, then Q is a necessary condition for P. But from this it does not follow that just because P produces Q (and is a sufficient condition) one can eliminate P by eliminating Q, e.g., one cannot eliminate magnetism by eliminating the movement of iron. Why is this? This is because there is a *causal arrow* from magnetism to the movement of iron; but not the other way around: moving pieces of iron do not cause an object to be magnetic. In short, the traditional account of causality as "necessary and sufficient condition" is not completely adequate.

Similarly, if P is a necessary condition for Q, then Q is a sufficient condition for P. But from this it does not follow that if a germ is a necessary condition for a disease, that one can produce the germ by producing the disease. Again, this is because the causal arrow goes from germ to disease, not from disease to germ. Again, the traditional account of causality is not adequate as a complete account of causality: it may specify a necessary set of conditions for causality but it does not specify a sufficient set. This is because there seems to be a directionality to causality.

Hume's Account of Causality

Many people have believed not only that there is a directionality to causality, that causes give rise to effects but effects do not give rise to causes, they have also believed the causal agent has a causal *power* to produce the effect. According to this view, what distinguishes causes from effects and what explains the directionality of causality is the fact that the causes *make* the effect occur by virtue of the power resident in the cause: there is a necessary connection between cause and effect such that given the cause, the effect has to occur, e.g., an unsupported object (heavier than air) has to fall.

If there were such a power in the cause, reasoned **David Hume**, then there would be a causal tie or necessitation between the cause and the effect. Furthermore, as a good empiricist, Hume believed that if there was such a causal necessity lodged in the cause, one ought to be able to observe it empirically, by means of the senses. But no matter how long you inspect the cause and the effect, Hume claimed, you will never verify the existence of such a causal necessity or tie. You will only observe that one event (the cause) is *regularly followed by* another event (the effect) and this occurs in (roughly) the same region of space and time. In short, causality is nothing more than *the constant conjunction of two events contiguous in space and time*. Hence, if you cannot verify the existence of this alleged causal power (necessitation) between cause and effect, it does not exist. Hence, it must be eliminated from one's legitimate conception of causality.

If Hume is correct, then any two events constantly conjoined would be related as cause and effect. But this is clearly not the case. For example (to give a well-worn example), night regularly follows day but day is not the cause of night. There seems to be something wrong with Hume's account. Similarly, as we pointed out in an earlier chapter, *correlation is not causation:* just because two variables are correlated, it does not mean that one is the cause of the other. This is another illustration that Hume's analysis of causation is inadequate since the constant conjunction of two events is a case of high correlation between two variable (events), but not necessarily causation.

However, there does not seem to be any way of empirically verifying the existence of a causal power (over and above empirically verifying the correlation between two sets of events). Hence, Hume seems to be right that the concept of causal necessity must be rejected (several contemporary thinkers have, nevertheless, defended the concept of causal *power*). Although there have been several important alternative accounts of the nature of causality, none have resulted in universal acceptance.

Mill's Methods

Although the issue of how to define causality is fraught with conceptual difficulties, perhaps enough has been said to introduce the reader to some of the relevant notions. Regardless of what an adequate definition of causality is, the task remains of identifying causes. How can one sort out causes from correlations? What methods should be followed in order to draw causal inferences from the data?

For certain kinds of phenomenon and under certain kinds of conditions, **John Stuart Mill** suggested a set of procedures that, he believed, successfully solved this problem--his famous "Five Methods of experimental inquiry": the method of agreement, the method of disagreement (difference), the joint method, the method of residues, and the method of concomitant variation.

In all of these methods, there is the observed effect--the *phenomenon* to be explained--and a set of *antecedent conditions* (the potential causes) of the phenomenon. Now, assuming one can reasonably identify the most likely causes of the effect, Mill's methods can be used to *eliminate* all the unlikely causes except one, with the result that what is left must be the cause of the effect.

When Mill advanced his Methods of Experimental Inquiry, he suggested these were forms of inductive inference importantly different from *induction by simple enumeration* (see Chapter 4). In this method, the reader will recall, one infers that since all examined things have a particular property, therefore everything of that class has that property; one argues from an observed sample to a conclusion about the entire population.

Mill believed this type of inductive inference was not adequate as a method of inference and that there was a more powerful form of inductive inference--induction by elimination or **eliminative induction**. This form of inference has the following structure:

Either A or B or C or . . . or N is the cause of e.
But B is not the cause of e and C is not the cause of e and . . .
  and N is not the cause of e.
Therefore, A is the cause of e.

Although Mill seems to have thought this form of induction by elimination was an example of a deductive inference (demonstrative induction), it is clear that if it is, it can't be an example of an inductive inference. Hence, it needs to be put into an inductive form, although how precisely to do this remains unclear.

It is clear, however, that we are using a form of negative reasoning--we are proceeding by the process of elimination. By eliminating all causes *except one*, we are showing that these other causes are *not* causes and hence the remaining one must be the cause. The crux of the methods lies in step (2), where we are attempting to show that the other antecedent factors aren't the causes.

How do we do this? We do this by showing that each candidate except one cannot be a cause because it lacks a certain property a cause must have: by showing it is *not* a necessary condition for the effect, or *not* a sufficient condition.

The Method of Agreement

To illustrate the first method--the method of agreement--suppose one is a psychotherapist in a university counseling center and that several students (clients) have come to you seeking relief from severe and chronic depression. As you discuss this problem with your client, several possible causes suggest themselves: low self-esteem (A), diet (B), income level (C), socio-economic background (D), satisfaction with ones major (E), early childhood experiences (F), etc. Let these factors be A, B, C, D, etc. and let the effects of these factors be a, b, c, d, etc. (For the sake of simplicity suppose we designate the effect under consideration as *e* and simply ignore all other effects a, b, c, etc.). Which of these factors is the cause? According to the method of agreement, *that factor is the cause which is present when the effect is present.* As John Stuart Mill (1872) put it:

> *If two or more instances of the phenomenon under investigation have only one circumstance in common, the circumstance in which alone all the instances agree is the cause (or effect) of the given phenomenon.* (p. 255)

Let us construct the following table to illustrate the method of agreement; if a letter appears in the table, this indicates it is present, but if it is not present, this is indicated by a --.

Table 10.1 Method of Agreement

Individuals	Antecedent Factors						Effect
#1	A	B	C	D	E	F	e
#2	A	B	C	D	E	-	e
#3	A	B	C	D	-	F	e

#4	A	B	C	-	E	F	e
#5	A	B	-	D	E	F	e
#6	A	-	C	D	E	F	e

Mill's first method says: if the effect is always present when one of the antecedent factors is always present, this common antecedent factor is the cause of the effect. In the above table, there is only one antecedent factor (low self-esteem) always present when the effect (depression) is present. Hence, low self-esteem is the cause of depression.

What is the rationale for this conclusion? Why couldn't another factor, say, early childhood experience (F) be the cause of depression? The answer is the following: assume a cause C must be a necessary condition for the effect e. That means: ~C ⊃ ~e. If early childhood experience is the cause of chronic depression, it must be a necessary condition for it. If so, the following must be true: if not-childhood experience, then no depression.

One can either verify this claim or falsify it. One can *verify* the claim that C is a necessary condition for e (~C ⊃ ~e) in the following way: find cases of the absence of childhood experience (~C) and cases of the absence of depression (~e). One can *falsify* this claim in the following way: find cases of an absence of C (~C) and cases of the presence of e (e). Looking back at our table, we will note that in case #2, factor F (childhood experience) is absent but e is present (~F ⊃ ~e). Hence, F is not a necessary condition of e. This same situation applies to the factors B, C, D, and E. But we have not shown this for A, since there are no cases of A being absent; therefore, we have no cases of ~A and e. We are, in short, ignorant about whether A is a necessary condition or not. But we have shown that none of the other factors (B, C, D, E, F) are necessary conditions, and if A or B or C or D or E or F must be the cause, it follows that A must be the cause (since it is the only one left). We have eliminated every factor except A. In short, the argument here is

Either A or B or C or D or E or F are necessary conditions for e.
But B and C and D and E and F are not necessary conditions for e.
Therefore, A is a necessary condition for e.
Therefore, A is the cause of e.

Note that if we wanted to show (positively) that A was a necessary condition, we would need a case in which A was absent and e was absent (~A and ~e), but we have no such case at present.[1] To show that, we need a different form of the Method of Agreement--the Negative Method of Agreement. (See below.)

If something is a cause of something else, it must be a *necessary condition* for the latter. But another property often ascribed to causality is that of being a *sufficient condition* for the effect. No information in our table shows that A (as opposed to B, C, D, E, F) is the sufficient condition for E. This is because it is true of every factor that whenever it is present, the effect is present.

Exercise 10.1

In each of the following arguments (1) identify the effect (e), (2) identify relevant antecedents circumstances (A, B, ...), and (3) identify the cause by means of the Method of Agreement. Use a table (as suggested).

1. Edward Jenner kept neatly detailed records of his work, noting how Sarah Portlock, Mary Barge and Elizabeth Wynne, and Simon Nichols, Joseph Merret and William Rodway, had "taken" cowpox and how they showed immunity when he inoculated them with smallpox. He repeated his observations with others, and years passed as he accumulated page upon page of records of cowpox and smallpox. Eventually he was satisfied. He was convinced that the people who had taken cowpox were without exception immune to small pox.

—A. L. Baron, *Man Against Germs*

2. Johnston compared the effects of smoking with those of hypodermically injected nicotine. The smokers almost invariably thought the sensation pleasant. Johnston, who gave himself 80 injections of 1.3 mg. of nicotine from three to four times a day, found that he preferred the hypodermic injections of nicotine to inhaling a cigarette. In his case, it would appear that nicotine was the major factor in the pleasant sensation due to smoking.

—Ernest Wynder, *The Biologic Effects of Tobacco*

3. It is interesting to note that one of the frequent symptoms of extreme combat anxiety cases is an interference with speech that may run from complete muteness to hesitation and stuttering. Similarly, the sufferer from stage fright is unable to speak. Many animals tend to stop vocalizing when frightened, and it is obvious that this tendency is adaptive in preventing them from attracting attention of their enemies. In the light of this evidence one might suspect that the drive to fear has an innate tendency to elicit the response of stopping vocal behavior.

—John Dollard and Neal Miller, *Personality and Psychotherapy*

4. An alarming number of malignant skin cancer cases have been discovered among members of the District of Columbia Police Department who were repeatedly exposed to tear gas during the riots and demonstrations of 1968-1971 here.

Dr. Robert F. Dyer, director of the D.C. Police and Firemen's Clinic, said Thursday that one chemical component of tear gas apparently causes cancer.

"Over the past five years I personally have collected a series of 12 patrolmen with malignant melanoma [a form of skin cancer], which in a group of 4,800 men is higher than I would expect," Dyer said.

All 12 officers reported being present at one or more of the many riots and demonstrations quelled with tear gas between 1968 and 1971, he said.

---Washington Star

5. A few years ago a small number of people living in various sections of the United States were inflicted with an identical disease. At about the same time the eyes of these individuals developed what the physicians call cataracts--small, irregular, opaque spots in the tissue of the lens. Cataracts interfere with the clear passage of light through the transparent medium of the eye lens. In severe cases, they may block vision, visual acuity is lost and the lens must be removed. It turned out that all individuals who developed these cataracts were physicists and that all of them had been connected with nuclear-energy projects during the war. While they worked with cyclotrons in atomic-energy laboratories they had been the targets of stray neutron rays. They were under medical supervision all during their work, but the density of the neutrons was thought to be entirely harmless. Several years later, however, they developed cataracts.

This case is one of the best examples of the insidiousness of nuclear radiation.

---Heinz Haber, *Man in Space*

The Joint Method of Agreement and Disagreement

In the method of agreement, which we can call the *positive method of agreement,* one can conclude that B, C, D, E and F are not necessary conditions (and hence can't be causes). It doesn't follow that A is a necessary condition, but by a process of elimination A looks like it is the cause (and hence is a necessary condition). Suppose one wanted more positive evidence that A is a necessary condition for e? What would one do? To show that A is a necessary condition for e, we would have to show that whenever ~A, then ~e. Although we had no such cases in our table, we could construct a variant of the method of agreement, which we could call the *negative method of agreement,* by means of which we could show that A is a necessary condition for e. It would look like this:

Table 10.2. Negative Method of Agreement

Individuals	Antecedent Factors						Effect
#1	-	B	C	D	E	F	-
#2	-	B	C	D	E	-	-
#3	-	B	C	D	-	F	-
#4	-	B	C	-	E	F	-
#5	-	B	-	D	E	F	-

#6	- - C D E F	-

By a process of reasoning similar to that above, it follows that A is a necessary condition for e because in all cases, whenever ~A then ~e. But now notice the same conclusion follows for all of the other factors (B, C, D, ...): they are also necessary factors because in each case it is true that whenever, say, ~B then ~e. Hence, although we have shown that A is a necessary condition for e, we have also shown that each of the other factors is also a necessary condition for e.

By examining the table, however, one can see it allows us to draw a conclusion about which factors are not *sufficient conditions* for e, namely, B, C, D, E, and F. But we have no evidence that this is true of A. Hence, we can reason

Either A or B or C or D or E or F are sufficient conditions for e.
But B and C and D and E and F are not sufficient conditions for e.
Therefore, A is a sufficient condition for e.
Therefore, A is the cause of e.

By means of the positive and negative methods of agreement, one can show that everything except one factor is not a necessary condition and everything except one factor is not a sufficient condition. It would seem to be logical, therefore, to combine these two methods.

The simplest way would be to combine the two methods into what is called the joint method of agreement and disagreement, which is really a combination of the positive and negative methods of agreement. Combining our earlier positive and negative methods of agreement (and shortening them somewhat) would give the following (where the top half of the table is the positive method of agreement and the bottom half is the negative method of agreement):

Table 10.3 Joint Method of Agreement and Difference

Individuals	Antecedent Factors			Effect
#1	A	B	C	e
#2	A	B	-	e
#3	A	-	C	e
#4	-	D	E	-
#5	-	D	-	-
#6	-	-	E	-

Mill characterized this method in the following way:

> *If two or more instances in which the phenomenon occurs have only one circumstance in common, while two or more instances in which it does not occur have nothing in common save the absence of that circumstance, the circumstance in which alone the two sets of instance differ is the effect or the cause, or an indispensable part of the cause, of the phenomenon.* (Mill, 1872, p. 259)

The method of agreement and difference allows us to compare what difference the presence of A makes to the production of the effect e. It allows us to eliminate all factors except A since the joint method allows us to conclude, first, that B, C, etc. are *not necessary* and second, that B, C, etc. are not sufficient. By a process of elimination that leaves only A. But more than this, the joint method allows us to say something positive about A, namely, that it is necessary (by the second method) and sufficient (by the first method). In the joint method, therefore, we have something more than merely negative knowledge; we have positive knowledge.

As an example of this method, consider the following:

> Christian Eijkman, a Dutch doctor, was concerned about discovering the cause of beriberi. He originally thought it was caused by bacteria. He noticed, however, that chickens in the courtyard of the prison where he worked acted very much as did the human prisoners who had beriberi. That prompted the idea that the chickens' illness and the prisoners' were the same—their food. The prisoners were fed almost exclusively on polished rice and the hens lived on what the prisoners threw out of their windows into the courtyard. Upon inquiring about the diet in other prisons, he learned that some prisons had scarcely any beriberi at all. In these cases, the prisoners were given unpolished rice. Thus, he paved the way for the famous discovery that beriberi (and similar problems) is caused by vitamin deficiency.

This can be represented in the following way. There are two groups: prisoners fed on polished rice (group A) and prisoners fed on unpolished rice (group B). For the sake of illustration, let us suppose there are three individuals in each group. The cause of beriberi was polished rice, whose absence (unpolished rice) resulted in no beriberi. Let *e* be beriberi and *A* be polished rice. What other antecedent factors were present? We are not told, but presumably there are numerous other factors (individual ones and those common to a particular group). These can be represented as B, C, etc. and which we presuppose to be specific to individuals. The first three cases show that B through D are not necessary conditions (and hence can't be the cause of e), thus leaving A; the second three cases show E through G are not sufficient conditions, thus leaving A. Together, they show that A is a necessary and sufficient condition for e. Because factors B, C, ... are possessed by individuals and are not common, we can therefore ignore them and focus on the presence versus the absence of factor A.

Individuals	Antecedent Factors				Effect
#1	A	B	C	-	e
#2	A	-	C	D	e
#3	A	B	-	D	e
#4	-	E	F	-	-
#5	-	-	F	G	-
#6	-	-	F	-	-

Method of Difference

A third method touted by Mill is what he calls the direct method of difference (or simply the method of difference). Looking back at the schematic outline for the joint method, one notices we could compare a single individual (e.g., #l) in two conditions: first, with the presence of A and second, with its absence:

Table 10.5 Method of Difference

Individual	Antecedent Factors			Effect
#1	A	B	C	e
#1	-	B	C	-

(Similarly with individuals #2, #3, etc.). Here, we would have an example of the method of difference.

This method is characterized by Mill in the following way:

> *If an instance in which the phenomenon under investigation occurs, and an instance in which it does not occur, have every circumstance in common save one, that one occurring only in the former; the circumstance in which alone the two instances differ is the effect, or the cause, or an indispensable part of the cause, of the phenomenon.* (1872, p. 256).

For example:

> In ancient times, men noticed that if a substance such as meat was kept around for too long, it became infested with maggots or other insects. From this, a belief grew up that these creatures were actually produced by the decaying meat. This theory of spontaneous generation--that living

organisms can be produced naturally by inanimate matter--was long accepted, until in the nineteenth century Pasteur decided to test it. He placed meat broth in a flask, boiled it to destroy all bacteria, and sealed the flask. After some time he examined the liquid and found it still free of bacteria. Then he left it exposed to the air for a time. In a few hours it contained bacteria and was beginning to decay. Pasteur concluded that the bacteria had entered the broth from the air, rather than being spontaneously generated by the broth itself." (Baum, 1975, p. 370)

This can be represented in the following way: let *e* be the presence of bacteria and A be exposure to air. B, C, etc. are additional factors not mentioned, but *present in both conditions*. Because this will prove to be significant we will continue to include these factors. By boiling the liquid, Pasteur eliminated bacteria, thus producing ~e (the absence of bacteria). By sealing the flask, he prevented exposure to the air, thus producing ~A. So, ~A is followed by ~e. Then, he exposed it to air (A), which was followed by e (presence of bacteria). In short, if all the conditions (B, C, etc.) were the same with the only difference being the absence versus the presence of the cause (A), then, since the effect still appeared, it must be due to the cause.

Individuals	Antecedent Factors			Effect
#1	-	B	C	-
#1	A	B	C	e

The argument here is the following: the first entry shows A is a necessary condition for e (and that B and C are not sufficient conditions); the second entry shows A is a sufficient condition for e. Hence, A is a necessary and sufficient condition for e. (Although the second entry also shows B and C are sufficient conditions for e, the first entry shows they are not; hence, they can be eliminated.)

Relation between These Two Methods

What exactly is the difference between the two methods? Most textbooks on logic misinterpret the difference between them. Both are methods of differentiating (distinguishing) the absence versus the presence of a potential cause.

(l) One difference is that in the joint method of agreement and difference one is using two *groups* whereas in the method of difference one is using two *individuals* (or the same individual twice). If one employs the joint method with just two cases, then it becomes a limiting case of the method of difference (under the assumption of common background factors A, B, C, ...). For example, suppose in employing the joint method, one found that instead of

Individuals	Antecedent Factors				Effect
#1	A	B	C	-	e
#2	A	-	C	D	e
#3	A	B	-	D	e
#4	-	E	F	-	-
#5	-	-	F	G	-
#6	-	-	F	-	-

all of the factors besides A were the same:

Individuals	Antecedent Factors			Effect
#1	A	B	C	e
#2	A	B	C	e
#3	A	B	C	e
#4	-	B	C	-
#5	-	B	C	-
#6	-	B	C	-

This would be logically indistinguishable from the method of difference and would be one example of the use of *control groups*. The only difference here would be whether one is focusing on a single individual or a group. But note, that in the above case, one could not conclude that B and C were not necessary conditions for e (since they were not absent) but that they are sufficient conditions. This is ruled out in the second three cases, however, so we have shown that they are not sufficient conditions. But have we shown they are not necessary conditions? No, since they were never absent. We can, in short, eliminate them as sufficient conditions but not as necessary conditions. At the same time, we can claim to have shown that A is both a necessary and sufficient condition for e.

(2) An important difference noted by Mill concerns the question of whether one is doing an *experiment* or making a set of *observations*. If one can perform an experiment in which one is comparing two conditions, the major difference being that the experimenter has introduced an experimental variable, then one can employ the method of difference. But if one cannot perform an experiment but must rest content making a series of observations concerning what antecedent factors accompany the absence or presence of an effect, one uses the joint method of agreement and difference. For example, let B, C, etc., be a set of constant background factors in the presence of which the

effect is not present. Then introduce the cause (A), while keeping these factors B, C, etc., constant, and determine if the effect occurs. If it does, one has an example of the method of difference:

$$-BC \supset \sim e$$
$$ABC \supset e$$

Suppose, however, one cannot experimentally introduce A. Then one can perform the following set of observations. Given the absence of e in numerous cases, observe what antecedent factors were absent; then observe numerous cases of the presence of e and observe what antecedent factors were present. You can use the joint method:

Exercise 10.2

In each of the following arguments (1) identify the effect (e), (2) identify relevant antecedent circumstances, and (3) identify the cause by means of the Joint Method of Agreement and Difference or the Method of Difference. Use a table.

1. The primitive brain, as we saw it in the planaria, served chiefly as a sensory relay--a center for receiving stimuli from the sense organs and then sending impulses down the nerve cord. This is also true of the nereis, for, if the brain is removed, the animal can still move in a coordinated way--and, in fact, it moves about more than usual. If it meets some obstacle, it does not withdraw and go off in a new direction but persists in its unsuccessful forward movements. This very unadaptive kind of behavior shows that in the normal nereis the brain has an important function which it did not have in flatworms--that of *inhibition* of movement in response to certain stimuli.

---Ralph Buchsbaum, *Animals Without Backbones*

2. We have recently obtained conclusive experimental evidence that there can be no tooth decay without bacteria and a food supply for them. In germ-free laboratories at the University of Notre Dame and the University of Chicago, animals innocent of oral micro-organisms do not develop cavities. Where animals in normal circumstances average more than four cavities each, the germ-free rats show no signs of caries. At the Harvard School of Dental Medicine we have demonstrated the other side of the coin: that food debris also must be present. Rats that have plenty of bacteria in their mouths but are fed by tube directly to the stomach do not develop cavities. In a pair of rats joined by surgery so that they share a common blood circulation, the one fed by mouth develops tooth decay, the one fed by tube does not.

---Reider Sognnaes, "Tooth Decay" (*Scientific American*, 1957)

3. Eijkman, we will recall, discovered there was an important difference between polished rice and unpolished rice and the presence versus the absence of beriberi. To test his ideas further, he took two groups of chickens. One group was fed exclusively on white, husked rice, which subsequently developed beriberi and died. The other group was fed unpolished, unhusked rice. None of them developed the disease and died.

The next question that occurred to him was: what is the difference between polished rice and unpolished rice? Unpolished rice has a silvery husk, which merchants removed to make the rice look more attractive and to fetch a higher price (this husk contained the germ of the rice-grain which contains vitamins), whereas polished rice had the husk removed. To test for the difference between these two types of rice, Eijkman gathered up the polishings (husks) from rice and fed them to other chickens with beriberi; in a short time they recovered.

4. An experiment by Greenspoon on the reinforcement of a response (1950) provides another clear example of direct, automatic, or, in other words, unconscious learning. He had his subjects sit facing away from him so that they could not see him. He asked them to say all the words they could think of, pronouncing them individually without using any sentences or phrases, and he recorded their responses on a tape recorder. The response he was reinforcing was that of saying plural nouns; he did this by saying "Mmmm-hmmm" immediately after the subject said a plural noun. IN this case the response was a highly generalized part of language habits and the reinforcing value of the stimulus "Mmmm-hmmm" must have been acquired as a part of social learning.

Greenspoon found that during the "training" period the experimental group, to whom he said "Mmmm-hmmm" after each plural noun, greatly increased the percentage of plural nouns spoken, while the control group to whom nothing was said after plural nouns, showed no such increase. Furthermore, this happened with subjects who on subsequent questioning showed that they had no idea what the purpose of the "Mmmm-hmmm" was and were completely unaware of the fact that they were increasing their percentage of plural nouns.

This clearly demonstrates that the effect of a reinforcement can be entirely unconscious and automatic. Somewhat similar experiments have been performed by Thorndike (1932) and Thorndike & Rock (1934). A great deal of human learning seems to be of this direct, unconscious kind. Apparently many attitudes, prejudices, emotions, motor skills, and mannerisms are acquired in this way.

---John Dollard & Neal Miller, *Personality and Psychotherapy*

5. It was assumed for a long time, by analogy with the mosquito and other blood-sucking vectors, that the virus of typhus was injected by the louse when sucking blood. But apparently this is not so. The infection is not in the saliva of the louse, as it probably is with the mosquito, but in the feces. The disease is thought to be spurred through the feces coming into contact with scratches or abrasions in the skin, and scratching and louse infection are generally inseparable. This fact was first suggested in 1922 by the two workers who fed infected lice to a monkey, while taking great care that no feces from the lice should come into contact with the monkey's skin. They found that the monkey remained healthy.

--K. Smith, *Beyond the Microscope*

6. In a study at Vipeholm, a mental institution in southern Sweden, 436 adult patients on a nutritionally adequate diet were observed for several years. They were found to develop caries [cavities] at a slow rate. Subsequently, the patients were divided into nine groups to compare the effect of various changes in their carbohydrate intake. Sucrose was included in the diet as toffee, chocolate, or caramel, in bread, or liquid form Caries increased significantly

when food containing sucrose was ingested between meals Not only the frequency but the form in which sucrose was ingested was important: sticky or adhesive forms were more cariogenic [cavity-producing] than forms which were rapidly cleared from the mouth. After two years on the test diets, the patients were again placed on the control diet, and the caries activity reverted to the pretest pattern.

---E. Newburn, *Science* (1982)

The Method of Residues

The Methods of Agreement, of Difference, and Agreement and Difference employ an argument form that is eliminative: they allow one to eliminate certain candidates as possible causes and, hence, to conclude that the remaining factor must be the cause. As we have seen, one can also draw a positive conclusion about the causal status of a certain factor, but the thrust of Mill's methods is its power to *eliminate* factors. The same logic applies to the next method we will discuss--the method of residues.

Suppose we know that factor B causes a particular effect (or part of an effect) e_1 and that factor C causes a particular effect (or part of an effect) e_2.

$B \rightarrow e_1$

$C \rightarrow e_2$

Suppose now that the total effect e is $e_1 + e_2 + e_3$ and that, in addition, factor A seems to be a likely candidate for a cause. Then one can reason in the following way:

A, B and C are antecedent factors of e (and hence are likely causes of e).
B caused e_1 and C caused e_2 but neither caused e_3.
Thus, A is the likely cause of e_3.

Here we are eliminating B and C on the assumption they caused other parts of e (e_1 and e_2) but not the remaining part--e_3. As Mill (1872) puts it:

> *Subduct from any phenomenon such part as is known by previous inductions to be the effect of certain antecedents, and the residue of the phenomenon is the effect of the remaining antecedents.* (p. 260)

As Mill points out, the method of residues is really a case of the method of difference.

A	B	C	e_1	e_2	e_3
-	B	C	e_1	e_2	-

Here, B and C are not sufficient conditions for e_3 (row 2), and hence can be eliminated as possible causes of e_3. That leaves A as the cause of e_3. Of course, if one wants more certain knowledge that A is the cause of the residue e_3, one would determine that B and C are not necessary conditions for e_3:

A	B	C	e_1	e_2	e_3
A	-	-	-	-	e_3

One could then combine the above into another table of Difference:

-	B	C	e_1	e_2	
A	-	-	.	-	e_3

Virtually all textbook writers cite the following example as employing the method of residues:

> In 1821, Bouvard of Paris published tables of the motions of a number of planets, including Uranus. In preparing the latter he had found great difficulty in making an orbit calculated on the basis of positions obtained in the years after 1800 agree with one calculated from observations taken in the years immediately following discovery. He finally disregarded the older observations entirely and based his tables on the newer observations. In a few years, however, the positions calculated from the tables disagreed with the observed position of the planet and by 1844 the discrepancy amounted to 2 minutes of arc. Since all the other known planets agreed in their motion with those calculated for them, the discrepancy in the case of Uranus aroused much discussion.
>
> In 1845, Leverrier, then a young man, attacked the problem. He checked Bouvard's calculations and found them essentially correct. Thereupon he felt that the only satisfactory explanation of the trouble lay in the presence of a planet somewhere beyond Uranus which was disturbing its motion. By the middle of 1846 he had finished his calculations. In September he wrote to Galle at Berlin and requested the latter to look for a new planet in a certain region of the sky for which some new star charts had just been prepared in Germany but of which Leverrier apparently had not as yet obtained copies. On the twenty-third of September Galle started the search and in less than an hour he found an object which was not on the chart. By the next night it had moved appreciably and the new planet, subsequently named Neptune, was discovered within 1° of the predicted place. This discovery ranks among the greatest achievements of mathematical astronomy. (Fath, 1934)

The effect being considered here is the movement of Uranus. Part of this movement (e_1), which took the shape of an ellipse, was known to be caused by B, the gravitational influence of the sun and the inner planets. The remainder (e_2)--the residue--was the perturbation of the orbit, a slight depression of the ellipse. The antecedent causal factor A was the unknown planet, hypothesized to exist and whose attraction pulled Uranus off course.

Exercise 10.3

In each of the following arguments (1) identify the effect, (2) the relevant antecedent circumstances, and (3) the cause by means of the Method of Residues. Use a table.

1. Is avarice a natural tendency or an acquired habit? Two Harvard psychologists have been investigating this question with rats. Louise C. Licklider and J. C. R. Licklider provided six rats with all the food they could eat and more. Their food after weaning consisted of pellets of Purina Laboratory Chow. Although none of the rats had ever experienced a food shortage, all immediately started hoarding pellets. Even after they had accumulated a hoard and the food-supply bin was empty, they kept coming back to hunt for more.

This behavior confirmed what previous investigators had found. But the Lickliders refined the experiment to try to unearth the rats' motives for hoarding. They covered half of the pellets with aluminum foil, thus eliminating their value as food. The experimenters discovered that four of the six avaricious rats actually preferred the worthless, inedible pellets in hoarding.

The rats were then put on short rations for six days. After this "deprivation period" they hoarded even more greedily and showed more interest in the plain food pellets, but some still hoarded foil-wrapped pellets and continued to prefer them.

The Lickliders conclude, in a report to the *Journal of Comparative and Clinical Psychology*: "The factors that lead to hoarding and that determine what is hoarded are by no means entirely alimentary. The initiation of hoarding seems to be for the rat, as for the human being, a complex motivational problem to which sensory and perceptual factors rather than blood chemistry, hold the key."

---Science and the Citizen (*Scientific American*, 1950).

2. The radioactivity of every pure uranium compound is proportional to its uranium content. The ores are, however, relatively four times as active. This fact led M. and Mm. Curie, just after 1896, to the discovery that the pitchblende residues, from which practically all of the uranium had been extracted, exhibited nevertheless considerable radioactivity. About a ton of the very complex residues having been separated laboriously into the compounds, it was found that a large part of the radioactivity remained with the sulfate of barium. From this a product free from barium, and at least one million times more active than uranium, was finally secured in the form of the bromide. The

nature of the spectrum and the chemical relations of the element, now named radium, placed it with the metals of the alkaline earths. The ratio by weight of chlorine to radium in the chloride is 35.46: 113, so that, on the assumption that the element is bivalent, its chloride is $RaCl_2$ and its atomic weight is 226. With this value it occupies a place formerly vacant in the periodic table.

---James Kendall, *Smith's College Chemistry*

3. In H. Davies' experiments on the decomposition of water by galvanism, it was found that besides the two components of water, oxygen and hydrogen, an acid and an alkali were developed at opposite poles of the machine. Since the theory of the analysis of water did not give reason to expect these products, their presence constituted a problem. Some chemists thought that electricity had the power of producing these substances of itself. Davies conjectured that there might be some hidden cause of this part of the effect--the glass might suffer decomposition, or some foreign matter might be in the water. He then proceeded to investigate whether or not the diminution or total elimination of possible causes would change or eliminate the effect in question. Substituting gold vessels for glass ones, he found no change in the effect and concluded that glass was not the cause. Using distilled water, he found a decrease in the quantity of acid and alkali involved, yet enough remained to show that the cause was still in operation. He inferred that impurity of the water was not the sole cause, but was a concurrent cause. He then suspected that perspiration from the hands might be the cause, as it would contain salt which would decompose into acid and alkali under electricity. By avoiding such contact, he reduced the quantity of the effect still further, till only slight traces remained. These might be due to some impurity of the atmosphere decomposed by the electricity. An experiment determined this. The machine was put under an exhausted receiver and when it was thus secured from atmospheric influences, no acid or alkali was produced.

---G. Gore, *The Art of Scientific Discovery*

4. It was not merely the amount of water in circulation which was influenced by temperature. . .It was the total amount of hemoglobin. The mystery was: "Whence came this outpouring of hemoglobin?" It was not credible that the bone-marrow could have provided the body with new corpuscles at the rate required. Moreover, there was no evidence of increase of immature corpuscles in circulation. . .

The question then was forced upon us: Has the body any considerable but hidden store of hemoglobin which can be drawn upon in case of emergency? . . .In searching for a locality which might fulfill such a condition one naturally seeks in the first instance for some place where the red blood corpuscles are outside the circulatory system--some backwater outside the arteries, capillaries, and veins. There is only one such place of any considerable size in the body--that place is the spleen.

---Joseph Barcroft, *The Lancet* (1925)

The Method of Concomitant Variation

The Method of Residues is a method that can be used where the phenomenon under investigation is a quantitative one, i.e., one in which the quantity of something (e.g., amount of hoarding) can vary. The same applies to the final method concomitant variation.

According to many individuals, if something is a cause, then it must be a necessary and sufficient condition for the effect. Furthermore, if the cause and the effect can vary in quantity, what additional fact must be true of them? If the effect increases (i.e., changes from one value e_1 to another value e_2), then there must be a corresponding change in the cause, say, from A_1 to A_2 for if there weren't any such change in the cause, then by employing the method of difference we could show that A is not the cause of e. On analogy with the method of residues, suppose we wish to explain e_1 and e_2 and suppose we know A_1 is the cause of e_1. Then the following is true:

A_1	e_1 --
A_1	-- e_2

That is to say, supposing e_1 changes to e_2 and A_1 remains the same. Then A_1 cannot be a sufficient condition for e_2. Furthermore, we can also show it is not a necessary condition by removing it:

A_1	e_1 --
--	-- e_2

Hence A_1 is neither a necessary nor a sufficient condition for e_2. A_1 cannot explain e_2 since, when e_1 changes to e_2, A_1 does not change to A_2 but remains the same. In short, the underlying assumption here is: *nothing is a cause if it varies when the effect is constant, or is constant when the effect varies or varies in no functional way with it.* As Mill put it:

> *Whatever phenomenon varies in any manner whenever another phenomenon varies in some particular manner, is either a cause or an effect of that phenomenon, or is connected with it through some fact of causation. (p. 262)*

We can thus represent the method of concomitant variation in the following way: let A, B, and C be antecedent factors and e be the effect. Assume that these can vary in amount, an increase being represented as a '+' and a decrease as a '-.' Then, one type of concomitant variation is the following:

Antecedent Factors	Effect
ABC	e
A+BC	e+
A++BC	e++

another is

Antecedent Factors	Effect
ABC	E
A-BC	e-
A--BC	e--

Both of these express relations between the cause and effect that is one of *direct proportionality,* i.e., as the one increases (decreases), the other increases (decreases). For example, in the gas laws (Boyle's law), as the pressure of a gas increases, its temperature increases.

Still another possibility is the following:

Antecedent Factors	Effect
ABC	e
A-BC	e+
A--BC	e++

In this case, there is a relation of *inverse proportionality* i.e., as one increases, the other decreases. Again, in the gas laws (Boyle's Law), as the volume of a gas increases, its pressure decreases.

Example: Consider the following example: In England, which imports most of its sugar, records of the last 100 years show a steady increase in per capita consumption of sugar, from about 20 pounds per year in 1820 to over 100 pounds per year today. This represents 15 to 20 percent of an individual's caloric requirement. Concomitant with this increased consumption of sugar has been an almost parallel rise in the prevalence of cavities. Conversely, surveys in Europe and Japan demonstrated that cavities were dramatically reduced during periods of wartime restrictions of sugar, syrup, and all sugar products.[2]

Here the effect e is amount of cavities, with the obvious cause, A, being amount of sugar. We are told that as the consumption of sugar increased in England, there was a corresponding increase in the number of cavities. We are also told that, as sugar consumption decreased, frequency of cavities also decreased. This can be represented as:

Group	Antecedent Factors	Effect
England	ABC	e
Europe	A+BC	e+
Europe	A++BC	e++
Japan	A-	e-

Hence, A (sugar) is the cause of e (cavities).

Exercise 10.4

In each of the following arguments (1) identify the effect, (2) identify relevant antecedent circumstances, and (3) identify the cause by means of the Method of Concomitant Variation.

1. Careful studies have been made of the incidence of leukemia in the survivors of the atomic bomb burst over Hiroshima and Nagasaki. These survivors received exposures ranging from a few roentgens to 1000 roentgens or more.

They are divided into four groups. . .The first group, A, consists of the estimated 1,870 survivors who were within 1 kilometer of the hypocenter (the point on the surface of the earth directly below the bomb when it exploded). There were very few survivors in this zone, and they received a large amount of radiation.

The second group, B, consists of the 13,730 survivors between 1.0 and 1.5 kilometers from the hypocenter, the third, C, of the 23,060 between 1.5 and 2.0 kilometers, and the fourth, D, of the 156,400 over 2.0 kilometers from the hypocenter.

The survivors of zones A, B, and C have been dying of leukemia during the period of careful study, the eight years from 1948 to 1955, at an average rate of about 9 per year. . .many more cases of leukemia occurred in the 15,600 survivors of zones A and B than in the 156,400 survivors of zone D, who received much less radiation. There is no doubt that the increased incidence is to be attributed to the exposure to radiation.

. . .The survivors of zone A received an estimated average of 650 reoentgens, those of zone B, 250; those of zone C, 25, and of zone D, 2.5. . .To within the reliability of the numbers, the incidence of leukemia in the three populations A, B, and C is proportional to the estimated dose of radiation, even for class C, in which the estimated dose is only 25 roentgens.

—Linus Pauling, *No More War*

2. In a *Harper's* article, the following observations on fuel-cancer relationships were attributed to Eugene J. Houdry, president of a firm that is developing a device for destroying harmful auto wastes: between 1940 and 1945 gasoline consumption in the United States dropped about 35% because of wartime rationing, and at the same period lung cancer in US white males dropped by approximately the same percentage; between 1914 and 1950 lung-cancer mortality increased nineteenfold and the rate of gasoline consumption also increased nineteenfold.

---*The New Yorker*, October 31, 1959

3. Even as Banting was slaying dogs to save men, Evans was achieving a startling discovery in this field with another mysterious gland, *hypophysis cerebi,* commonly called the *pituitary*. This is a bit of an organ safely housed in a small pocket of bone attached to the base of the brain. Both Galen and Vesalius knew of this gland and thought it supplied the body with spit (in Latin, *sputus*). It is one of the most inaccessible glands in the living body. For many years, there appeared to be some connection between body growth and the functioning of this gland. In 1783 John Hunter had bargained with an undertaker for the body of an Irish giant of eight feet, four inches--Charles O'Brien, who had died at the age of twenty-two. The physician finally bought the body for twenty-five hundred dollars, and found a pituitary almost as large as a hen's egg. That of a normal adult man weighs hardly more than half a gram. A century later, *acromegaly*, an enlargement of the hands, feet, nose, lips, and the jaw, was declared to be due to a tumor of the pituitary. The pituitary glands of dwarfs, some of them only eighteen inches high, all showed relatively small development or partial atrophy.

---Bernard Jaffe, *Outposts of Science*

4. In the period from the early 1920s to 1960 the consumption of tobacco products in the United States rose about 30 percent. Significantly, the rise was due to the use of cigarettes. The use of tobacco in other forms actually declined. During this same period death rates from all infectious diseases declined rapidly except for the death rates from lung cancer. Deaths from lung cancer in the United States climbed from 4,000 in 1935 to 36,000 in 1960. Painstaking studies have clearly shown that the increase in lung cancer is real and not attributable to improved diagnosis.

It was known for some time that lung cancer could result from prolonged and heavy occupational exposure to certain industrial dusts and vapors. This led to the hypothesis that the increase in lung cancer was due to increased exposure of the human population to air contamination of some sort. The factor involved had to be widespread and not confined to any particular occupational group. (In all countries with adequate mortality statistics lung cancer was found to have increased.) Three factors that met the requirements were: fumes from the combustion of solid and liquid fuels, dust from asphalt roads and the tires of motor vehicles, and cigarette smoking.

A number of studies were made comparing the smoking habits of lung cancer patients with the smoking habits of individuals free of the disease. The findings in all the investigations were remarkably similar. Lung cancer is an extremely rare cause of death among nonsmokers, except for those who had had prolonged and heavy occupational exposure to certain dusts and fumes. The death rate from lung cancer was very low for men who had never smoked, it increased with the amount of cigarette smoking, and it was very high for men who smoked two or more packs of cigarettes a day. For those who smoked 40 or more cigarettes per day, the death rate from lung cancer per 100,000 man-years was 214. 143.9 was the figure for those who smoked 20 to 39 cigarettes a day, and 3.4 for those who never smoked. It was also found that the relative death rate from all causes increases with the degree of inhalation of cigarettes.

--- E. Hammond, "The Effects of Smoking" (Scientific American, 1962)

5. In a four-year study of 560,000 female and 440,000 male subjects in 25 states, the following data were obtained regarding mortality from lung cancer.

Males		Females	
No. Cigarettes Smoked per day	Mortality Rate	No. of Cigarettes Smoked per day	Mortality Rate
None	1.0	None	1.0
1-9	4.8		
10-19	6.4	1-19	2.1
20-39	11.6		
Over 40	12.6	Over 20	4.4

H. Diehl, *Tobacco and your Health* (1969)

Exercise 10.5

In each of the following arguments (1) identify the effect, (2) identify relevant antecedent circumstances, (3) identify the cause by means of one of Mill's methods (indicate which of Mill's Methods is being used).

1. It was Wolfgang Pauli who in 1930 postulated the existence of the neutrino in order to reconcile an apparent contradiction between the quantum-mechanical model of the phenomenon of nuclear beta decay and the observed products of this decay. (The term "beta decay" is used by physicists to describe the spontaneous transformation of a neutron in an unstable atomic nucleus into a proton and an electron; it is also applied to the transformation of a proton into a neutron and a positron, or positive electron.) When experimenters measured the total energy of the nuclear system both before and after beta decay, they invariably found the reaction not only varied in total energy from measurement to measurement: they were poorer in energy than the original nuclear system. Faced with the disturbing alternative of declaring a failure in the law of the conservation of energy, Pauli was moved to suggest a less radical solution, namely to postulate the existence of an unobserved particle that carries away the missing energy.

—Frederick Reines & J.P.F. Sellschop, "Neutrinos from the Atmosphere and Beyond" (*Scientific American*, 1966).

2. Alzheimer's disease is a serious form of senility that afflicts between 5 and 10 percent of all persons over age 65. Recently, researchers discovered that patients who die of Alzheimer's disease have much smaller amounts of a particular enzyme in the cortexes of their brains than persons the same age who do not suffer from the disease. The loss of this enzyme activity has been suspected as a cause of the lesions found on the brains of Alzheimer's patients. In postmortem examinations of five Alzheimer's patients and five people without the disease, it was found that all the diseased patients had lost neurons from the nucleus basalis (a tiny area found deep in the brain; its function is not understood by neurologists), while all the people without the disease had the normal number of cells in this area. Scientists who conducted this study believe that the loss of neurons from the nucleus basalis may be responsible for the decreased activity in the cortex.

3. On the 31st of August, 1909, Paul Ehrlich and Hata stood before a cage in which sat an excellent buck rabbit. Flourishing in every way was this rabbit, excepting for the tender skin of his scrotum, which was disfigured with two terrible ulcers, each bigger than a twenty-five cent piece. These sores were caused by the gnawing of the pale spirochete of the disease that is the reward of sin. They had been put under the skin of the rabbit by S. Hata a month before. Under the microscope--it was a special one built for spying just such a thin rogue as that pale microbe--under this lens Hata put a wee drop of the fluid from these ugly sores. Against the blackness of the dark field of this special microscope, gleaming in a powerful beam of light that hit them sidewise, shooting backwards and forwards like ten thousand silver drills and augurs, played myriads of these pale spirochetes. It was a pretty picture, to hold you there for hours, but it was sinister--for what living things can bring worse plague and sorrow to men?

Hata leaned aside. Paul Ehrlich looked down the shiny tube. Then he looked at Hata, and then at the rabbit.

"Make the injection," said Paul Ehrlich. And into the ear-vein of that rabbit went the clear yellow fluid of the solution of 606, for the first time to do battle with the disease of the loathsome name.

Next day there was not one of those spiral devils to be found in the scrotum of that rabbit. His ulcers? They were drying already! Good clean scabs were forming on them. In less than a month there was nothing to be seen but tiny scabs--it was like a cure of Bible times--no less! And a little while after that Paul Ehrlich could write:

"It is evident from these experiments that, if a large enough dose is given, the spirochetes can be destroyed *absolutely and immediately with a single injection.*"

---Paul de Kruif, *Microbe Hunters*

4. Over in Denmark Johannes Fibriger, a pathologist of the University of Copenhagen, had been working thirteen years on the problem of tuberculosis among laboratory animals. During a series of postmortem examinations of tubercular rats, he found three had suffered from stomach cancers. Fibiger knew enough about cancer to realize that he had come across a singular phenomenon. Rats rarely suffered from tumors of the stomach.

Fibiger made a visit to the dealer who had been supplying him with these rats, and on questioning found that those sent to his laboratory had all come from a sugar refinery. Was there anything peculiar about this refinery which could account for the unusually large percentage of stomach-cancerous rats from this spot? He investigated the place and found nothing unusual except a high infestation with cockroaches which formed a fairly large part of the diet of its rats. Could he find some connection between roaches, rats, and cancer? Cancer as a disease of filth had been spoken about for years, and vermin were said to be responsible for the so-called "cancer houses," private homes from which emerged many a human cancer victim of the same family.

Fibiger planned a controlled experiment. He collected thousands of the refinery roaches and fed them to rats from another breeding establishment. The rats enjoyed this strange treatment, and for three years--that was the normal life span of his rodents--Fibiger remained skeptical. Then they died, and one by one he opened them up. To his astonishment, he found many stomach cancers. Fibiger made a careful microscopic study of the growths. He discovered that in every case they had formed around a parasitic worm to which the roach had been host before it was fed to the rat. The larva of the worm coiled up in the muscles of the rat, later developing into an adult worm in the animal's stomach. Around this the tumorous growth had appeared. Fibiger had actually for the first time produced artificial cancer in a laboratory animal.

---Bernard Jaffe, *Outposts of Science*

5. A group of psychologists at the University of Florida have worked on a curious riddle concerning the role of age in tolerating sleep loss. In the early 1960s, Wilse B. Webb and his associates at Gainesville placed a group of young rats on a mesh water wheel which rotated slowly and continuously. If the animals fell asleep, they toppled into cold water. The researchers kept expecting the rats to topple, but some of them managed to stay on the water wheel for 27 days. However, when older rats were placed on the wheel they fell off in 3 to 4 days. By testing rats of intermediate ages, the experimenters saw that the staying power of the animal was directly correlated with its age. Delicate EEG techniques revealed that the young animals were snatching brief naps of 10 to 15 seconds as they rode the wheel. They would sleep for a few seconds, then move to the end of the wheel, and sleep as they rode forward.

---G. Luce & J. Segal, *Insomnia*

6. The French historian Roustan asserts that the *philosophes* had a direct and strong influence upon the revolutionary development that exploded in 1789. Poverty, despair, and mistreatment inflamed the masses but without the intellectuals there would have been no revolution: the men of 1789 were not merely escaping starvation which had been an integral part of the national tradition. In 1753, 800 people died of hunger in one small community alone--more died in other cities. There were riots, but no revolution--the military won the day. It took almost 40 years of education and propaganda on the part of the philosophers before the French could stand up and demand their rights and human dignity: "The spirit of the *philosophes* was the spirit of the Revolution."

G. Remmling, *Road to Suspicion*

Summary of Mill's Methods

Method of Agreement

Individuals	Antecedent Factors						Effect
#1	A	B	C	D	E	F	e
#2	A	B	C	D	E	-	e
#3	A	B	C	D	-	F	e
#4	A	B	C	-	E	F	e
#5	A	B	-	D	E	F	e
#6	A	-	C	D	E	F	e

Joint Method of Agreement and Difference

Individuals	Antecedent Factors				Effect
#1	A	B	C	-	e
#2	A	-	C	D	e
#3	A	B	-	D	e
#4	-	E	F	-	-
#5	-	-	F	G	-
#6	-	-	F	-	-

Method of Difference

Individuals	Antecedent Factors			Effect
#1	A	B	C	e
	A	B	C	e
	A	B	C	e
#2	-	B	C	-
	-	B	C	-
	-	B	C	-

Method of Residues

Antecedent Factors			Effect:		
A B C			$e_1 + e_2 + e_3$		
A	--	--	e_1	--	--
--	B	--	--	e_2	--
--	--	C	--	--	e_3

Method of Concomitant Variation

Antecedent Factors	Effect
A+BC	e+
A++BC	e++

For Further Reading

The literature on scientific laws and causality is vast. For a good introduction to the former see Hempel (1966, ch.5) or Walters (1967), followed by Mehlberg (1958/1971) and then Jobe (1967/1971). For a comprehensive philosophical introduction to *causality* see Taylor (1967) and then (in the context of science) Cook and Campbell (1979, ch. 1). An excellent anthology is Beauchamp (1974).

Hume's (1748/1960) important critique of causal necessity is reprinted in countless anthologies. For contemporary defenses of Hume's account and the causal powers account, see parts 2 and 3 of Beauchamp (respectively).

The issue of correlation vs. causation in the context of statistics presupposes a basic knowledge of statistics. Discussions tend to be rather technical and not very philosophical.

Mill's methods seem archaic to many (and not widely discussed); others believe they are relevant to the scientific method (and also not discussed). See, however, Mackie (1967) and von Wright (1951) (which is more advanced).

Notes

[1]Strictly speaking, this is false because of the following: $(\sim A \supset \sim e) = (e \supset A)$. Since we do have an example of the latter in the case of A, i.e., whenever e is present, A is present, we have shown A is a necessary condition for e. But for the sake of illustration I have ignored this point.

[2]From E. Newburn, "Sugar and Dental Caries: A Review of Human Studies" (Science, 1982), reported in Salmon (1967), p. 120.

11

Analogy

Introduction

Scientists often reason by analogy: a conclusion about a particular individual is made based upon a similarity (resemblance) between it and other particular instances. For example, when tar from cigarettes is smeared on mice, they get cancer; therefore, cigarette smoking probably causes lung cancer in humans. The similarity between the two groups is not a perfect resemblance in all respects; it is a particular kind of resemblance called an analogy—mice are analogous to humans. Here, a class A of individuals is like a class B of individuals in certain respects (P, Q, R), and B has an additional property S; therefore, it is likely that A also has S.

The use of such analogies in science is widespread: the human mind is analogous to a computer; human society is analogous to an organism; the activity of science is analogous to a business, etc. These analogies are known as **descriptive analogies**.

On the basis of these descriptive analogies, further conclusions are made. This kind of inductive inference is an example of *reasoning by analogy*, an *analogical argument:*

Mice are analogous to humans
Mice get cancer from cigarette nicotine
Therefore, humans will get cancer from cigarette nicotine

When scientists classify data, they often do so by analogy. x is classified as an A because x has properties P, Q and R, and members of A have properties P and Q. For example, a whale is classified as a mammal and not a fish because whales are like mammals (in certain key respects) although unlike them in other respects. Biologists (engaged in *Systematics*, the science of classification) have traditionally employed analogies (sometimes called *homologies*) when they group organisms together: organisms such as dog, bird, whale, and human may be classed together because of their vertebrate forelimbs even though they differ in numerous other ways, or they may be grouped together because underlying structures have similar functions (in which case the analogies are called *analogues*).

Scientists often invent new and original theoretical ideas involving things like electrons, genes, etc. to explain certain data. These theoretical concepts are not to be found in the data but are invented to explain the data. How does this creative leap come about?

On the basis of his experience and knowledge, a scientist comes to "see" relations between things. What is seen is an analogy between items of his experience and knowledge. In this case, scientists are often said to be creating and employing a *model* to explain the data. For example, because there is an analogy between the behavior of the planetary system and the behavior of an electron (both are particles, both travel in roughly circular orbits about a center, etc.), the planetary system became a model for the electron—the Bohr model—in which there is a force of attraction between the electron and the neutron like the force of attraction between the earth and the sun, and this force of attraction decreases as the square of its distance, etc. But a model appears to involve a certain kind of analogy. According to several individuals, therefore, creativity in science occurs by analogy.

Analogy: The Basic Model

Every student is familiar with examples of reasoning by analogy. Such examples abound both in everyday life and in science. The following is an example:

My neighbor has a VW rabbit and it gets 50 miles per gallon
I have a VW rabbit

Thus, my VW rabbit should get 50 miles per gallon

This example is an example of reasoning in which a conclusion is drawn based upon premises of a certain kind. In the premises, there is contained, either explicitly or implicitly, an analogy in which one thing is said to be *like* another thing: one VW automobile is like another VW automobile, humans are like rats, humans are like machines, etc. This is called the descriptive analogy. What it means to say they are "like" each other is that there is an analogy between them. An analogy is of the form:

A is to B as C is to D

or

A:B::C:D

The underlying logic of such an analogy is the following:

Individual or class a has properties P and Q and R
Individual or class b has properties P and Q

Thus, individual or class b (probably) has property R

or

```
a | P & Q & R
 b| P  & Q
==+===========
b |          R?
```

Why does it seem likely that b has property R? Because both a and b have properties P and Q and a also has R. Since a and b share certain properties, P and Q, it appears likely that they should also share property R. Clearly, the more individuals or classes that share P, Q and R, the more likely the conclusion.

a, b, c, d have properties P and Q [the resemblance]
a, b, c also have property R [the known property]

Thus, d also has property R [the expected or predicted property]

or

```
a| P & Q & R
b| P & Q & R
c| P & Q & R
=============
d| P & Q & ?
```

The key element here is the underlying resemblance or similarity between a, b, c and d—the properties they share and the importance of these common properties.

A particularly useful way of representing this situation has been given by Max Black (1952). Following Black's model, we can represent the first analogy in the following way:

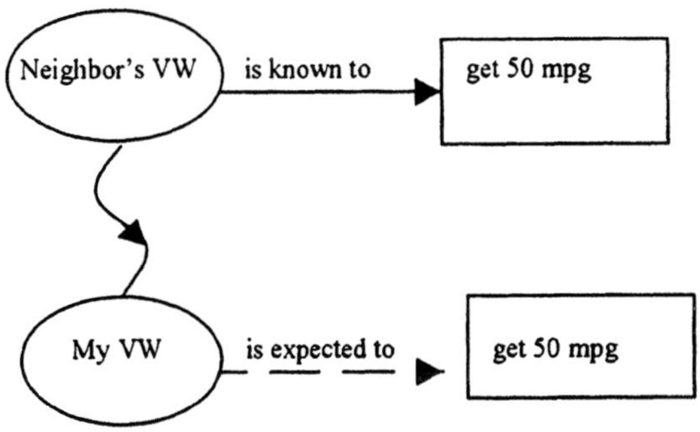

where the solid line represents what is known and the dashed line represents what is not known but expected. The general outline of this would be:

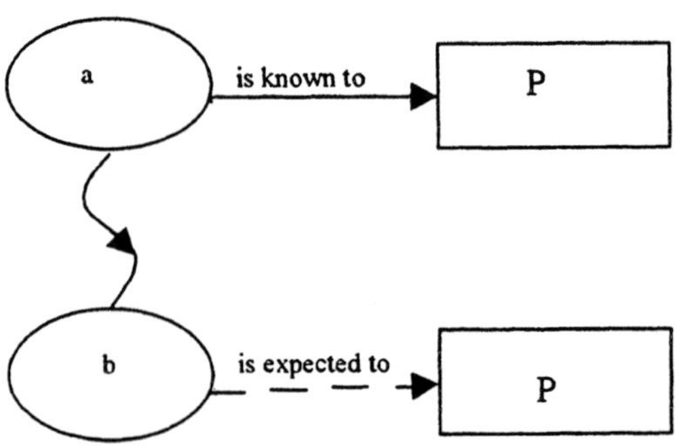

The key ingredient here is the resemblance between a and b--the curved arrow. Again, following Black, we can say that in such a case there is what can be called the **basis of resemblance** between a and b, which is the set of properties a and b are *known* to have in common. For example, in the case of the VW's with similar internal combustion engines, both are known to be VW rabbits; in the case of the rats and humans, both are known to be biological organisms with similar physiology; and in the case of the humans and the computer, both are known to be information-processing machines.

One key issue here is the following: Is the basis of resemblance *connected* with the property P? For example, is the property of being an internal combustion engine of a certain kind connected to its gas mileage? Yes. Is the property of being a particular color connected to gas mileage? No. In so far as there is a connection between the basis of resemblance and the predicted property, the strength of the analogy is increased.

Let us represent the preceding in the following way: let R be the properties making up the basis of resemblance. Then:

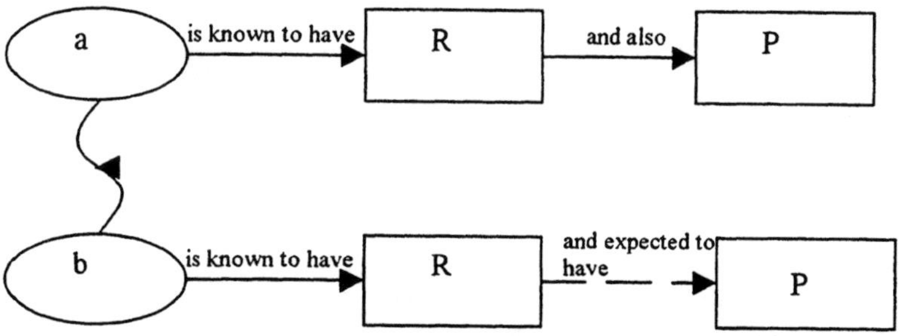

In such an argument, a key question concerns the relation between R and P. In particular, is it the case that *whenever R occurs, P also occurs*? Does there exist, in other words, what Black calls a linking generalization, a generalization linking R and P? If so, we can apply this linking generalization to the case of b and hence expect b to have property P.

One key question concerning the linking generalization is: how well-confirmed is the connection between R and P. If we have substantial scientific evidence that there was a linking generalization between R and P, say, in the form of a *scientific law*, then the strength of the analogy would be substantial and our confidence in our conclusion would be warranted. In the case of our VW, there seem to be scientific laws linking the properties of internal combustion engines with gas mileage but no laws linking color to gas mileage. The relevant laws in this case would be the laws of physics and chemistry. Thus, the strength and adequacy of reasoning by analogy is itself based upon the existence of linking generalizations and upon the empirical (and theoretical) evidence supporting such linking generalizations.

It is important to distinguish passages in which there is just a descriptive analogy from passages in which the descriptive analogy is being used as a premise for drawing a subsequent conclusion. Sometimes, individuals will present a descriptive analogy without giving an argument in support of the claim. For example:

> Knowledge is a comfortable and necessary retreat and shelter for us in an advanced age

Or

> God is our heavenly father

are descriptive analogies for which no reasons are given. If reasons were given, then one could cast these into the form of an argument, with the descriptive analogy as the conclusion:

> God and human fathers create their offspring
> God and human fathers love their offspring
> God's abode is heaven
> Therefore, God is our heavenly father

Sometimes, a descriptive analogy is presented as one of the premises in an argument in which a further proposition is alleged to follow from it:

> Knowledge is a comfortable and necessary retreat and shelter for us in an advanced age; and if we do not plant it while young, it will give us no shade when we grow old.

Here, a conclusion is drawn on the basis of the analogy. The same occurs in the following passage:

> God is our heavenly father.
> Hence, we must obey him completely and without question.

In identifying an analogical argument, therefore, one needs to identify the premises (containing the descriptive analogy) and the conclusion. Not surprisingly, there are analogy indicator words that aid in this process.

Analogy Indicator Words: Words that indicate that what follows is the conclusion of an analogy argument include:

same. . .
in comparison. . .
alike. . .
resembles. . .
as (in). . .
analogously. . .
similar(ly). . .
like. . .
just as. . .
is a kind of . . .

Exercise 11.1

In each of the following passages, circle the analogy indicator word and then indicate what is being claimed to be analogous to what.

1. Perhaps the most startling discovery made in astronomy this century is that the universe is populated by billions of galaxies and that they are systematically receding from one another, like raisins in an expanding pudding.
--Martin Rees & Joseph Silk, "The Origin of Galaxies"

2. The brain secretes thought as the liver secretes bile; the kidneys urine.

3. One of the pleasures of science is to see two distant and apparently unrelated pieces of information suddenly come together. In a flash what one knows doubles or triples in size. It is like working on two large but separate sections of a jigsaw puzzle and, almost without realizing it until it happens, finding that they fit into one.
---John Bonner, "Hormones in Social Amoebae and Mammals"

4. A city is a kind of social animal. It has its heart, and its vital arteries. It is born; it grows; it ages; and eventually it dies, leaving its accomplishments to others to take and develop.

Exercise 11.2

In each of the following passages, underline the premises and bracket the conclusion of the analogical argument. Indicate the basis of resemblance and the linking generalization by use of the diagram suggested in the text. How strong is the linking generalization?

1. Artistic energy and its periodic release in a burst of creativity can be compared to blowing up a balloon until it pops. The building up of energy can be compared to the filling of the balloon. The pop of the balloon, like creativity, appears to be sudden but it has really been preceded by a storing up of vital energy.

2. Larger animals tend to be slower than smaller animals. However, the larger animals are not less dangerous. It may take them longer to get moving, but once they are moving, no power can easily stop them. It is the same way

with nations. Big nations make decisions slower than smaller nations, but once a decision is made, it is nearly irrevocable.

3 "That the universe is formed by a fortuitous concourse of atoms, I will no more believe than that the accidental jumbling of the alphabet would fall into a most ingenious treatise of philosophy."

--Jonathan Swift

4. We have said that normal persons have little motivation to prompt special efforts at self-study. The same thing is true of arithmetic. If motivation were not supplied from parents and school pressure, there would be little learning of mathematics. By analogy, it seems possible that children could be motivated and trained to use their mental skills to solve *emotional* problems. They get almost no training in this important skill at the present time.

--John Dollard & Neal Miller, *Personality and Psychotherapy*

5. It seemed to us that the synchronous behavior of malaria parasites, all coming to cell division at the same time every 24 hours or a multiple thereof, was remarkably like the behavior of microfilariae, all entering or leaving the peripheral blood at the same time every 24 hours. Microfilariae are carried from one patient to another by mosquitoes, which suck blood mostly at night, and the swarming of the microfilariae in the peripheral blood is arranged to coincide with this time of sucking blood. The biological purpose of the cycle of the microfilariae is clearly to help them encounter mosquitoes and so get transmitted to new patients.

Since malaria is also carried from one person to another by mosquitoes, it seemed to us that the periodic behavior of the malaria parasites in the blood was similarly designed somehow or other to facilitate transmission by mosquitoes.

--Frank Hawking, "The Clock of the Malaria Parasite"

6. Photographs that have actually been taken from space reveal in a dramatic way that our planet is very much a spacecraft, turning on its axis while making its endlessly repeated trip around the sun. This comparison allows us to apprehend the importance of keeping the ship in good repair and functioning properly. We are adrift in the limitless sea of space, and there is no port where we can ever expect to put it. All that we shall ever have is aboard. If we destroy our life-support system—the atmosphere, the lakes, the rivers, and oceans, the forests and swamps—then we will eventually destroy ourselves. If we allow the ship to become more crowded than our necessarily limited supplies can support, then we can expect competition, bloodshed and almost endless suffering. Thus, we must put an end to pollution and see that our population is limited if we are to save ourselves and future generations.

---R. Munson, *The Way of Words*

Criteria for Assessing the Adequacy of Analogical Arguments

An analogical argument is based upon the presumed existence of an analogy between two or more things. The adequacy of this analogical argument depends upon how good (how strong) the underlying analogy is. But how does one determine whether the underlying analogy is sufficiently strong or not? Several criteria have been given for assessing the strength of an analogy and for judging the strength of an analogical argument. We will mention six such criteria: (1) the strength of the conclusion, (2) the number of entities compared, (3) the diversity of the instances, (4) the similarity between them, (5) the dissimilarity between them, and (6) the relevance of the similarities and dissimilarities.

1. The Strength of the Conclusion

The first criterion to be considered is how strong a conclusion is being drawn? *In general the stronger the conclusion, the less likely it is to be true, and the weaker the premises support the conclusion*; on the other hand: *the weaker the conclusion, the more likely it is to be true and the more support the premises give to the conclusion.* This general point is worth remembering in the context of analogical arguments. It clearly makes a difference, for example, whether the VW argument concluded that "*my* VW rabbit is *likely* to get *approximately* the same gas mileage as my neighbor's VW versus a conclusion that *all* other VW rabbits will *certainly* get *precisely* the same gas mileage as my neighbor's."

As the above example shows, there are several ways in which conclusions may vary in strength: in their *scope* (do they apply to *all* or *some* or *one?)*; in their *modality(*in the certainty claimed by the indicator word: *certainly, probably, possibly*, etc.); and in the *precision* of the predicted property (approximately 50 mpg, exactly 50 mpg, etc.). Clearly the more precise the predicted relationship, the more the conclusion is likely to be false and the weaker the analogical argument. On the other hand, the more general the conclusion, the vaguer the prediction and hence the less valuable the analogy. It would be relatively useless to be told, for example, that the computer and the human will have some property in common since, such a vague claim is likely to be true of any two things in the world.

2. Number of Entities

Our first criterion concerned a feature of the conclusion--how strong it was. The second criterion concerns a numerical feature of our premises: how *many* entities are being compared, i.e., the size of our sample. *Other things being equal, the greater the number of individuals mentioned in the premises, the more likely they will have other properties in common; hence, the stronger the analogy.* Thus, if instead of mentioning only my neighbor's VW rabbit, I also mentioned several other VW rabbits that also got 50 mpg, the analogy would be stronger:

My neighbor has a VW rabbit and gets 50 mpg
My sister has a VW rabbit and gets 50 mpg
My grandmother has a VW rabbit and gets 50 mpg

Thus, my VW rabbit should also get 50 mpg

In short, the larger the sample the better the analogy.

3. Diversity of Instances

Another important property to consider concerns how diverse or heterogeneous are the instances mentioned in the evidence. Consider, for example, the following argument:

My neighbor's 1969 gas VW bug gets 50 mpg
My sister's 1986 gas VW golf gets 50 mpg
My grandmother's 1982 diesel VW rabbit gets 50 mpg

Therefore, my 1984 VW rabbit should get 50 mpg

Here, the conclusion would be relatively strong because the evidence cited in the premises contain information about a great diversity of different kinds of VW's (bugs, Golfs, Rabbits), fuel systems (gas and diesel), years of manufacture (1969, 1982, 1986), and manner of driving them. Now, if in such a diversity of instances they all get 50 mpg, then it seems reasonable to expect my VW to get the same mileage. Why? Obviously, gas mileage has little to do with kind of fuel, year, model or even driving record; rather, it has something to do with the set of properties common to all these diverse VW's. Such properties can be expected to be present in any VW, including mine. In the same way, if I were to offer the following analogical argument

I took a Philosophy class and liked it
I took another Philosophy class and liked it
I took a third Philosophy class and liked it

Thus, if I take another Philosophy class I will like it

such an argument would be weaker than the following one

I took a Philosophy class in Ethics and liked it
I took a Philosophy class in Religion and liked it
I took a Philosophy class in Logic and liked it

Thus, if I take another Philosophy class, I will like it

The second case gives a better reason for the conclusion than the first because the evidence cited concerns a greater diversity of types of Philosophy classes. By contrast, in the first case it may have been that all the Philosophy classes were in Religion and that the next Philosophy class under consideration might be one in Logic. It may be that the individual only likes Philosophy classes in Religion (but this information is masked in the premises). When the premises mention a diversity of types of Philosophy classes, our confidence that it must be Philosophy classes in general that the student likes is increased.

The next two criteria concern the similarity (analogy) and dissimilarity (disanalogy) between the cases mentioned in the premises and that mentioned in the conclusion. Following Mary Hesse (1965), we can call the known similarity the **positive analogy**, the known dissimilarity the **negative analogy**, and the remainder the **neutral analogy**. The positive analogy consists of all those properties known to be possessed in common; the negative analogy consists of all those properties known not to be possessed in common; and the neutral analogy consists of all those properties such that we do not know whether they are possessed in common or not.

4. Similarity (Positive Analogy)

Clearly, the strength of an analogical argument depends upon the existence of an important similarity or positive analogy between two things. The greater this similarity the stronger the argument. In the case of our VW rabbits, for example, the criterion of similarity would involve the question of *how similar* my neighbor's VW rabbit was to mine: Were they manufactured in the same year?, Were they made in the same factory?, Were they driven in the same way? Etc. Clearly, two 1979 VW rabbits produced in Wolfburg, Germany, driven in the same way would be more similar than a 1979 VW rabbit produced in Wolfburg, Germany driven by a little old lady from Pasadena and a 1985 VW rabbit produced in Pennsylvania driven by my reckless 16 year old son.

It may appear that this criterion conflicts with the third criterion, the diversity of the instances mentioned in the sample, but this is really not so. Criterion 3 mentioned *diversity within the evidential sample,* whereas the present one mentions *similarity between the evidential sample and the conclusion*. If an analogical relationship holds when the sample contains a great deal of diversity, this indicates a *general* (generic) property possessed by all these diverse cases, which was the basis for the analogy, namely something like: any VW will get 50 mpg. However, in the case where there is a great deal of similarity between premises and conclusion and not much diversity, we suspect there is some very *specific* or narrowly limited property possessed by the instances mentioned in the premises. Consider the following argument:

my neighbor's 1979 VW diesel rabbit gets 50 mpg
my sister's 1979 VW diesel rabbit gets 50 mpg
my grandmother's 1979 VW diesel rabbit gets 50 mpg

Thus, my 1979 VW diesel rabbit should get 50 mpg

Here, the similarity between premises and conclusion points to a more specific property that appears to account for the 50 mpg--something peculiar to 1979 VW diesel rabbits, not (presumably) present in gas models, other years, other models, etc. Suppose, on the other hand, we drew a different conclusion from the same premises, namely:

my neighbor's 1979 VW diesel rabbit gets 50 mpg
my sister's 1979 VW diesel rabbit gets 50 mpg
my grandmother's 1979 VW diesel rabbit gets 50 mpg

Thus, my 1986 VW gas Golf should get 50 mpg

Clearly, the argument is weaker precisely because of a lack of similarity between premises and conclusion.

In short, *diversity of instances within the premises* points to a general property, whereas *similarity between premises and conclusion* points to a specific property. In evaluating analogical arguments, therefore, both criteria of similarity and diversity must be used and (as it were) weighed against each other.

5. Dissimilarity (Negative Analogy)

In the case of the VW's, we would expect the positive analogy to be very strong; whereas in the case of the analogy between a human and a computer, one might be inclined to say the analogy is weaker. Clearly, for example, a computer and a human have some known features in common--they share a positive analogy to a certain degree—but there are many features they obviously do not have in common—a *negative analogy*. For example, one is made of protoplasm, whereas the other is made of circuits and transistors, one was made by a human, whereas the other one was not, one has free will and creativity, whereas the other does not, etc. In such a case one might want to argue that there is a strong disanalogy between the computer and humans. In the case of an analogical argument, therefore, the stronger the negative analogy, the weaker the analogical argument.

Like the positive analogy, the negative analogy concerns the relation between premises and conclusion. We can represent these three criteria, therefore, in the following way:

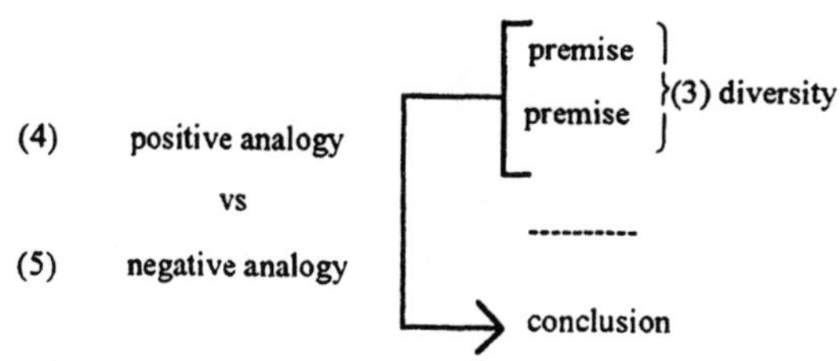

6. Relevance of Similarities and Dissimilarities

The final criterion to be considered concerns the underlying relevance of the analogy. No doubt there are many analogies one could draw between any two things chosen, but many of these analogies would be insignificant. A shoe, for example, is analogous in some ways to a house and a house to a whale but these analogies seem to be insignificant. In our example of the VW's and gas mileage, certain factors such as construction of the engine, driving style, and type of fuel used were given as the basis for the assertion that one car would likely get the same gas mileage as the other ones because these factors are relevant to how much gas mileage a car gets. But suppose one argued that my VW should get the same gas mileage as three other VWs because the owners' names all begin with D, they all have blond hair, their cars all have the same type of bumper stickers, their cars are all pink, etc. Here, although there is indeed a positive analogy between the cars, the factors mentioned seem to be irrelevant. So the existence of a positive analogy by itself is not decisive in evaluating the strength of an analogical argument; more important than number of common factors is the relevance of these common factors.

Many times a single relevant factor will provide a stronger basis for an analogical argument than a dozen irrelevant positive similarities. For example, suppose a doctor were to conclude that Hansel's illness will be improved by taking a particular kind of medicine because Hansel and Gretel both have the same illness, they have the same type of bacteria present in their kidneys, and the medicine in question helped Gretel. Suppose this same doctor were to draw the conclusion that Hansel's illness is likely to be abetted by the medicine in question because Gertrude, Otto, and Hermann were all helped by this medicine, they and Hansel are Lutherans, all voted for the Social Democratic party in the last election, all were students at the same Gymnasium, all wear Lederhose, and all are Gemini. These factors, although making up a substantial positive analogy, are fundamentally irrelevant. In short, when someone suggests that because certain individuals have properties P, Q and R and another individual has P and Q, that this individual also has R, the question is whether P and Q are relevant to R.

As we have already pointed out, the question of relevance is related to the concept of linking generalization. A linking generalization is a generalization linking P and Q (the basis of the resemblance) to R (the predicted property). In fact, the question of relevance *is* the question of whether there is a linking generalization and if so, how strong it is. There is a linking generalization if the occurrence of P and Q is accompanied by the presence of R. If so, then P and Q are relevant to R.

There are various kinds of linking generalizations and each has various degrees of strength. They may vary in strength from *whenever P and Q, then R* to *(sometimes) if P and Q, then R*. Likewise, a linking generalization may be an association of these properties without our knowing why they go together (a correlation), or it may be that P and Q are the *causes* of R (causal generalization). We expect, for example, that the structure of the internal combustion engine is related to gas mileage just as we expect there is a relationship between physiological makeup and rate of learning, and between biochemistry and lung cancer. These factors are thought to be relevant largely because they appear to be causally related to the expected property.

Although the linking generalization between P and Q, and R (and the associated concept of relevance) is often thought to be a causal relation, there are other kinds of relationships also possible: P and Q might be the effect of R; P and Q and R might be effects of some hidden cause, etc. Thus, P and Q might be the cause of R in the case where P and Q are factors such as the structure of the internal combustion engine and type of fuel used and R is gas mileage. P and Q might be the effect of R in the case where P is body temperature, Q is spots on the skin, and R is the presence of a virus. An analogical argument using this form would be the following:

> several patients have symptoms P and Q and also the virus R
> another patient also has symptoms P and Q
> --
> Thus, the patient also has the virus R

One can also argue that because P, Q, and R are effects of a common cause, the connection between P, Q, and R can be used to infer the presence of R. For example, P, Q, and R might very well be three symptoms of the same underlying disease entity, with P being high temperature, Q being red spots on the skin, and R being enlarged glands. All three are the presumed effects of an underlying causal agent--the disease, which (let us again suppose) is a virus. Then one can argue that because P, Q, and R have occurred together in other patients, and because P and Q are present in the patient under consideration, this patient is also likely to have R--enlarged glands. One set of symptoms doesn't cause the other symptom; rather, they are all effects of one underlying cause. But from the presence of some symptoms, which are correlated with the presence of another symptom, one can infer by analogy the latter's likely presence.

Exercise 11.3

A. For each of the following groups, list the statements in order of increasing precision (decreasing vagueness). How does this relate to the question of the probable truth of the statement? Why?

1. a) Aunt Audrey will arrive tomorrow.
   b) Aunt Audrey will arrive tomorrow at 3 PM.
   c) Aunt Audrey will arrive tomorrow afternoon.
   d) Aunt Audrey will come for a visit.
   e) Aunt Audrey will come this week.
2. a) Helmut Schmidt will be re-elected.
   b) Something will happen to a major politician.
   c) Schmidt will be re-elected by a wide margin.

d) A major politician will be re-elected.
e) Something will happen to someone.

B. Each of the following arguments by analogy has six additional premises suggested for it. For each of these alternative premises, decide whether its addition would make the resulting argument more or less probable or, whether it is irrelevant to the strength of the conclusion. Why?

1. A psychologist is experimenting with rats to determine the strength of sexual privation in determining rat behavior. He has worked with five albino male rats and one albino female. The male rats were sexually deprived through isolation for one month. When placed in a maze with the female rat as the bait, the five rats found their way through the maze much more quickly than they had under normal conditions. The psychologist now postulates that the next five albino rats, treated similarly, will also run faster through the maze after sexual deprivation than under normal circumstances.

a. The first five male rats and the second five male rats were all young rats.

b. The first five male rats were given all the other comforts--food, shelter, freedom to roam. The second five were kept isolated in a very small cage.

c. Instead of using albino rats, the psychologist used ordinary gray rats on the second round.

d. The second five rats were also deprived of food during the last two days before being placed in the maze.

e. The color of the maze was changed.

2. Mrs. Carpenter has been investing in insurance stocks for over the past 15 years. The overall value of his stocks has increased about 4 percent per year, and the companies have paid a good and steady dividend each year. She concludes that it is now time to buy another 200 shares of insurance stocks and that he will not only increase the value of her purchase but gain modest divided earnings.

a. The stock she intends to purchase is stock of one of the companies of which he already owns stock.

b. She is purchasing stock in a relatively new insurance company, whereas her previous stock has all been stock of old and established insurance companies.

c. Rather than the usual 4 percent increases in appreciation and the usual 5 percent dividend, her insurance stocks this past year increased its appreciation by 8 percent and passed a 9 percent stock dividend.

d. The insurance company in which Mrs. Carpenter bought the new stock is heavily into providing medical malpractice insurance for physicians, whereas the other insurance companies in which she has invested are primarily in life insurance.

e. The oil company stocks have recently decreased in value.

f. Several of the state governments pass laws regulating medical malpractice suits and insurance premiums charges for medical malpractice.

Exercise 11.4

For each of the following arguments, identify the premises and the conclusion of the argument. Evaluate them in terms of the six criteria that have been explained. What general conclusion do you draw about the strength of the analogical argument?

1. Every species of plant or animal is determined by a pool of germ plasma that has been most carefully selected over a period of hundreds of millions of years.

We can understand now why it is that mutations in these carefully selected organisms almost invariably are detrimental. The situation can be suggested by a statement made by Dr. J. B. S. Haldane: my clock is not keeping perfect time. It is conceivable that it will run better if I shoot a bullet through it; but it is much more probable that it will stop altogether. Professor George Beadle, in this connection, has asked: "What is the chance that a typographical error would improve *Hamlet*?"

--Linus Pauling, *No More War*

2. In the United States, especially, atom-smashing equipment, such as the Van de Graaff generators and cyclotrons, had been constructed. These machines were already capable of accelerating certain particles used as 'projectiles' up to the enormous energy of nine million volts. Nevertheless, even they had only damaged, without breaking into, the protective walls with which Nature in her wisdom had encircled the atomic nucleus and the tremendous stores of energy it contained. The idea that neutrons, which carried no electrical charge at all, might have been able to accomplish what could not be done with such heavily charged projectiles, was too fantastic to be credited. It was as though one were to suggest to troops which had been vainly shelling an underground shelter with guns of the heaviest caliber for a long time that they should start trying their luck with Ping-Pong balls.

--Robert Jung, *Brighter than a Thousand Suns*

3. While he was attending a group of drug addicts at a sanitarium in Berlin in 1927, it occurred to Dr. Manfred Sakel to try insulin on them. This hormone promotes the utilization of sugar in the body, and on theoretical grounds he believed its effect should relieve the paradox by which a slave of the drug habit requires larger and larger doses of what is essentially a poison. He hoped that though the insulin he might free the victim of dependence on morphine.

. . . .

Some of the men reacted to the insulin with convulsions, but most of them broke into perspiration and lapsed into deep sleep. When they came out of their seizure, or were awakened after a few hours of coma, their conduct surprised the doctor. He noticed that the morbid fears and anxieties which habitually oppress addicts had diminished. Odd notions of persecution, jumpy nerves, and other psychotic symptoms were gone.

This unexpected outcome set Dr. Sakel to thinking. If insulin improved the mental climate of the drug-crazed men, what would it do for the frankly insane?

--George Gray, *The Advancing Front of Medicine*

4. The discovery of this remarkable weapon against disease dates back to 1929. It was purely accidental. Dr. Alexander Fleming, in St. Mary's Hospital, London, was growing colonies of bacteria on glass plates for certain bacteriological researches. One morning he noticed that a spot of mold had germinated on one of the plates. Such contaminations are not unusual, but for some reason, instead of discarding the impurity and starting fresh, Dr. Fleming decided to allow it to remain. He continued to culture the plate, and soon an interesting drama unfolded beneath his eyes. The area occupied by the bacteria was decreasing, that occupied by the mold was increasing, and presently the bacteria had vanished.

Dr. Fleming now took up this fungus for study on its own account. He recognized it as of the penicillium genus, and by deliberately introducing a particle into culture mediums where bacteria were growing, he found that quite a number of species wouldn't grow in its presence. . .In his laboratory, whenever he wanted to get rid of a growth of gram-positive bacteria, Fleming would implant a little penicillium, and after that the microbes disappeared. . . So the medical scientists began to speculate. Since the mold destroyed gram-positive organisms on a culture plate, could it be used to destroy gram-positive disease germs in the living body?

–George Gray, *Science at War*

5. Indeed, the most compelling similarity between science and myth is the drama each derives from living in the twilight zone between fantasy and reality. Physics lives there just as assuredly as do the ancient stories of the gods and goddesses and demons, for physics is never unrestrictedly true. It is only temporarily true, and in a broader context, it is always false for sometime in the future each of our theories will most likely be disproved by a new discovery. . . Further, like myth, the power of modern science is derived from the metaphorical linkage between facts, not by the facts themselves. Thus, science, the actor, emerges from the darkened winds of the human mind into the bright lights of shared experience, pretending in spite of the mask and the costume that he is real, his world is real, and that he is more than an actor. Yet all the time he knows that he is only pretending and that he will be judged not by the quality of his truth but by the quality of his illusion. Truth slips from our hand when we insist that science be true.

—Kim Malvill, *A Feather for Daedalus*

6. An electron is no more (and no less) hypothetical than a star. Nowadays we count electrons one by one in a Geiger counter, as we count the stars one by one on a photographic plate. In what sense can an electron be called more unobservable than a star? I am not sure whether I ought to say that I have seen an electron; but I have just the same doubt whether I have seen a star. If I have seen one, I have seen the other. I have seen a small disc of light surrounded by diffraction rings which has not the least resemblance to what a star is supposed to be; but the name "star" is given to the object in the physical world which some hundreds of years ago started a chain of causation which has resulted in this particular light-pattern. Similarly in a Wilson cloud chamber I have seen a trail not in the least resembling what an electron is supposed to be; but the name "electron" is given to the object in the physical world which has caused

this trail to appear. How can it possibly be maintained that a hypothesis is introduced in one case and not in the other?

–Sir Arthur Eddington, *New Pathways in Science*

Analogy and Scientific Models

Reasoning by analogy is an important type of inductive inference, one central to much of scientific reasoning. It is based upon certain scientific facts about a class of objects and hence depends upon scientific observation. But more than this, the adequacy of such analogical arguments depends upon linking generalizations which can be construed as scientific laws and generalizations pointing to (or explicitly involving) underlying causal relations. But scientific generalizations themselves are, in turn, rooted in facts, many of which are obtained by scientific observation. In short, one cannot engage in a scientific analogical argument without a sufficient amount of evidence to support it.

But analogies, like other forms of inductive inferences, are not certain; in fact, it is abundantly clear that when a scientist engages in analogical reasoning, the conclusion drawn is always a risky one, with the degree of risk varying from case to case and the amount of evidence relevant to each. It is risky not only because the conclusion goes beyond the information contained in the premises but, more importantly, because the inference is based upon an *analogy* between two or more things and by its very nature such analogies may be misleading or even mistaken. As we have seen, there may be a positive analogy between two things and yet the underlying resemblance may be absent. In such a case, the analogy would be a questionable one.

However, although analogical reasoning is a form of inference susceptible to error and misuse, it remains a fundamental way of thinking in science. Indeed, some authors have gone so far as to claim that all really creative thinking, not only in science but in other areas as well, depends upon thinking in terms of analogies, of seeing similarities and relationships between things not noticed by others and by exploring these similarities in some detail. When we see relationships between diverse and apparently unrelated things, our understanding of things is often increased. Part of the reason for this is the fact that the human mind seems to understand something when it can relate it to other things, especially when it can be fit into a larger pattern of relationships. When connections between things are noticed, previously unrelated things are brought together into larger conceptual patterns.

Thus analogy, together with the related notions of metaphor, simile, model, etc., underlies creative thinking at its best by pointing out new relationships between things. This is likewise true of science and scientific thinking; in particular, it is true of one dominant kind of scientific theorizing running consistently throughout the history of science: theorizing in terms of models. Indeed, whenever a scientist advances a theoretical model, there is an underlying analogy which is the basis for the model. In short, scientific hypotheses often come from metaphors and analogies, which constitute one fruitful source for generating new scientific knowledge. To the question: How do scientists discover new theories?, the answer often is: *by using analogies.*

Two Brief Case Studies: Light and Motivation

The history of optics is largely a history of a debate between two theories of light: the corpuscular (atomic) theory of light and the wave theory of light. Sir Isaac Newton championed the first account and his contemporary **Christian Huygens** the second. Initially, the corpuscular theory was the dominant one largely because of the influence of Newton, but in the 19th century things were reversed largely because of the work of Thomas Young.

What is important for our purposes is Huygens' defense of the wave theory of light based upon an analogy. In his famous work, he begins with a resume of widely accepted facts about light: light consists in the motion of some kind of matter, light has an extremely great velocity, light is always propagated in straight lines, light is reflected from bodies, etc. To explain all of this, Huygens proposed a model of light based upon an analogy: *light spreads the way sound spreads, namely as a wave.* This conjecture, in turn, is based upon an analogy between sound waves and our ordinary experience of water waves:

> I call them [i.e., light movement] waves from their resemblance to those which are seen to be formed in water when a stone is thrown into it, and which presents a successive spreading as circles, though these arise from another cause, and are only in a flat surface" (1690/1964, p. 170).

He proceeds to suggest that in both kinds of wave, there is a gradual movement of matter—a spreading wave—and that this account can explain the recognized properties of light.

Here, Huygens is suggesting that light is like a wave, that there is an analogy between light and water waves. Hence, a wave is a *model* for the nature and behavior of light: Both light and waves exhibit the same type of movement, a successive spreading out in circles. Here is a positive analogy between the two. Of course, there is a negative analogy; for example, they are made of different types of material and the medium is different, but the positive analogy is sufficiently strong to suggest that one can use a wave as a model for light.

What is involved in using one thing as a **model** for something else? First of all, of course, there must be a positive analogy between them. More importantly, however, is the use of what has been called the *neutral analogy*. This is the set of properties possessed by the one thing, in this case, a wave but which is *not known* to be possessed by the other thing--light. By using the neutral analogy, one is able to make new predictions about light simply by suggesting that these neutral properties of waves are also possessed by light. For example, it is known that water (and sound) waves manifest the properties of *reflection* (waves "bounce back") and *refraction* (waves bend as they pass from one medium to another)—and so it was hypothesized that light also should reflect and refract. Similarly, since water waves and sound waves travel through a medium--water and air--so (it was hypothesized) light should also travel through a medium--in this case a special light medium called the *ether*, a transparent and elastic medium. (But as the famous Michelson and Morely experiment of 1887 showed, there is no such ether. Hence, this prediction was falsified and the way was paved for Einstein's Special Theory of Relativity of 1905.)

We can represent this by using the following diagram:

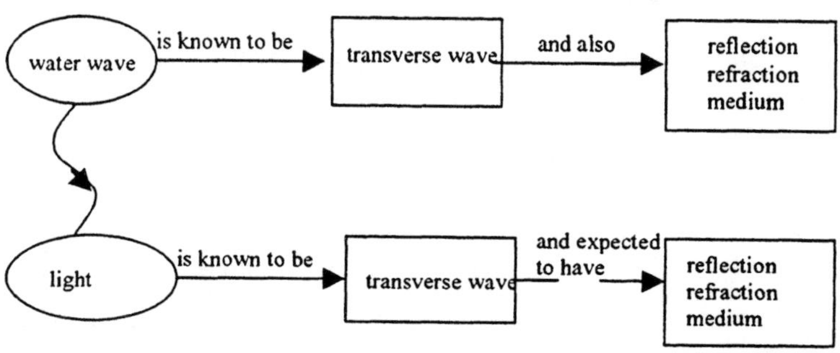

Another famous case of using an analogy to construct a theoretical model can be found in psychology and the behavioral sciences--the hydraulic model of motivation (drive reduction) of the psychoanalysist Sigmund Freud and the ethologists Niko Tinbergen and Konrad Lorenz. Drives and motives, e.g., the sexual urge, have several features in common with mechanical systems of water works (e.g., the toilet): water gradually gets higher and higher in a water reservoir as its source continues to produce water (just as the energy behind a drive increases over time if not released); as the level of the water increases, there is an increase in the water pressure (just as there is an increase in tension as the urge increases); flushing the toilet empties the reservoir of water and reduces the pressure (just as releasing the drive eliminates the tension that has built up); etc. Here, then, we have a positive analogy between the two and this suggested that a hydraulic system could be the *model* for motivation. What about the neutral analogy? As we have seen, the neutral analogy is the basis for making predictions about the thing modeled and for one's explanations of puzzling phenomena. For example, in a hydraulic system there are overflow pipes so that if the water builds up and cannot be released down its normal channel, it will overflow and traverse a different channel. One might suggest, therefore, that the same thing is true of motivational drives, that as sexual drive increases and cannot be discharged along normal sexual channels, this sexual energy may "overflow" into a nontypical channel, with deviant sexual behavior resulting. Thus, the hydraulic model would explain certain puzzling (aberrant) motivational phenomena.

The wave model of light and the hydraulic model of motivation are just two examples of many models to be found in the history of science. The following is a brief list of some of these models (Leatherdale, 1974):

The Modeled is analogous to	*The Model*
Bohr-Rutherford atom	Planetary system
Light Waves	Water Waves
Kinetic theory of gases	Container of billiard balls
Nuclear Fission	Division of a liquid drop
Electric field	Incompressible fluid flowing through tubes of variable section
Society	Organism
Natural Selection	Malthusian struggle in society
Animal and Human body	Machine
Atmosphere	Ocean
Brain	Computer
Atom	Vortex
Electric current	Fluid
Social Evolution	Biological Evolution
Ideas in the Mind	Billiard Balls

Ideas in the Mind Billiard Balls

Exercise 11.5

Below are some examples of some of the models cited above. Identify which is the underlying model being used. Point out the positive and negative analogies, and indicate how the neutral analogy might lead to new predictions or new explanations.

1. Let us consider the atom of hydrogen with its nucleus and one electron. There will be a force of attraction between the two, proportional to the product of their charges and inversely proportional to the square of the distance between them; in fact, the law of attraction is of exactly the same form as that between a planet and the sun. The electron will thus fall toward the nucleus, unless it already possesses a velocity in some other direction, just as the earth would fall into the sun if its velocity in its orbit were suddenly checked. We must consider then that the electron describes an orbit around its nucleus just as a planet describes an orbit around the sun; and since the law of attraction is the same in the two cases, this orbit will also be an ellipse, or in special cases a circle. Thus we may speak of the astronomy of the atom.

---Discovery, V, p. 64

2. Just as the bottom of a bucket containing water is pressed more heavily by the weight of the water when it is full than when it is half empty, and the more heavily the deeper the water is, similarly the high places of the earth, such as the summits of mountains, are less heavily pressed than the lowlands are by the weight of the mass of the air. This is because there is more air above the lowlands than above the mountain tops; for all the air along a mountain side presses upon the lowlands but not upon the summit, being above the one but below the other.

---Blaise Pascal, *Treatise on the Weight of the Mass of the Air*

3. Thus the nervous system and the automatic machine are fundamentally alike in that they are devices which make decisions on the basis of decisions they have made in the past. The simplest mechanical devices will make decisions between two alternatives, such as the closing or opening of a switch. In the nervous systems, the individual nerve fiber also decides between carrying an impulse or not. In both the machine and the nerve, there is a special apparatus for making future decisions depend on past decisions, and in the nervous system a large part of this task is done at those complicated points called 'synapses' where a number of incoming nerve fibers connect with a single outgoing nerve fiber. In many cases it is possible to state the basis of these decisions as a threshold of action of the synapse, or in other words, by telling how many incoming fibers should fire in order that the outgoing fibers may fire. [Further], the machine, like the living organism is, as I have said, a device which locally and temporarily seems to resist the general tendency for the increase of entropy.

--Norbert Weiner, *The Human Use of Human Beings*

4. Benjamin Franklin developed and refined the one fluid theory. Each unelectrified body contains a normal, or equilibrium, amount which produces

no observable effects. The process of electrification, he supposed, consists in taking some of the electrical fluid from one body and giving it to another. Thus, the fluid is not created, but is merely transferred. When an unelectrified body is rubbed, it either gains electrical fluid and thus reaches a positive state (plus), or loses some of its natural amount, leaving it in a negative state. In the process of electrification by conduction he assumed that a normal body will receive additional fluid from a positively charged body, and will give up fluid to a negatively charged body. Thus, the 'direction of flow' of electrical fluid is from a positively electrified body to one that is either a normal or negatively electrified body, and from a normal body to one that is negatively electrified. This convention of determining the flow of electricity (electric current) is still in use.

--Charles Kegley & Jacquelyn Kegley, *Introduction to Logic*

5. It must be supposed that a gas contains an enormous number of tiny molecules in rapid motion, which like billiard balls exert only contact force on each other and can travel only in a straight line. These are subject to the conservation laws of classical mechanics. Their diameters are small compared to the mean-path of the molecules; they collide elastically with each other and the walls of the container; and the absolute temperature of the gas is a function of the mean kinetic energy of the molecules.

6. Society is to be considered an organism, similar in many ways to a biological organism. A biological organism has as its basic unit, cells, while a social organism has as its basic unit, individual human beings. Thus, when you talk about the structure of a biological organism, you talk about the relations between cells; whereas, when you talk about the structure of the social organism, you talk about the relations between human beings. If you talk about 'activities' of a biological organism, you speak about the activities of cells and systems of cells; if you are concerned about the 'activities' of a social organism, you are concerned with the observed behavior of individual human beings and groups of human beings. Further, just as all biological organisms are concerned with maintenance and adaptation, so also are all social organisms. Thus, it is proper to speak of *functions* in both cases. In the social context, *functions* are 'those observed consequences which make for the adaptations or adjustment of a given system and *dysfunctions* are those observed consequences which lessen the adaptation or adjustment of the system.

7. In the next chapter, the Struggle for Existence, amongst all organic beings throughout the world, which inevitably follows from their high geometrical powers of increase, will be treated of. This is the doctrine of Malthus, applied to the whole animal and vegetable kingdoms. As many more individuals of each species are born than can possibly survive; and, as consequently, there is frequently recurring struggle for existence, it follows that any being, if it varies, however, slightly in any manner profitable to itself, under the complex and sometimes varying conditions of life, will have a better chance of surviving, and thus be *naturally selected*. From the strong principle of inheritance, any selected variety will tend to propagate its new and modified form.

--Charles Darwin, *The Origin of the Species*

For Further Reading

Virtually all books on logic include a chapter on analogical reasoning.

A comprehensive overview of the entire field of metaphors and model can be found in Leatherdale (1974). Hesse's (1965) work on models has been very influential, as has Black (1962) work on metaphors. On the connection between metaphors and human cognition, see Lakoff & Johnson (1980).

12

Hypotheses and Hypothesis-Testing

The Hypothetico-Deductive Method

Introduction

Historically, there have been two influential answers to the question, What is the nature of scientific method? One is inductivism and the other is hypothetico-deductivism. We have already discussed two forms of inductivism—naive and sophisticated inductivism--and we have indicated some of their strengths and weaknesses. Several kinds of scientific activities can be characterized as versions of sophisticated inductivism, scientific inferences in which data are collected and then inductive inferences drawn from them. Such inferences include inferences to a scientific law, complex processes of qualitative and/or quantitative classification and measurement together with the construction of a mathematical equation to describe the quantitative data, and so forth. In still other cases, there are more complex forms of reasoning by analogy in which hypotheses are invented on the basis of analogical connections concerning earlier experience and knowledge.

Although sophisticated inductivism describes certain types of scientific activities and, hence, is a partial account of scientific method, it is inadequate as a complete account. It does not adequately characterize several important kinds of scientific activity, in particular, those methods involving the creation and testing of hypotheses involving theoretical entities. Inductivism is often linked to the epistemological view of empiricism; and although some forms of empiricism may be tenable, other interpretations are not. The making of observations and the drawing of inductive inferences typically involve very complex cognitive activities which are conjectural in nature, and any theory that claims such activities are absolutely certain must be judged to be inadequate.

The fundamental problem with narrow versions of inductivism and empiricism, therefore, is that they cannot account for one crucial kind of scientific method—the postulation of **theoretical entities.** These hypothetical entities (for example, genes, electrons, fields, black holes, viruses, unconscious mechanisms, social structure, etc.) cannot be observed by the unaided, naked eye, and often cannot even be observed with sophisticated scientific instruments. Inductivism does not appear to be able to handle this form of scientific reasoning since, in this kind of activity, the scientific mind is engaged in a complex process that is creative, an activity some individuals characterize as being both resistant to scientific explanation, unpredictable, and mysterious. Even if we reject the latter interpretation, the fact remains that the construction of scientific theories, such as Einstein's theory of relativity, is a process in which initial empirical data play a relatively minor role and where the inference in question is very conjectural. For these types of activities, we need a method other than inductivism.

As an account of scientific method, inductivism does not cover cases of what can be called *the indirect testing* of hypotheses and theories. When scientists test for the presence of some entity, they can do so in two ways, either by a direct test or by an indirect test. A direct test for the existence of an entity involves the direct observation of the thing in question, but in an indirect test, one is not observing the thing itself but a consequence of its presumed existence and nature. In a direct test, an experimental arrangement will allow one to decide whether the thing in question is present or not, for example, individuals may look into a microscope to see if a particular bacterium is present. But this type of testing will not be available when the entity in question is hidden from our direct gaze. In this case, we must perform an indirect test: we must test for its presence by determining if one of its hypothetical consequences is present. For example, if we observe one of the effects of an alleged cause, our confidence is increased that the cause has been detected; and if we do not observe its predicted effects, we have reason to believe the thing does not exist. Inductivism may adequately characterize the first type of detection—direct testing—but it can not characterize the second type--indirect testing. In order to do this, we need a different method--the hypothetico-deductive method.

The General Hypothetico-Deductive Model

A Case Study: Semmelweis[1]

To begin our discussion of the hypothetico-deductive model, let us look at a clear cut example of this model in action—the famous case of Ignaz Semmelweis and the discovery of puerperal fever. After describing this discovery, we will use it as an illustration of the various steps involved in the hypothetico-deductive model. (In the following passage, numbers in brackets refer to the corresponding steps in the hypothetico-deductive method.)

> In 1858 Semmelweis, while advocating chemical treatment of the hands as a prevention of the spread of puerperal fever, received a letter from one of his students describing the conditions at Gratz: "Infection of all sorts occurs at Gratz lying-in-hospital. . . The dissection-room is the only place where the students can meet and pass the time when waiting for their midwifery cases, and they often devote their attention to dissecting or studying and manipulating preparations. (Be it noted parenthetically that the cadavers of the dissecting-room at that time were not embalmed in antiseptics as they are today, and in Germany, where an autopsy occurred invariably after each death in the hospital, the body, afflicted with whatever disease, furnished material for the medical student.) When they (the students) are summoned to the lying-in hospital, which is just across the street, they do not make any pretense at disinfection; some of them do not even wash their hands. . . The patients might as well be delivered in the dissecting-room. As it is, the students cross the street with hands wet and bloody from dissecting; they dry their hands in the air, and stick them a few times into their pockets, and at once proceed to make examinations. It is no longer a riddle to me why, after a clinical meeting, the medical officer of Gratz exclaimed: 'The lying-in hospitals are really nothing but murder institutions!' "
>
> Of the lying-in hospital of Budapest in 1850, it is said that the patients' view from the window was the burying-ground, varied on the other side by glimpses of the dissecting-room, with underneath the privies and an open sewer. In 1860 the hospital was moved to a new building, and of this the following is written in a publication of that time:

"While it is not to be denied that the institutions have obtained the advantages of more room, it must be admitted also that the internal fittings (furniture, beds, etc.) are in the old wretched condition; the broken tables and the ragged and worn-out bed clothes, all brought from the old hospitals. Especially the lying-in clinic is in an indescribably pitiable condition; there poor lying-in women are to be found, some of them partly on straw spread on the floor, some of them on wooden benches, others crouching in any corner of the room, weary and worn out; only to a few is it vouchsafed to find a regular bed on which to stretch their weary limbs. Everywhere you find dirty bed linen, with bedclothes old and worn and almost in rags. . ."

The bed linen referred to in the quotation above as being almost in rags suffered also from other difficulties. In 1860 graft was known in hospital administration and Semmelweis encountered and ended, for his wards, one of the most loathsome varieties. The mortality among his patients remained high in spite of his efforts, and he examined the instruments, the utensils, and the bedding. He made a shocking discovery. He found that patients in labor were laid upon filthy sheets, which actually stank of decomposed blood and uterine discharge. These had been received and accepted as clean by the head nurse from the laundry contractor, who had accepted the contract at an especially low rate. All the circumstances pointed to dishonesty, which extended all the way from the superintendent to the pupil-midwives . . .

With a picture in mind of the conditions existing in the maternity hospitals of the middle of the last century, we turn again to Semmelweis and his problem. The maternity hospital of Vienna, where Semmelweis started his work, was in two divisions; in one the medical students were instructed and in the other the women who were to become midwives were trained. In the First Division, that of the medical students, there were from 68 to 158 deaths in each 1,000 births, with an average, over a period of six years, of 99. In the Second Division, that of the midwives, the average for the same period was 33 deaths in 1,000 births. [1] The difference in these figures started the active mind of Semmelweis to puzzling over the reason: [2] he passed in review all the possible causes, which had been suggested. [3] Some physicians had confidently attributed the frightful devastation wrought by childbirth fever in the First Division to epidemic influences, which in those days signified some not exactly defined state of the weather which often spread over whole districts of the country and caused fever in women bearing children. [4] But Semmelweis dismissed this theory [5] on the ground that the same atmospheric conditions existed in both divisions of the hospital. [6] He recognized that the cause of the disease lay within the hospital. He said: "A very common and successful method of diminishing the ravages of a puerperal-fever epidemic is to close the lying-in hospital. The hospital is not closed in the expectation that the patients will die somewhere else, but because of the belief that if they remain in the hospital they will die, owing to the epidemic influences at work within the hospital; whereas if they are confined outside, away from the hospital, they will remain well."

Semmelweis makes point after point in tracing the course by which he established to his own satisfaction that the disease arose from some unknown cause within the hospital. It may be added that only a dreadful dilemma saved the lying-in hospitals from destruction. The patients were charity cases and most of the children born were retained under the care of the state. Due to the hospitals, many women in the bloom of life were carried off by fever. Without the hospitals a large number of women would remain well, but because of the anxiety over the maintenance of themselves and their infants they would resort to the crimes of abortion, child exposure, desertion, and murder. The hospitals were tolerated as the lesser of the two evils.

[7] Semmelweis demonstrated that crowding could make little difference in the distribution of deaths between the two divisions of this hospital, [8] because each was crowded. [9] Fear was declared by some physicians to be the cause of the fever, for the dread of going to the First Division was very great; so great, indeed, that it was thought the women sickened and died for that reason. In this connection Semmelweis says: "That they were afraid of the First Division there was abundant evidence. Many heartrending scenes occurred when patients found out that they had entered the First Division by mistake. They knelt down, wrung their hands, and begged that they might be discharged. Lying-in patients with uncountable pulse, meteoric abdomen, and dry tongue, only a few hours before their death would protest that they were quite well in order to avoid medical treatment, for they believed the doctor's interference was always the precursor of death." [10] Semmelweis dismissed this theory that fear was the cause of the difference in mortality, [11] because this difference in mortality would of necessity have to exist before the fear could develop.

[12] Certain religious observances were also accused of increasing the mortality. The chapel of the lying-in hospital was so situated that the priests, bearing the last sacrament to the dying, could, in the Second Division reach the ward where the patient lay without passing through the other wards, but in the First Division they must pass through five wards before reaching the sick-room beyond. It was usual for the priests, arrayed in their robes, with an attendant marching before them ringing a bell according to Catholic ritual, to proceed to the sick women to administer the sacrament. According to ordinary arrangement this should be done only once in twenty-four hours, but in childbed fever twenty-four hours is a long time, and the priests had to be sent for occasionally in a few hours after their regular visit. It is easy to imagine the impression, which the fateful tolling of the bell would produce upon the lying-in women. "Even to me myself," says Semmelweis, "it had a strange effect upon my nerves when I heard the bell hurried past my door; a sigh would escape my heart for the victim that once more was claimed by an unknown power. This bell was a painful exhortation to me to search for this unknown cause with all my

might. During my first term of office I appealed to the sense of humanity of the servant of God, and without difficulty [13] it was arranged that for the future the priests would take a roundabout route, without ringing the bell, so as to reach the sick-chamber in silence and unobserved. The two divisions were made similar in this respect, but [14] the difference in their mortality still remained."

Other and even more absurd reasons were given for the difference in mortality. [15] It was alleged that the reason for the great mortality was because patients were unmarried women of the most hopeless class of the community, accustomed to earn their bread in want and misery and amid conditions which produced great and constant depression of spirits. [16] If this had been the cause of the mortality it would have been as great in the Second Division, for to it exactly the same class of patients was admitted. [17] The higher mortality of the First Division was ascribed to the wounded modesty of the poor women going through the process of parturition in the presence of men. Most of the patients in the First Division certainly suffered from fear, but [18] not many were troubled with a sense of shame. Truly it shows with what want of thought the whole question of the etiology of puerperal fever has been discussed when the persons who at times are depicted as the most abandoned of the population, have attributed to them in the next sentence a tenderness of modesty such as the upper and highest classes of the community do not claim. Among the upper and even the highest ranks of society labor is conducted by physicians, and their patients do not die of puerperal fever in consequence of wounded modesty in the same proportion as is alleged of the inmates of the lying-in hospitals who, for the sake of argument, are often depicted as the most loose and abandoned of the community."

[19] Semmelweis goes on step by step to eliminate ventilation, dirty laundry, and improper diet [20] on the grounds that these things were the same in the two divisions. Yet as he eliminated these possible causes he knew that the real cause lay undiscovered before him in the hospital. This fact was proved by the observation that women who were overtaken with labor on the street while making their way to the hospital were not affected with child fever. . .

The restless energy of Semmelweis in his work in the hospital, his sympathy with suffering women, and his constant criticism of the old orthodox opinions of the causation of childbed fever, probably reached the ears of the director, Kline, in exaggerated form. Semmelweis was demoted from his position of assistant in charge of the First Division, to that of provisional assistant: another physician was put in his place. After some six months of depressing inactivity Semmelweis resumed his former position. At the same time the death of Dr. Kolletschka occurred at the hospital under circumstances which greatly impressed Semmelweis. Kolletschka, while performing an autopsy, received a puncture wound on the finger from the knife of one of his pupils who was assisting him. In consequence of this slight wound he sickened and died. [21] The general symptoms of his malady were those of childbed

> fever. Of this occurrence Semmelweis says: "In the excited condition which I then was, it rushed into my mind with irresistible clearness that the disease from which Kolletschka had died was identical with that from which I had seen so many hundreds of lying-in women die."
>
> "Day and night the vision of Kolletschka's malady haunted me, and with every-increasing conviction [22] I recognized the identity of the disease from which Kolletschka died with the malady which I had observed to carry off so many lying-women."
>
> [23] Semmelweis was on the verge of his great discovery that childbed fever was wound infection, blood-poisoning, which was transmitted to women by the unclean hands of the physicians and medical students who examined them during their childbirth.
>
> [23] "In the case of Kolletschka the cause of the disease was cadaveric material carried into the vascular system; I must, therefore, put this question to myself: Did, then, the individuals whom I had seen die from an identical disease also have cadaveric matter carried into the vascular system? [24] To this question I must answer, Yes."
>
> All of the physicians and students attending the First Division of the hospital had frequent occasion to come into contact with and work upon the bodies of those who had died at the hospital. According to the usual method of washing the hands, merely with soap and water, particles adhering to the hands were never completely removed, a fact demonstrated by the odor which the hands retained. In the examination of the women in the wards, any of the students were entitled to make internal examination for their instruction; the raw surfaces left at delivery were infected; childbed fever followed. Moreover, when one woman was infected, the hands of the examiner, unwashed between examinations, carried the infection to the next woman examined.
>
> [25] With his discovery made, Semmelweis at once required each student to wash his hands in a solution of chloride of lime before making examinations. [26] At that time in his division there were 120 deaths in 1,000 births; in the next seven months the deaths fell to 12 in 1,000 and for the first time in the history of the hospital were below that of the Second Division. In that year also there were two months in which not one single death occurred among the patients of the First Division.

The scientific community (1848) had not yet recognized the role of microorganisms in infection. Although his hypothesis received clear support from his experimental test, the medical community was less than receptive to his novel ideas. His position was not renewed at Vienna. He then went to Budapest, where he introduced his new ideas without any more success. His subsequent career ended in failure and he died (1865) in a lunatic asylum—from an infected wound in a finger received during a gynecological operation (the same kind of infection he had spent his entire life combating). It was only after more than a decade, and due to the work of individuals such as Pasteur, that the medical community began to accept Semmelweis' hypothesis, and then only after they began to accept the existence of bacteria.

Fundamental Steps in the General Model

From the above case study, one can abstract the following steps involved in the hypothetico-deductive model. Although not all examples of the hypothetico-deductive model exemplify all of these steps these steps can be seen to be an *ideal type*, which all cases exemplify to one degree or another.

I. Background. In any example of the hypothetico-deductive method, one must always consider the background, for it is always the background that generates the problem requiring a solution. The common background or context typically includes a certain body of knowledge, which everyone takes for granted. This background knowledge includes at least two types of things: (1) accepted scientific knowledge including scientific facts, theory, etc. and (2) assumptions about the nature of reality, the nature of knowledge, etc.—something we can call their accepted "world view."

In the case of Semmelweis, this background consisted of those facts concerning medical treatment of the women, the administration of the hospital, the religious beliefs and practices of the people, the social and economic conditions of the times, beliefs about the non-existence of bacteria, etc.

II. Problem. After becoming clear about the relevant background conditions, the first real step of the hypothetico-deductive method begins--the formulation of the problem. In many cases, the problem will appear to be so clear that it does not require precise formulation, but in many other cases, as John Dewey (1938) points out, the initial problem may be a vague sense of uneasiness or puzzlement, in which case it is necessary to formulate the problem in more precise terms so that an answer can be given to it. As many examples in the history of science show, it is often the case that a major scientific discovery or breakthrough occurs only after the problem-situation has been clarified or re-defined. In fact, what is a problem for one tradition (or theory) may not necessarily be a problem for another tradition (or theory). Hence, it is often necessary to conceptualize the problem in a different way so that a solution is forthcoming. As several individuals have pointed out, scientific discovery is fundamentally a problem-solving activity. But before a solution can be found, it is often necessary to think about the problem in a different way, to reject assumptions that may appear to be self-evident, and so forth; only when one's entire way of approaching the problem-situation is changed is it often times possible for a solution sometime to emerge. Hence, one cannot underestimate the importance of carefully thinking about the problem and formulating it in an adequate way.

In the case of Semmelweis, however, we have an easier task, for the problem appears to be relatively straight-forward: how can one explain the difference between the incidence of childbed fever between the two hospital divisions—medical students versus midwives—a difference of 9.9% and 3.3%? (Number [1] in the text.)

III. Collection of Facts. The next step in the hypothetico-deductive method can be described in different ways. Sometimes, new observations will be needed; sometimes, new facts or data must be collected. Sometimes, however, such relevant facts have already been discovered by other scientists (and may form part of one's knowledge background) and instead of duplicating these discoveries, a scientist need only perform a survey of the existing scientific literature to see if such relevant observations exist. This is one of the major reasons scientists will ordinarily do a search of the literature prior to the creation of a hypothesis.

In other situations, as in the case of Semmelweis, one will not need to collect more data, but instead one will immediately proceed to the construction of a hypothesis in order to explain the problem; in some cases, therefore, this third step will be missing. Whether this step is necessary or not will depend upon the particular situation at hand; no hard and fast rules can be given at this point. What is needed will depend upon the intelligence, the background, and the training of the scientist, and no explicit list of rules can be given. We include this as a third step, however, because there are numerous cases, which exemplify this pattern.

IV. Tentative Hypothesis. Whether one goes through the third step or not, a scientist, upon encountering a problem, will propose a tentative hypothetical solution to the problem. This initial hypothesis will often require elaboration, modification, and development. Sometimes, however, this step will be missing in the procedures used by scientists. A scientist such as Semmelweis may proceed immediately to the next step, that of the evaluation of alternative hypotheses. If that is so, this step will occur after step V. In either case, however, this step is necessary to the scientific method.

V. Evaluation of Hypotheses. In our case study of Semmelweis, steps III and IV are missing. Instead, Semmelweis immediately proceeds to the step of the evaluation of possible hypotheses. The reason for this is clear: before advancing one's own hypothesis, it is often valuable to see what other hypotheses are available, hypotheses which might be able to explain the problem in question. The human mind often finds it valuable to survey the existing theories to see if they can solve a problem before taking the time and energy to construct a new one. This is precisely what Semmelweis did--"he passed in review all the possible causes which had been suggested" ([2]). One reason for this, illustrated in the present case, is that the consideration of alternative hypotheses often results in a conclusion that, in turn, will lead to one's preliminary hypothesis.

The first hypothesis considered was

(1) Epidemic influences ([3])

But Semmelweis subjects this possible hypothesis to a searching examination and rejects it [5] as an adequate explanation:

> Epidemic influences could not be the cause because epidemic influences existed in both divisions ([5])

From this critical examination he concluded [6] that the cause must be in the hospital because there was a difference between the frequency of puerperal fever in the hospital and outside of it, and the frequency could be decreased if one simply closed the hospital. This conclusion constitutes an *intermediate conclusion*, one, which ultimately led him to his own hypothesis.

Next, he proceeded to consider eight additional hypotheses. A second possible hypothesis [7], Crowding, was rejected because it existed in both divisions [8]. [9] Fear was rejected [10] as a hypothesis because [11] the cause of childbed fever had to exist before the fear did. That is to say, something had to cause childbed fever, which in turn caused the fear. Fear, therefore, could not be the cause since a cause must exist before (or at least simultaneously with) its effects.

The next hypothesis [12], religious observances, was subjected to empirical test to determine if it could have this effect [13]: the two divisions were made similar with respect to the absence of overt religious observances but [14] there still was a difference between the two divisions.

[15] Depression due to their socio-economic condition was rejected because [16] this condition also existed in the other division. [18]

"Wounded modesty" was rejected because [19] women in the First Division did not have such "wounded modesty," only fear. [19] Ventilation, dirty laundry, and improper diet were rejected because [20] they were present in both divisions.

In all, Semmelweis rejects nine possible hypotheses. He does so on the grounds that they could not possibly be the explanation of the phenomenon in question. It is important to note that he employs two distinct types of evaluation of competing hypotheses, what we will call **a priori** (or logico-conceptual) **evaluation** and *a posteriori* (or *empirical) evaluation.* The reader will note that he actually subjects only one hypothesis--that of religious observance--to an empirical (experimental) test. All the other hypotheses he rejected without subjecting them to actual experimental test. Why? Because he could reject them on purely logical and conceptual grounds: All of them violated certain basic assumptions about causality mentioned in our discussion of Mill's methods, namely: *the presence of a cause (of some phenomenon) versus the absence of this cause must make a difference to the production versus the absence of the effect.* If the effect is produced whether the cause is present or not, then the alleged cause cannot be the cause of it. A cause, in short, must make a difference to the phenomenon. But in all the cases considered by Semmelweis except two, the hypothetical cause was present in both divisions. Hence, it could not be the cause.

The other case considered by Semmelweis and rejected by him on further logical grounds was case [9]—fear—where it was rejected because of the principle: a cause must exist before the effect. Fear could not have caused childbed fever because the fear was fear about childbed fever, which must have existed before fear began.

What we see, therefore, is that a scientist may evaluate the logical adequacy of a hypothesis on purely conceptual or a priori grounds. He may reject a hypothesis if it is untenable. In such a case, it is not necessary actually to carry out an empirical test of the theory.

Such logical (or a priori) evaluation may be based on numerous criteria: Is a hypothesis compatible with accepted scientific knowledge? If it is not, then scientists may reject such a hypothesis as unworthy of further investigation. Is a hypothesis clear? If not, it may be impossible to decide what test would be relevant to deciding its adequacy. Is the hypothesis testable? Are there ways in which it is possible to make a test which would determine if the hypothesis is true or false? Again, there are many hypotheses that are not testable and, hence, simply cannot be subsequently tested. We will discuss these and other criteria for the evaluation of a hypothesis below.

Although this step may contain a component of a priori evaluation of competing hypotheses, it may also contain a slightly different step. If a hypothesis has already been advanced in step IV, this step may involve what we will call an elaboration of this initial hypothesis. Such elaboration may involve several of the steps already mentioned. For example, it may turn out that the initial hypothesis is not something that can be put to empirical test and, hence, is in need of revision so that it can be tested. Again, the hypothesis may be vague or unclear and this may require considerable conceptual revision before it is judged *worthy of further pursuit.* Of course, it may happen that, after considerable thought, the initial hypothesis is declared to be a non-serious candidate, and a scientist will turn to another hypothesis or create a new one. There are, therefore, four possible decisions that can occur at this stage: (1) a hypothesis (whether a competing one or an initially proposed one) may be accepted as plausible and worthy of test, (2) rejected as implausible, (3) modified, or (4) a new hypothesis may be created.

In the case of Semmelweis, after he subjected various hypotheses to critical evaluation and rejection, he then came up with his own hypothesis. This occurred as a result of observing the death of Kolletschka [21], an event which suddenly suggested to him that Kolletschka's symptoms were the same as those of childbed fever. Here, Semmelweis was able to see "unity in diversity," i.e., that there was a similarity or analogy between Kolletschka's death and the death of the women and, in fact, that they were analogous (See Chapter 11). He reasoned by analogy:

> Kolletschka's symptoms were the same as those of puerperal fever;
> Kolletschka's symptoms were caused by a wound infection [22];
> thus, it is likely that childbed fever is caused by a wound infection.

How could the women have received a wound infection, asked Semmelweis, and what was the cause of it?

At this point, Semmelweis [23] introduces a hypothetical entity—*cadaveric matter*—as a theoretical cause. This is really just a name for something in the cadaver that caused the disease. Semmelweis really knew very little about it but postulated its existence nonetheless. Suppose, he reasoned, there is something called "cadaveric matter" that causes the disease of Kolletschka and the women. This is Semmelweis' hypothesis [24].

Before testing this hypothesis, Semmelweis subjects it to critical examination to see if it could explain the disease and if so how. This description occurs between steps [23] and [24]. He reasoned that such cadaveric matter would be present when doctors came into direct contact with the cadavers, e.g., in performing an autopsy on them. Of course, the doctors washed their hands, Semmelweis conceded, but apparently mere soap and water would not kill the cadaveric matter, a point which had confirming evidence from the resulting odor of the hands. The doctors then would come into contact with the women's vascular system by means of an internal examination and cadaveric matter would be transmitted, thereby causing childbed fever. In short, this hypothesis was a plausible one and could explain the disease.

VI. Deduction of Implications of the Hypothesis. Although the above description of the possible cause of childbed fever was a very plausible hypothesis, Semmelweis was not content with this. Instead, he performed an empirical test of the hypothesis, an indirect test of the hypothesis since a direct test was not possible.

The logic of this test is one with which the reader is quite familiar: it is simply deductive logic. Suppose the hypothesis H were true. Then a consequence would follow deductively from it as an effect follows from a cause. This consequence would be an implication of the hypothesis—a test implication I. The reasoning is thus:

If H is true, then an implication I should follow

The arrow (the 'if-then') is a purely deductive arrow. In this step, therefore, one is employing deductive logic and nothing more. The test implication (I) has the following *prediction*: if one kills the cadaveric matter, then there would be no puerperal fever.

VII. Testing of Implication. Step VI consists of the formal deduction of I from H. But one needs to test this implication I empirically. How will one do this? How does one kill the cadaveric matter? Soap and water won't do but chlorinated lime might. Thus, a test of Semmelweis' hypothesis involved the following testing procedure [25]: before performing an examination, students must wash their hands in chlorinated lime, which presumably would kill the cadaveric matter and consequently no childbed fever would result. That was the prediction Semmelweis made.

The actual testing of an implication I drawn from a hypothesis H also has the form of a conditional; namely, if one sets up certain test conditions T (before performing an examination, students must wash their hands in chlorinated lime), then one will get a particular result R (no childbed fever will result). Thus, the testing of a hypothesis has this form: $H \supset I$, and I, in turn, entails that if a certain Test T is performed, then a certain Result R should occur. So $I \supset (T \supset R)$. Hence:

$$H \supset (T \supset R)$$

This stage of actually testing the test implication I of H will, thus, normally but not always involve the establishment of an experiment, a subject we will turn to in a subsequent chapter.

Semmelweis's prediction that there would be no childbed fever was correct [26] since the number of deaths fell from 12% to 1.2% and in fact during two months there were 0 deaths. Semmelweis' hypothesis thus seemed correct.

In this step, therefore, one either actually observes I (as predicted by H) or one observes not-I. If one observes not-I (the reader will recall from an earlier chapter), then the hypothesis has been falsified by the logical principle of *Modus Tollens*. If one observes I, then the hypothesis has been confirmed (although it has not been proven to be true, since to make such an inference would be committing the *fallacy of affirming the consequence*).

The logic involved here is the following:

$$\begin{array}{l} H \supset I \\ \quad \sim I \\ \hline \sim H \end{array}$$

or

$$\begin{array}{l} H \supset I \\ \quad I \\ \hline\hline \quad H \end{array}$$

VII. Decision: Accept or Reject the Hypothesis. In the final stage, what occurs is a decision by the scientist to either accept the hypothesis (tentatively) or to reject it. If one accepts the hypothesis, then one will ordinarily want to subject it to additional tests and perhaps to develop it even further. If one rejects the hypothesis, one will want to come up with another hypothesis and subject it to subsequent test. In either case, however, one will want to repeat certain steps of the hypothetico-deductive method, e.g., steps 5 and 6 or step 3, until one comes up with a workable hypothesis.

Exercise 12.1

In each of the following passages, point out the eight steps involved in hypothesis formulation and testing as outlined above. [In Case 1 numbers have been supplied in order to help locate the eight steps.]

1. [1]In the second decade of the twentieth century, a disease called "pellagra" had become epidemic in the southern states. The disease was also endemic (i.e., its occurrence was confined to some localities and did not occur in others). One village might be affected; a neighboring one not. Gastric disturbances, skin eruptions, and nervous derangement characterize pellagra. It was the opinion of the world of medicine that the disease was caused by an unknown microbe. The U. S. Public Health Service decided to send Dr. Joseph Goldberger to Mississippi to try to find a way of eliminating the disease.

[2] In tackling the problem of the cause of pellagra, [3] Goldberger observed the fact that the sufferers were always in contact with other victims. Since this is the consequence that would be expected if the disease was caused by microbes, [4] he adopted the tentative hypothesis that the disease was caused in this manner. [6] He then reasoned that if microbes were the cause, persons in close contact with the victims should catch the disease. [7] But in the first hospital he visited, he found that the orderlies, nurses, and doctors who were in close contact with the victims never caught the disease. [8] This led him to discard the hypothesis that the disease was caused by microbes. He now needed another hypothesis, so he continued his observations. [3] He found that only the poor suffered from the disease. He also noted that the victims' diet consisted of corn-meal mush, hominy grits, and similar foods, but practically no milk or fresh meat. On one occasion, while visiting an orphanage, he found that only the children in the six to twelve age group had the disease. In investigating this further, he found that the children under six were given milk and the children over twelve were fed meat. The orphanage's funds were limited, so they skimped on the food of the six to twelve year old group, who were too young to work (and so did not "deserve" meat) and who were beyond the "baby" stage, the stage when milk was regarded as essential. [4] This fact suggested to Goldberger the hypothesis that a diet lacking in milk and fresh meat was the cause of disease. [5] He then reasoned out the consequences of this hypothesis, namely, [6] that remedying these deficiencies in diet would cure the disease. [7] He secured a sufficient amount of milk and fresh meat for the children at the orphanage. The result: All cases of pellagra disappeared. [8] The cause of pellagra had been found.

2. Benjamin Franklin, it will be remembered, became convinced that lightning is nothing more than electricity passing from one cloud to another. This hypothesis was suggested to him by a consideration of the many respects in which lightning resembled electricity. Lightning sets things on fire; so does electricity. Lightning melts metals, kills people, causes blindness; so does electricity. Lightning is attracted to pointed bodies, e.g., spires, trees, mountaintops; so is electricity. His famous experiment with the kite was designed to confirm a prediction based on the hypothesis suggested by these many analogies. If lightning is electricity, it ought to be possible to get a spark from it and to charge a Leyden jar. So he sent a kite up in a storm. To the end

of the kite string he tied a key. To the key he tied a silk ribbon by means of which he controlled the kite. When the rain had dampened the string, sparks were drawn from the key, and being attached to the key charged a Leyden jar. Franklin thus believed his hypothesis to be confirmed.

3. Columbus had arrived at the hypothesis that there was land to the west, long before he made his first voyage. This hypothesis accounts for a wide range of facts that were almost enigmatical on any other hypothesis. The earth was known to be spherical, and two-thirds of its surface had been explored. The remaining third lay between the West Coast of Europe and the East Coast of Asia. Many geographers, both ancient and modern claimed a westward passage to India was a genuine possibility. Columbus knew that a Portuguese navigator, sailing some 450 leagues west of Cape St. Vincent, had picked up a piece of carved wood which had not been worked with any metal instrument. As the wind had drifted it from the west, there was a chance that it came from land to the west. He knew, also, that reeds of an immense size had floated to the island of Porto Santo, from the west; and that those reeds were unlike any European flora. He knew, too, that a westerly wind had deposited on the shores of the Azores Islands trunks of huge pine trees of a kind that did not grow on those islands, and the bodies of two dead men whose faces differed from any known face of people. To these facts he was able to add the report of a mariner blown far westward while bound for Ireland, that he had sighted land away to the west which he had taken to be some extreme part of Tartary, i.e., Asia. Upon these facts, Columbus erected the hypothesis that there was land to the west, that it was attainable, that it was fertile, and that it was inhabited. His voyage confirmed the prediction.

4. In the south of England there are many large shallow ponds, on hilltops, known as "dew-ponds" because it is believed, locally, that they are replenished by dew. This local hypothesis is supported by the belief that a heavy layer of straw, placed upon the basin of these ponds when they are being constructed, acts as a non-conductor of heat, thus keeping the basin cooler than the surrounding earth, thus causing dew to be deposited, since dew is only deposited on something which is cooler than the dew itself.

In 1911 the Royal Geographical Society sent one of their Fellows to investigate these peculiar ponds. His analysis of the case is interesting and instructive logically, whatever it may be geographically. There were, he reasoned, only three sources from which the water could come, ignoring springs: dew, rain, or some hitherto unascertained source.

The dew hypothesis he eliminated on three accounts: the straw rots during the first winter and becomes useless as a non-conductor. The least likely place for a dew deposit would be the surface of the water in the pond, because in water the hottest part always rises to the top and forms the surface; and on a warm summer's night there is not sufficient time for this tepid surface water to fall below the dew point and it could, therefore, receive no dew. The investigator carefully checked the night temperature of ground, air, and water, and found that the water surface remained the warmest. The maximum rate of dew-fall for all England is one and one-half inches per annum. That quantity would be required every night to compensate for what cattle drink from the pond each day.

The rain hypothesis was eliminated by the fact that 1911 was a dry year and practically no rain had fallen. In spite of these facts, the ponds did not dry up. Upon what remaining hypothesis could this fact be accounted for? "It is not until one stays all night by one of these ponds that the explanation is found, namely, that the ponds are replenished by the early morning sea mists which roll up over the hills and saturate their tops." This hypothesis explains "why these self-filling ponds are not to be found in valleys, but only on hilltops," since the mists roll up the hills, never stopping in the valleys. If this hypothesis is true, he proceeded to argue, then the pond water should contain a slight amount of salt, since the sea mists are evaporated sea water and by mechanical action a slight amount of salt would be taken up with the moisture. He analyzed the pond water and found that it did, as a matter of fact, contain a slight amount of salt.

5. William James put forth the hypothesis that bodily changes follow directly upon the perception of the exciting fact and that our feeling of these changes—we feel sorry because we cry, we are angry because we strike, we are afraid because we tremble. To begin with, particular perceptions certainly do produce widespread bodily effects by a sort of immediate physical influence, antecedent to the arousal of an emotion or emotional idea. If we abruptly see a dark form in the woods, we catch our breath instantly and before any articulate idea of danger can arise. In listening to poetry, we are often surprised at the cutaneous shiver which like a sudden wave flows over us. In hearing music, the same is even more strikingly true. If the hypothesis that emotions follow internal bodily changes is true, argued James, then a person with pathological visceral disturbances (in whom the bodily changes are constantly taking place) should feel an emotion even though there is no external occasion for that emotion. He found such a patient in an asylum. The patient suffered pathological disturbances of the internal glands and would laugh or cry for no reason at all.

James then reasoned that if his hypothesis were true, a person with an anesthetic viscera (i.e., one from which he could get no sensations due to destruction of sensory nerves) should feel no emotions. He found a war veteran whose sensory visceral nerves had been destroyed, and the man claimed that he never felt any emotions.

James further argued that if we fancy some strong emotion and then try to abstract from our consciousness of it all the feelings of its bodily symptoms, we find we have nothing left. If we imagine away every feeling of laughter and of tendency to laugh from our consciousness of the ludicrousness of an object, we cannot imagine that the object would be ludicrous anymore. If the theory is true, then any voluntary arousal of the so-called bodily manifestations of an emotion should give us the emotion itself. The more you sit in a moping posture, sigh, and reply to everything with a dismal voice, the more your melancholy lingers. However, James had a number of actors go through all the bodily motions connected with some specific emotion and asked them whether they felt the emotion--the reply was "no."

Sherrington, a physiologist, severed the spinal cord of a dog, thus preventing any sensations from the body from reaching the brain. He then brought a cat into the room, and the dog showed all the emotions associated with such an occasion.

Dr. Cannon introduced electrodes into the body of a cat which cause the same changes to occur as occur in the case of an emotion. When he turned on the electric current and these bodily changes took place within the cat, the animal showed no emotion but continued to purr complacently.

The Logic of the Hypothetico-Deductive Method

In the preceding discussion, we have proceeded as if the underlying logic of testing a hypothesis was a simple case of deducing a test implication I from a hypothesis H and determining whether I is correct or not. Although the above account may do as a schematic account of hypothesis testing, the logic underlying this account is too simple in fact, it is simplistic. Earlier, we discussed the difference between Modus Tollens and merely confirming a hypothesis. If one concludes that the hypothesis is true (or verified), one commits the Fallacy of Affirming the Consequent. We suggested a better way of viewing the situation: there are two appropriate kinds of test of a hypothesis:

$H \supset I$	$H \supset I$
$\sim I$	I
--------	====
$\sim H$	H
Falsification by Modus Tollens (Deductive Logic)	Confirmation by Affirming the Consequent (Inductive Logic)

We suggested, furthermore, that the above schema represented a fundamental asymmetry between falsification and confirmation in that falsification was certain, whereas confirmation was only probable. Thus, we suggested one case of confirmation couldn't be counted as equivalent to one case of falsification, since there is this asymmetry between them. In fact, one case of falsification is sufficient to refute the hypothesis no matter how many cases of confirmation there are. In short, no amount of confirmation can ever equal falsification.

After having said all of this, however, we must seriously qualify these remarks. The basic reason for this qualification concerns what is called auxiliary assumptions.

Auxiliary Assumptions

Auxiliary assumptions are those assumptions one makes that are necessary to derive the test implication from the hypothesis but which are not actually part of the hypothesis being tested. For example, in the case of Semmelweis, in order to derive this test implication, he had to make the auxiliary assumption *chlorinated lime kills cadaveric matter*. Without this assumption, he would not have been able to derive his prediction. So, auxiliary assumptions are necessary for one's test implication, but these assumptions are ones that are accepted as being true and as not requiring proof. Because they are not part of the hypothesis being tested, they are called auxiliary or supplementary assumptions.

If we make these auxiliary assumptions explicit, however, and include them in our deduction of the test implication, then we have the following schema:

> If our Hypothesis H is true and if our Auxiliary Assumptions A are true, then Test Implication I will also be true, i.e., $(H \& A) \supset I$.

Here, the entire antecedent of the conditional is (H and A) and not merely H. Now, suppose I turns out to be false. By Modus Tollens, this leads to:

$$\begin{array}{r} (H \& A) \supset I \\ \sim I \\ \hline \end{array}$$
$$\sim(H \& A)$$

which is logically equivalent to

$$\sim H \vee \sim A$$

Thus, if our test implication is false, we know either that our hypothesis is false or our auxiliary assumption is false. *But not-I does not tell us which one is false.* At this point, deductive logic is no longer of help, since we can either reject H or we can reject A, but logic does not tell us which one to reject. That is a matter of choice on our part. We could choose to reject A and still keep H (i.e., not-A and H) or we could choose to reject H and keep A. But if we do the first, the falsity of our test implication has nothing to do with falsifying our hypothesis or (to put it somewhat differently) we have failed to falsify our hypothesis. In fact, it is, logically speaking, impossible to falsify a hypothesis since one may always save the hypothesis by abandoning one's auxiliary assumptions. This is now known as the **Duhem-Quine thesis** (after two famous philosophers). Thus, suppose the prediction by Semmelweis had turned out to be false and washing in chlorinated lime did not result in a decrease in childbed fever. Would this have falsified Semmelweis' hypothesis? Not at all, since a failure of the test implication shows either that his hypothesis is false or that chlorinated lime does not kill cadaveric matter, but it doesn't show which.

Things are not quite so bad as Duhem and Quine make it out to be, however. Formal deductive logic does not dictate that we must reject our hypothesis; as far as formal deductive logic is concerned all we have to do is either to abandon H or to abandon A. A scientist may choose to abandon A and still accept H (even though he was testing H). But whether he will abandon H or A will not be a matter of arbitrary whim; instead, his decision will be based on the question: is not-H or not-A more likely to be true? This, in turn, will be decided by inductive logic, by the overall evidence in favor of H or A. This evidence will normally consist of *background knowledge.* How likely is it that chlorinated lime doesn't kill cadaveric matter? To answer this question we would have to rely on current scientific knowledge about chemistry, biochemistry, micro-biology, etc. Such knowledge might indicate that this auxiliary assumption was indeed substantially well supported by contemporary science; in which case this would lead us to reject H. On the other hand, current scientific knowledge might indicate the evidence in favor of H was substantial (if we had accumulated other evidence supporting it). If our current scientific knowledge did not substantiate the likelihood of A, then it would perhaps be more reasonable to adopt H and to reject A.

In short, additional evidence may assist us in deciding whether to reject H or to reject A—an inductive decision. Suppose the evidence supporting A was sufficiently strong so that A seemed to be correct. Then we could conclude that H was false. Furthermore, if the test implication I turns out to be true, this would confirm the conjunction (H and A), which in turn would confirm both H and A. This is what we would expect from the case of Semmelweis: a correct prediction about handwashing confirms both his hypothesis about cadaveric matter and the auxiliary assumption that chlorinated lime kills cadaveric matter. (See Appendix A on Hempel's criteria of confirmation.)

Crucial Experiments

So far, we have been talking about the steps involved in testing a single hypothesis. But there are many examples in the history of science in which we have two competing hypotheses, that is, two hypotheses that offer alternative explanations of the same phenomenon. For example, in astronomy, the geocentric view of Ptolemy was opposed to the heliocentric view of Copernicus; in physics, Galilean kinematics was opposed to Aristotelian kinematics; in twentieth century psychology, the learning theory of Clark Hull was opposed to that of Edwin Tolman; in biology, Darwin's theory of evolution was opposed to Lamarck's, etc.

Perhaps the most famous example of such competing hypotheses concerns the question, What is the nature of light? According to the corpuscular theory, light consists of corpuscles, which were analogous to the atoms making up ordinary material objects. According to the wave theory, light consists of disturbances in an intervening medium, analogous to the spread of water waves. Here were two competing theories, which battled back and forth for centuries, from the 17th century until the 20th century.

The hypothetico-deductive method can be applied to the testing of two competing hypotheses. In this case, one often speaks of a **crucial experiment**, an experiment that decisively decides the case between two competing hypotheses. The logic of a crucial experiment is the following. Suppose we have two hypotheses (H_1 and H_2) that are *contraries*, i.e., such that not both can be true but both might be false. This would be the case, for example, with the corpuscular versus the wave theory of light. Suppose, furthermore, that from one hypothesis H_1 one may deduce a test implication I such that the other hypothesis H_2 would predict that I is false.

$$H_1 \supset I \qquad H_2 \supset \sim I$$

Now upon performing an experiment, one will either conclude that I is true or I is false. If I is true, then H_2 would be falsified (and H_1 would be confirmed), whereas if ~I were true, then H_1 would be falsified (and H_2 would be confirmed). In either case, however, we would have a crucial experiment, since we would be able to decide conclusively that one hypothesis was false and the other one was not.

For example, one consequence of the corpuscular theory of light was the prediction that light should travel faster in water than in air, whereas the wave theory predicted that light should travel faster in air than in water. Here, then, we would have a crucial experiment since (assuming light would not travel the same speed in water as in air) we can perform a test to see which of these two predictions is correct. Foucault performed such a test in 1850, which supported the wave theory of light against the corpuscular theory of light. The result was that the corpuscular theory of light seemed to have suffered a "fatal blow" and was judged by most scientists to be false.

What does a crucial experiment establish? According to some, it can only establish (at most) that one of the hypotheses is false. According to others, however, a crucial experiment can establish that one of the competing hypotheses is true. How can a crucial experiment do the latter? If, following the logic of Mill's eliminative induction, one could list all the hypotheses that could explain a particular phenomenon and if one could then falsify all of them except one, then the remaining one must be true. This assumes, of course, that there is a single theoretical explanation for a particular phenomenon, that one can falsify a hypothesis, and that it is possible to list all of the possible hypotheses. Several of these assumptions are obviously questionable.

One especially problematic assumption concerns the possibility of falsifying a competing hypothesis. As we have already mentioned, no hypothesis by itself entails a particular observational prediction; it always requires auxiliary assumptions. Thus, our model for a crucial experiment would have to be modified in the following way:

$$(H_1 \;\&\; A_1) \supset I \qquad (H_2 \;\&\; A_2) \supset {\sim}I$$

Suppose we find I is true. Then we have:

$$\begin{array}{r} (H_2 \;\&\; A_2) \supset {\sim}I \\ I \\ \hline {\sim}(H_2 \;\&\; A_2) \end{array}$$

$$\text{Thus } {\sim}H_2 \vee {\sim}A_2$$

Likewise if we found ~I:

$$\begin{array}{r} (H_1 \;\&\; A_1) \supset I \\ {\sim}I \\ \hline {\sim}(H_1 \;\&\; A_1) \end{array}$$

$${\sim}H_1 \vee {\sim}A_1$$

Once more it appears to be logically impossible for there to be a crucial experiment since the experimental result—I or not-I—can never conclusively decide in favor of (or against) the competing hypotheses. But, once again, if we include background knowledge regarding the probability of these auxiliary assumptions—A_1 and A_2—we may be able to have a crucial experiment. These points can be illustrated in the following case study.

The Case Study of Atmospheric Pressure

For centuries individuals have known that in order to open a keg of beer one had to have an opening on the top as well as on the bottom. Likewise, if one sucks up a liquid in a straw and closes the top with one's finger, the liquid won't run out. Why was this? According to one hypothesis, passed down from Aristotle to the Middle Ages, there can be no vacuum in nature (H_1). Thus, H_1 could explain both of these phenomena: if water were to flow out of the bottom of the keg without there being a hole in the top (through which air could enter), a vacuum would result in the keg and this is impossible. Hence, one needs an opening in order to let the air in.

In the 16th century, Galileo pointed out that a suction pump would raise water only a certain distance, namely, 34 feet so that in order for such a pump to work it had to be less than 34 feet above the water. Galileo's pupil, Evangelista Torricelli (in 1644), came up with an alternative hypothesis. Reasoning by analogy, he suggested that we live "at the bottom of an ocean of air" and that just as water exerts pressure on us when we dive to the bottom of a lake, so the "ocean of air" above us exerts pressure on everything below it. He reasoned that if the earth was surrounded by an "ocean of air" and if air had weight, there would be air pressure on all objects submerged in this ocean of air: $(H_1 \,\&\, A_1) \supset I$.

From I, Torricelli deduced three predictions:

P_1: since the atmosphere supports a column of water 34 feet high and since mercury is 13.6 times as dense as water, it follows that the atmosphere should support a column of Mercury equal only to 34/13.6 = 30 inches

P_2: such a tube filled with mercury could be used to measure fluctuations in atmospheric pressure, i.e., a mercury tube is a barometer

P_3: if this mercury tube were transported to a place where the air was more rarefied, e.g., on the top of a mountain, then the column of mercury would descend, since there would be less atmospheric pressure there.

These three predictions were subsequently tested and confirmed. It is the third prediction that is especially relevant, for it can be used to decide if "nature abhors a vacuum" or if Torricelli's theory of atmospheric pressure is true. Blaise Pascal (1647) reasoned (again by analogy) that P_3 could be a crucial test between these two hypotheses: From H_2 (Torricelli's hypothesis) it follows that there should be a difference between the pressure at the foot of a mountain and at the top of the mountain, whereas H_1 would predict there should be no difference. A test bore this out. Thus:

If H_2 then Difference   If H_1 then No Difference
Difference

================== ----------------------------------

H_2 is confirmed H_1 is falsified

But clearly there are auxiliary assumptions in each hypothesis. For example, one auxiliary assumption made by H_2 but not by H_1 was that "nature does not abhor a vacuum more at the foot of the mountain than at its summit." Another such assumption was: a mercury column was not shorter at the top of a mountain than below. But the proponents of H_1 might have denied this auxiliary assumption: "the relative densities of mercury and air might change; the measuring stick might alter in length on being moved upward several thousand feet," etc. To have a convincing crucial experiment one would have to rule out these (and other possibilities); one would have to show that "nature does not abhor a vacuum more at the foot of the mountain than at the top," the relative densities of mercury and air don't change, the measuring stick doesn't alter in length on being transported upward, etc. If all of these auxiliary assumptions were verified, then one would have a more convincing case of a crucial test.

Exercise 12.2

In the following passages, identify the competing hypotheses H_1 and H_2, their respective implications, and the crucial test between them. Indicate which hypothesis was established and which one was falsified. If no crucial test was possible, indicate why.

1. In the early nineteenth century the German geologist Abraham Werner propounded the geological theory called "Neptunism." The theory was offered as an explanation of rock formations. Werner held that at one time an ocean covered the earth. Processes of crystallization, chemical precipitation, or mechanical sedimentation had deposited all the rock strata.

"First of all came the primitive rocks, such as granite, which had crystallized out of the primeval ocean: these were entirely devoid of fossils. Then came the transitional rocks such as the micas and the slates, containing a few fossils, which had been precipitated from the ocean. Next there were the sedimentary rocks, richer in fossils, such as coal and limestone, formed by the deposition of solids from the waters. Finally there were the derivative rocks, such as sand and clay, which were derived from the others by a process of weathering. Werner thought that volcanoes were due to coal catching fire underground, the heat generated melting the neighboring rocks, and forcing the eruption of volcanic lava from time to time. Thus, for Werner, heat was not an important geological force: volcanic action due to burning coal was a late and subsidiary rock-forming agency, appearing only after the main strata had been laid down."

One of the shortcomings of this theory was the lack of an explanation for the disappearance of the ocean after the rock strata had been formed. James Hutton, an amateur scientist of Edinburgh, put forward a theory, which employed only geological forces, which are in operation today. The interior of the earth, Hutton held, was composed of molten lava, the solid surface of the earth serving as a containing vessel, which was closed apart from the volcanoes that served as safety valves. From time to time, he thought, the molten rock escaped through cracks just beneath the earth's surface and tilted up the overlying sedimentary strata. The molten rock then solidified to form the crystalline rocks, such as basalt and granite, thus giving the mountains with their crystalline cores and sedimentary sides.. . .

"Werner's followers argued against Hutton, first, that molten rock would not become crystalline on solidification but would be glassy like lava, and secondly, that some rocks, like limestone, would decompose if subject to heat. Hall observed in a glass factory at Leith that if molten glass were allowed to cool very slowly it became crystalline and opaque, whilst if it were cooled more quickly it became glassy and transparent. He presumed that molten rock would behave in a similar way, and accordingly he obtained some lava from Vesuvius and Etna, and melted it in the blast furnace of an iron works. As he had expected, the molten rock became crystalline, like basalt, when allowed to cool slowly, and glassy, like lava, when cooled rapidly. Hall showed further that if limestone were heated in a closed vessel, it did not decompose as the Neptunists thought, but melted and became marble on cooling as Hutton suggested."

During the period between 1790 and 1830 the study of fossils in rock was begun. Fossil remains of land animals of different ages were found in different rock layers. This gave new support to Hutton's theory. Some Neptunists responded by claiming that the agents of these formations were a series of catastrophic floods.

Around 1819 Sedgwick and Murchison studied the early rocks containing no fossils in coal mines and the like. They concluded that this rock was formed by the solidification of molten rock, not by crystallization from water.

—S. F. Mason, *Main Currents of Scientific Thought*

2. Those who believed that microorganisms could arise de novo without parents, did in fact accept that life was continually being created anew from inanimate matter. This belief came to be known as the doctrine of *spontaneous generation*. . .

Among the many other types of experiments that Pasteur designed to rule out spontaneous generation, one is worth some emphasis by virtue of its very simplicity and decisiveness and because it finally silenced his opponents and settled the issue--at least for the time being. A fermentable fluid was put into a flask, the long neck of which was then heated and drawn into the form of an S tube (hence the name "swan-neck flask"). When the liquid was boiled, the vapor forced the air out through the orifice of the neck. As the fluid became cool again, the air slowly returned to the flask, but was washed in the moisture that condensed in the curves of the neck after heating was interrupted. Under these conditions, any dust or particle carried by the air was trapped in the neck, and the fluid in the flask remained clear, sterile. However, when the neck of the flask was broken, and the unwashed air allowed to come into contact with the fluid, then microscopic life immediately began to develop.

Despite the spectacular success of these experiments, there were still unforeseen difficulties to overcome. They arose from the fact, then unknown but now well understood, that certain species of bacteria form heat-resistant spores. In some of the early experiments these spores persisted in the fluid that was presumed to have been sterilized by heating, and when they germinated, they gave rise to bacterial growth even though access to outside air had been prevented. These difficulties arising from the presence of heat-resistant spores were eventually overcome, and Pasteur was able to prepare his swan-neck flasks in such a manner that the broth remained sterile in them all.

In Pasteur's words, "Never will the doctrine of spontaneous generation recover from the mortal blow of this simple experiment."

---Rene Dubos, *Pasteur and Modern Science*

3. When something, say, a match burns, it appears that something is released from the match. According to a theory of combustion formulated by two German chemists, Becher and Stahl, in the 17th and 18th centuries, when a substance was burnt "*phlogiston*" was said to escape in the form of fire and flame.

One application of the "phlogiston" theory was eventually to lead chemists into much confusion and to help to bring about its downfall. It arose in this way. When a metal, such as copper or lead, is sufficiently heated, it turns into a powdery substance and its metallic properties are lost. (The same thing happens in the familiar rusting of iron, but there without the application of heat.) The chemists of that time explained this by saying that a metal, when heated, lost its "phlogiston," leaving the powdery residue, which they called a calx. They knew that if this calx was heated afresh with charcoal, it was converted back again into metal; and charcoal, since it would burn away almost entirely, was held to be very rich in "phlogiston." The heating of the calx with charcoal had therefore restored enough "phlogiston" to the calx to reconstitute the original metal. Thus a metal was a compound of its calx and "phlogiston"; and the process of heating a metal to give it calx, called calcination, was a decomposition, a kind of combustion in which "phlogiston" escaped from the metal.

It was known, on the other hand, that, when a metal was calcined, the weight of the residual calx or powder was greater than the original weight of the metal taken. But how could the weight increase, since something material, namely "phlogiston," had been lost from the substance of the metal? In answer to this, some of the chemists who accepted the "phlogiston" theory were driven to suppose that "phlogiston" did not gravitate as other matter, but levitated--that it naturally rose upwards to the heavens whereas other substances naturally tended to fall to the earth—that it had a negative weight, as we might say.

Antoine Lavoisier, the founder of modern chemistry, believed that air plays an important part in combustion. He believed that in combustion things combine with air and that this explains why the calx has greater weight. Lavoisier in a famous experiment succeeded in separating hydrogen and oxygen from the air. He noted that things would burn in the presence of oxygen but not with hydrogen. He then reasoned that when hydrogen burns it should, according to his theory, combine with oxygen. All attempts by Lavoisier and others failed until Cavendish found that when hydrogen burns, water is produced. After this experiment, the phlogiston theory was gradually abandoned.

There were attempts to modify the phlogiston theory in the face of these experiments. According to the modified theory, when metal is heated it turns to pure earth and its phlogiston goes into the air. The earth then combines with water from the air to form calx.

---Douglas McKie, "The Birth of Modern Chemistry"

4. When the moon is on the horizon it looks bigger than when it is high in the sky (zenith moon). However in photographs its image is the same size. This is also true of the image in the eye.

Two explanations have been advanced for this: the apparent-distance theory and the angle-of-regard theory. According to the first theory, the moon looks bigger because it seems further away. Any object seen through filled space, such as the horizon moon, is perceived as being more distant than an object just as far away but seen through empty space, such as the zenith moon. It is well known that an observer perceiving two equal images and receiving sensory information that one object is farther away than the other, correctly sees the farther to be larger. According to the second theory the moon looks smaller because the viewer raises his eyes or head to look at it.

Some years ago Edwin G. Boring subjected the apparent-distance theory to what he considered a critical test. He asked people to judge the relative distance of the zenith and horizon moons. Most said the horizon moon seemed nearer. He then had subjects match moons of the same size on a screen. Those they saw at eye level they said were larger.

In 1957 Lloyd Kaufman and Irvin Rock investigated the phenomenon. They devised an optical apparatus, which simulated the moon against the sky. With the use of this instrument they found that the horizon moon looks larger whether or not the eyes are raised. The moon in the same region of the sky was the same whether or not the eyes were elevated. They concluded that Boring's findings on eye elevation were peculiar to the methods he employed.

In Kaufman and Rock's words: "We therefore turned to the apparent-distance theory. Boring had rejected it because his subjects said that the horizon moon appeared to be nearer than the zenith moon. But, we wondered, did they really see the horizon moon as nearer? Or were they judging it to be nearer precisely because it looked bigger, effectively turning the reasoning upside down? In that case the reported distance would be a secondary phenomenon, an artifact of the very illusion it was supposed to test. To check this possibility we showed our subjects pairs of artificial moons of different diameters and instructed them to compare their relative distances. Whenever the zenith moon was larger, the subjects said it was nearer than the horizon moon; when it was smaller, they said it was farther away."

Kaufman and Rock then made additional experiments to confirm that the horizon moon looks larger only because it was seen over terrain. After this they showed the moon looks larger when it is seen over a more distant horizon, and that the illusion increases with the degree of cloudiness. When the horizon was inverted, the moon looked smaller. They also found that the framing-effect one gets in cities enhances the size of the horizon moon. Last, they showed that color has no effect on the illusion.

Kaufman and Rock concluded form all this that they had "tested" the apparent-distance theory and "provided evidence that it is correct."

—from Lloyd Kaufman and Irvin Rock, "The Moon Illusion"

"Ad Hoc" Hypotheses

The preceding discussion pointed out that, in the face of embarrassing evidence, a proponent of a particular hypothesis often refuses to let the embarrassing evidence refute his hypothesis. He often does this by pointing to hidden auxiliary assumptions his opponent was taking for granted but which he does not share. By pointing out the crucial role of these auxiliary assumptions, he has saved his hypothesis from refutation. But when is such an appeal legitimate and when is it unjustified? What we tacitly suggested was that when one appeals to an auxiliary assumption A to save one's hypothesis, this auxiliary assumption must be evaluated to determine if it is correct or not. If one refuses to subject these auxiliary assumptions to test, one is flirting with disaster.

The importance of subjecting auxiliary assumptions to subsequent test—typically but not always empirical test—is a point that also occurs in a slightly different context. Suppose, for example, that one encounters some embarrassing evidence E. Suppose, furthermore, that one wants to keep one's basic hypothesis H. One way to explain E is to cite an auxiliary assumption (or some aspect of one's original hypothesis) and then proceed to modify it. By modifying H to H' (or by modifying A to A'), one can then account for the embarrassing fact.

Sometimes these modifications are said to be **ad hoc**, that is, these changes were made just for the sake of saving the hypothesis in the face of this embarrassing evidence and for no other reason. Often such ad hoc modifications are scientifically suspect and violations of the canons of good scientific methodology. We will offer the following as a definition of when such a modification is ad hoc:

> A hypothesis H' is *ad hoc* in relation to evidence E = H' is introduced for the sole purpose of saving a hypothesis seriously threatened by the negative evidence E.

An example of such an ad hoc hypothesis has already been mentioned in the preceding exercises (number 3). In the history of chemistry, a recurrent problem concerned the phenomena of combustion. In the 18th century, there arose a popular theory—the theory of phlogiston—to explain certain facts of combustion. Phlogiston was said to be a substance that all bodies possess. When a body was heated, phlogiston was used up and was given off to the surrounding air. Phlogiston could explain numerous chemical facts concerning combustion, e.g., that earths (oxides) when heated yield metals, that some metals when heated yield earths (oxides), etc. When a metal is transformed into a calx (or oxide), this was explained by saying that the metal gave up its phlogiston to the environment and was transformed into a calx so that metal - phlogiston = calx. Naturally, one would expect that the calx would weigh less than the metal since some of the phlogiston contained in it had been given off. But when the calx was weighed afterwards, it was found to weigh more than the original metal.

Here, we have a case in which the phlogiston theory encountered some embarrassing evidence. Clearly, it had to explain this fact. Some proponents of the phlogiston introduced what has subsequently been thought to be a purely ad hoc modification: they claimed that phlogiston was unusual in that, unlike other elements which had positive weight, phlogiston had "negative weight," that is, the more phlogiston an object had, the more "negative weight" it had and hence the less it weighed. As this phlogiston was eliminated from an object, there was less "negative weight" and hence the object weighed more. Hence, one was able to explain the embarrassing transformation by claiming that as phlogiston is given off by a metal, the weight of the calx is increased because there was less "negative weight." This modification would be said to be ad hoc if it were made solely for the purposes of accounting for the embarrassing evidence and nothing more. Indeed, most individuals seem to think this was the case.

In the history of science, theories have sometimes been modified in the light of embarrassing evidence but such modifications have not been thought to be ad hoc but rather genuine or legitimate modifications. For example, Newtonian mechanics encountered a particularly embarrassing fact in the late 18th century: there was a deviation (or perturbation) in the orbit of Uranus. The orbit of a planet should be an ellipse (according to Kepler and Newton) but the orbit of

Uranus was not exactly one. This was embarrassing, perhaps even threatening, evidence. It certainly was an anomaly that was puzzling and one that Newtonians could hardly avoid.

Newtonians responded to the perturbation of Uranus by introducing an additional hypothesis to save Newtonian mechanics from refutation. They postulated that there was a hidden planet that pulled Uranus off course, accounting for its perturbation. Was this hypothesis ad hoc? Most individuals would say, "no" because of the following feature: the postulation of the existence of this invisible planet led to the prediction that one should be able to observe it at a certain time of the year. In fact, after several attempts, Adams and Leverrier did observe the new planet Neptune. Hence, this was not an ad hoc hypothesis but a legitimate one.

How can we distinguish between ad hoc modifications and genuine ones? This is a very controversial and often debated issue. But according to one influential account, the key difference is the fact that a new prediction is made in the case of a legitimate modification, whereas this does not occur in the case of an ad hoc change. Thus, we can say according to this account:

> A hypothesis H' is *ad hoc* in relation to evidence E = (1) H' does not explain any phenomena other than E, (2) there is no evidence for H' other than E, or (3) H' makes no novel predictions. In order for a modification of a hypothesis (or the introduction of a hypothesis) not to be ad hoc, this modification must lead to a new prediction.

The following, therefore, is a more detailed sketch of the underlying logic:

(H & A) ⊃ I
~I

~(H & A)

~H v ~A

A is probably true

Thus, ~H

H must be modified: H ⇒ H' so that

$(H' \& A) \supset \sim I$

and

$(H' \& A) \Rightarrow$ novel implications

The Sources of Hypotheses

According to the hypothetico-deductive (HD) method of testing, one begins with a hypothesis from which one deduces one or more consequences of the hypothesis. As we have seen, there is a logic underlying this method involving the deduction of observational predictions from a hypothesis. But a question that must be briefly discussed concerns the source of the hypothesis in question, Where do the hypotheses come from? What methods and procedures should the scientist follow in generating a hypothesis? How, in short, are hypotheses discovered?

Of all the subjects in scientific methodology, this one is shrouded in mystery and controversy. It is unclear what is being asked when someone asks if there is a method for generating novel hypotheses. If the individual is asking for a set of mechanical rules or **algorithms** for generating a hypothesis, a set of rules a computer could follow that would result in the creation of a hypothesis, then it is not clear any such set of rules exists. Some philosophers, such as Karl Popper and Carl Hempel have argued against the very possibility of such a set of algorithms for discovery, and most philosophers of science would appear to agree with them. On the other hand, some computer scientists (e.g., Herbert Simon) have argued that there is a logic of discovery, that there are rules a computer could follow to generate a hypothesis and hence a computer could be programmed to create novel hypotheses. However, Simon's computer program does not consist of just mechanical rules--algorithms--but also what are called **heuristics,** that are not mechanical algorithms but rather useful rules of thumb, helpful hints, plausible ideas, analogies, etc., for creating new hypotheses. According to these individuals, they have created computer programs that do make scientific discoveries. Others have disputed this, but even if they have succeeded in doing this, it seems to be the heuristics of the program that are doing the real work and not the algorithms. If this is the case, then is it still correct to say that scientific discovery is a mechanical process or does it involve creativity--creativity on the part of the computer? What is the nature of scientific creativity? Is it a solecism to assert that a computer can be creative? These questions remain unclear at best.

Let us suppose there were a set of heuristics for scientific discovery. Would such a set of heuristics constitute what is called a "logic" of scientific discovery, i.e., a set of *rational* procedures for generating hypotheses? According to many individuals (e.g., Popper, Hempel), it would not constitute a logic because (for them) to have a logic of scientific discovery would be to have a set of mechanical rules for making such discoveries and they believe (with justification) that there are no such rules. But, as we have pointed out, several individuals believe that, although there is no set of algorithms for scientific creativity, there is a set of heuristics for it. Although such a set of procedures would not guarantee or entail a solution, i.e., the creation of a new hypothesis that is a solution to a scientific problem, such a set would nonetheless increase the likelihood of one, i.e., the creation of a new hypothesis that was a likely or *plausible* solution. If there were a set of heuristics for generating plausible solutions, clearly it would be rational to follow such heuristics. Hence, such a set of heuristics would be a set of rational procedures for generating a hypothesis, for (after all) it is part of the meaning of rationality that if one wants to solve a problem and if a set of heuristics would increase the likelihood of obtaining such a solution, it would be rational to follow such a set of rules. It this sense, therefore, there is a logic of discovery.

But now suppose such a set of heuristics did exist. What are they? On this question, there is some agreement but not much. In fact, most of the received wisdom about such procedures comes from personal anecdotes from scientists who have made important discoveries (e.g., Poincaré, Helmholtz, Einstein, Kekulé). As of the present, there have been few really good scientific studies of this question. So, the following set of heuristics (useful hints) must be taken with a grain of salt, stemming from their impressionistic source.

1. Scientific discovery involves creativity. Scientific discovery is, after all, not a blind, mindless, mechanical task. It is one that involves ingenuity and the construction of original ideas. Hence, one must beware of procedures, attitudes, and activities that interferes with the creative process and decrease the likelihood of original problem solving.

2. Creative ideas often come suddenly, in a "flash." Many scientists have given examples from their own work illustrating this point, e.g., Newton's discovery of gravity by noticing an apple falling from a tree, Archimedes' sudden insight (his "Eureka" experience) in the bath tub, Kekulé dream. In short, one never knows when a creative idea will appear; one must always be on guard for the unexpected.

3. *Chance* often plays a role in scientific discovery. Scientists often hit upon an idea or make a significant discovery by accident, when they are least expecting it. Often this is expressed by saying it occurs by **serendipity** (being in the right place at the right time by accident). Galvani's discovery of current electricity, Fleming's discovery of penicillin, Röntgen's discovery of X-rays all involved this element of chance or accident.

4. It seems to be a well-confirmed generalization that scientists make discoveries only after they have been intensely working on a problem for a long time--after they have been "saturated" with it. It appears to be necessary, therefore, that a scientist be engrossed or even fixated on a problem before a solution emerges.

5. Scientific discoveries are often made when scientists are rested, taking a leisurely stroll and in some general way relaxed and stress-free. Creative ideas often appear when one is not consciously thinking about the problem at all but doing something else.

6. Scientific discoveries are often accompanied by visual images. This occurred, for example, in the famous case of Kekulé who, puzzling over the structure of benzene, drifted off to sleep, and dreamed of the image of a snake with its tail in its mouth. Several other scientists (e.g., Einstein) have documented the important role that images have played in their scientific discoveries.

7. Finally, a crucial element in scientific discovery is analogy and metaphor. Scientists often hit upon a new and original idea by seeing relations between things, e.g., by conjecturing (or "seeing") that light is like water waves, that the electron is like a planet in a solar system, or that the air surrounding us is like a sea. Sometimes, this is expressed by saying that scientific discoveries occur when there is a gap in one's understanding and it is filled in by seeing a larger, complete pattern into which data fit.

A widely cited account of the art of thinking captures many of these features just mentioned. According to the model of Graham Wallas (1924), there are four stages of scientific thinking leading to a discovery: preparation, incubation, illumination and verification. (1) The stage of *preparation* consists of informing oneself of all the relevant facts about the problem at hand, collecting the relevant data, doing a literature search, reflecting upon the problem, taking notes, discussing the problem with colleagues, etc. This is the stage of being saturated with the problem. (2) During the period of *incubation*, the problem is temporarily forgotten or ignored. We cease contemplating the problem--at least consciously: we relax, take a walk, go to the movies, sleep, and let the subconscious (or as we would phrase it perhaps today "the cognitive unconscious") take over the problem and the materials, and work on a solution. During this process "ideas are percolating" in the unconscious. (3) If we are lucky, a solution appears suddenly, in a flash of insight or intuition during the period of *illumination.* This is the "eureka" experience of Archimedes or Newton, when one sees how pieces of the puzzle fit together. (4) Finally, in the stage of *verification,* the proposed solution is tested by standard scientific means and evaluated accordingly.

Wallas' model is a descriptive model of the sequence of events occurring in scientific creativity. It does not tell us what is happening on the unconscious level during this period, and it is this phase that is obviously of importance. No doubt, what is occurring at the level of cognitive unconscious is that information obtained from the collection of data and information recalled from one's background knowledge is being processed by means of certain strategies and tactics of problem solving. At least something like this is the account proposed by Herbert Simon in his model of scientific discovery, the *problem solving approach.* By writing a computer program which employs such strategies (e.g., means-ends analysis, hill-climbing), Simon claims to have been able to program a computer to make scientific discoveries. By analogy, he suggests, scientists must be employing these and analogous kinds of heuristics. Whether this approach successfully accounts for the process of scientific creativity must obviously await the verdict of history.

Criteria for a Good Hypothesis

Although we have discussed the steps involved in empirically testing a hypothesis, a preliminary step in the scientific method typically involves what we have called a priori evaluation of a hypothesis. This occurred, for example, in the case of Semmelweis in which he subjected several hypotheses to a searching examination on logical and conceptual grounds and consequently rejected them as unworthy of subsequent empirical test. We have not said, however, what is involved in such logical and conceptual evaluation. How does one decide that a hypothesis is not worthy of empirical test? How does one decide that a hypothesis has potential and should be subjected to subsequent empirical test? In the following, we will suggest several criteria for the evaluation of the adequacy of hypotheses. In order for a hypothesis to be worthy of tentative scientific acceptance and subsequent empirical testing, it must be (1) logically possible, (2) relevant, (3) testable, (4) compatible with existing scientific knowledge, (5) simple, (6) adequate scope, and (7) depth.

(l) Logically Possible. The first criterion a hypothesis must satisfy is so obvious as to not merit any extended discussion: it must be logically possible. A hypothesis H cannot be internally inconsistent or incoherent. Some hypotheses are logically impossible because their logical structure is such that they appear to have inconsistent principles: they assert both that P is the case and that P is not the case. Other hypotheses are logically incoherent in more subtle ways; their logical faults may not be obvious at first sight but as one explores their principles more carefully, one sees there are conceptual and logical problems preventing such a hypothesis from being a serious candidate. These conceptual deficiencies are not, strictly speaking, ones involving logical inconsistency but are "logical" deficiencies in a broader sense. For example, hypotheses may be so obscure and vague that it is difficult to know what they assert. In any case, the fact that a hypothesis is logically impossible seems sufficient to disqualify it as a serious candidate.

(2) Relevance. Secondly, any proposed hypothesis H must be relevant to the phenomenon in question: the proposed hypothesis must actually explain the phenomenon in the sense of showing that if the hypothesis were true, then the phenomenon would be expected (one could understand why the phenomenon occurs). If the hypothesis is not relevant to the issue under discussion, the hypothesis can be dismissed. In our example of Semmelweis, any hypothesis advanced had to explain the fact that there was a difference between the two divisions in the Hospital; hence, if one cited a factor present in both hospital divisions, this would not be considered relevant to the phenomenon in question.

(3) Testability. In order for a hypothesis to be worthy of consideration, it must be testable: The hypothesis must logically entail observable predictions that are disconfirmable or confirmable. There must be some way to decide empirically if the hypothesis is true, false, probable, unlikely, etc., and this means there must be some way to test the hypothesis. To test it empirically, one must be able to deduce from the hypothesis some statement about what one would observe under certain kinds of conditions.

Some hypotheses are completely untestable. For example, a theoretical entity that exists outside of space and time and has no causal properties cannot be tested empirically. So is a hypothetical entity that is invisible, inaudible, intangible, undetectable by any scientific instrument, etc. Such a hypothesis is untestable because there are no observable consequences of this hypothesis.

(4) Compatibility with Existing Scientific Knowledge. A further criterion concerns the question, Is the hypothesis in question consistent with current scientific knowledge? We can imagine that currently accepted scientific knowledge, which consists of facts, laws, theories, etc., forms a network of interconnected elements and that any new scientific knowledge must fit into this network. The underlying assumption here is that scientific knowledge forms a unity and that no real scientific knowledge can exist as a separate item of knowledge that cannot be integrated into this body of accepted knowledge. If a hypothesis were advanced that was inconsistent with this accepted body of knowledge, then such a hypothesis can be looked upon with skepticism. This does not mean, of course, that such a hypothesis is false but only that its antecedent credentials are in serious question. Any scientist who continues to hold such a hypothesis does so at a risk, and any such individual should have other reasons for holding the hypothesis, reasons that outweigh or counteract this failure to agree with the rest of science.

A well-known example of such lack of compatibility with current scientific knowledge is extra-sensory perception, para-psychology, psycho-kinesis, and other para-normal phenomena. There has been a great deal of experimental evidence cited in support of the authenticity of such phenomena. But many scientists continue to reject them because they do not agree with the rest of our scientific knowledge. Whenever a hypothesis lacks this coherence with the remainder of our scientific knowledge, this constitutes grounds for judging it to be unacceptable.

If a hypothesis is not consistent with current scientific knowledge, this does not mean that it should be rejected out of hand. But it does mean that this fact counts against the hypothesis; it renders it less worthy of pursuit, less worthy of actual empirical test. Consequently, the hypothesis is less probable.

(5) Simplicity. A good hypothesis should be one that is clear. It should not be vague, imprecise, with terms not clearly defined. In addition to being clear, however, it is widely thought that good hypotheses should be as simple as possible. Although no one has yet come up with a precise sense of 'simplicity,' it is still a concept extremely useful in science.

Some hypotheses are simple (to a certain degree), whereas others are more complex. Other things being equal, simple hypotheses are better than complex ones. According to what is known as **Occam's Razor** (named for the 12th century philosopher William of Occam), *entities should not be multiplied beyond necessity.*

What exactly simplicity means is not clear, although simplicity seems to concern matters such as the internal coherence and structure of the theory, the number of basic types of entities (categories) in the theory, the aesthetic elegance of the theory, etc. Mathematicians, for example, frequently refer to mathematical proofs as "elegant" or "beautiful" and the same judgments apply to scientific theories. Einstein's theory of relativity, for example, had just this kind of elegance and simplicity. Indeed, Einstein himself was convinced that his theory was correct long before there was sufficient empirical evidence in favor of the theory largely because of the structural features of the theory.

One of the most famous examples of the importance of simplicity in science occurred in 16th century astronomy; there were two competing accounts--the Copernican (sun-centered) account and the Ptolemaic (earth-centered) account. To explain various types of astronomical phenomena, it was supposed that planets travel in epicycles—circles around circles—about other planets. The Ptolemaic theory required 84 such epicycles, whereas the Copernican account needed only 34. Hence, it is said, the Copernican theory was simpler--numerically simpler. In addition, the Copernican theory was qualitatively simpler since, like the Ptolemaic theory it employed the concept of eccentrics and epicycles, but did not employ the concept of an equant. Hence, it employed one less category of entities.

(6) Scope. A hypothesis is worthy of pursuit if it explains not only the phenomenon in question but additional things. The quantity of phenomena a hypothesis can explain is its scope or explanatory power. One hypothesis has more explanatory scope or power than another hypothesis if it can explain more things than the latter hypothesis can, if it can explain a greater variety of phenomena, etc. Newton's theory of mechanics, for example, had a great deal of explanatory power because it could explain not only the movement of the planets and the movement of ordinary objects like arrows and cannonballs, but also the tides, the movement of a pendulum, the flattening of the earth's poles, etc. Its scope was extensive. When a hypothesis does this, it gives a patterned explanation, finding "unity in diversity"; it discovers that underneath a great diversity of rather distinct and differently appearing phenomena, there is an internally coherent set of unified principles.

(7) Depth. Finally, hypotheses vary in terms of whether they give us a deep explanation of some phenomenon under investigation or merely a superficial explanation. Although scientists use this concept of "depth" in a variety of ways and contexts, it is difficult to say what they mean by it. One common way of interpreting depth is in terms of the distinction between macro-level phenomena and micro-level mechanisms. Sometimes one has an explanation of some observable phenomenon involving laws—sometimes called phenomenological laws--describing variables of the same type as the phenomenon under investigation (e.g., patterns of phenotypic traits in plants). This would be a macro-level explanation. A deeper explanation of the data would be one in which the phenomena is explained by processes occurring on a lower or deeper level, one in terms of hidden mechanisms (e.g., the genetic structure of the plants).

The underlying assumption here is that there are levels of reality and that the more elementary or basic an explanation is, e.g., in terms of atomic structure, the more powerful it is. For example, one can explain the behavior of gases by reference to the macro-level gas laws of Boyle and Mariotte involving pressure, temperature and volume. These variables occur on the macro-level. But these gas laws have been shown to be in slight error and to be limiting cases of what is known as Van der Waal's equation, which is more complex than Boyle's Laws. Van der Waals' equation in turn can be explained by the kinetic theory of gases, which is an explanation occurring at a deeper level than the gas laws. Hence, the kinetic theory of gases give us a deeper explanation of the behavior of a gas than the phenomenological gas laws do.

Exercise 12.3

In the following passages, indicate what data are to be explained and what hypotheses are proposed to explain them. Using the criteria for a good hypothesis, evaluate how good each hypothesis is.

1. The distant universe looks quite different from the nearby universe. To explain this, astronomers adopt an evolution hypothesis. Distant objects, they believe, are young objects. The signals we get from them, delivered here at the speed of light, were emitted when the objects were in an earlier stage of development than we see in the local region. If, for example, one of MacAlpine's survey objects [quasars—quasi-stellar objects whose light has traveled some billions of years to reach us] were one and one-half billion parsecs away, its light would have originated at a time when our own solar system was still forming and our sun quite young. What was our galaxy like at that time? It is hard to know except to infer that perhaps it looked then like quasars look to us today.

--I. Copi, *Introduction to Logic*

2. Like multiple sclerosis, poliomyelitis in its paralytic form was a disease of the more advanced nations rather than of the less advanced ones, and of

economically better off people rather than of the poor. It occurred in northern Europe and North America much more frequently than in southern Europe or the countries of Africa, Asia or South America. Immigrants to South Africa from northern Europe ran twice the risk of contracting paralytic poliomyelitis than South-African-born whites ran, and the South-African-born whites ran a much greater risk than nonwhites did. Among the Bantu of South Africa paralytic poliomyelitis was rarely an adult disease. During World War II in North Africa cases of paralytic poliomyelitis were commoner among officers in the British and American forces than among men in the other ranks. At the time various wild hypotheses for the difference were proposed; it was even suggested that it arose from the fact that the officers drank whiskey whereas men in the other ranks drank beer!

We now understand very well the reason for the strong distribution of paralytic poliomyelitis. Until this century poliomyelitis was a universal infection of infancy and infants hardly ever suffered paralysis from it. The fact that they were occasionally so affected is what gave the disease the name "infantile paralysis." With the improvement of hygiene in the advancing countries of the world more and more people missed infection in early childhood and contracted the disease for the first time at a later age, when the risk that the infection will cause paralysis is much greater.

This explains why the first epidemics of poliomyelitis did not occur until this century and then only in the economically advanced countries.

—G. Dean, "The Multiple Sclerosis Problem" (*Scientific American,* 1970)

3. While walking one night with Dr. Frink, we accidentally met a colleague, Dr. P., whom I had not seen for years, and of whose private life I knew nothing. We were naturally very pleased to meet again, and on my invitation, he accompanied us to a cafe, where we spent about two hours in pleasant conversation. To my question as to whether he was married, he gave negative answer, and added, "Why should a man like me marry?"

On leaving the cafe, he suddenly turned to me and said: "I should like to know what you do in a case like this: I know a nurse who was named as a co-respondent in a divorce case. The wife sued the husband for divorce and named her as co-respondent, and *he* got the divorce." I interrupted him saying, "You mean *she* got the divorce." He immediately corrected himself saying, "Yes, she got the divorce," and continued to tell how the excitement of the trial had affected this nurse to such an extent that she became nervous and took to drink. He wanted me to advise him how to treat her.

As soon as I had corrected his mistake, I asked him to explain it, but, as is usually the case, he was surprised at my question. He wanted to know whether a person had no right to make mistakes in talking. I explained to him that there is a reason for every mistake, and that if he had not told me that he was unmarried, I should say that he was the hero of the divorce case in question, and that the mistake showed that he wished he had obtained the divorce instead of his wife, so as not to be obliged to pay alimony and to be permitted to marry again in New York State.

He stoutly denied my interpretation, but his emotional agitation, followed by loud laughter, only strengthened my suspicions. To my appeal that he should tell the truth "for science' sake," he said, "Unless you wish me to lie, you must believe that I was never married, and hence, your psychoanalytic interpretation is all wrong." He, however, added that it was dangerous to be with a person who paid attention to such little things. Then he suddenly remembered that he had another appointment and left us.

Both Dr. Frink and I were convinced that my interpretation of his *lapsus linguae* was correct, and I decided to corroborate or disprove it by further investigation. The next day, I found a neighbor and old friend of Dr. P., who confirmed my interpretation in every particular. The divorce was granted to Dr. P.'s wife a few weeks before, and a nurse was named as co-respondent.

--A. Brill, *Psychoanalysis: Its Theoretical and Practical Applications*

4. Tonight I want to tell you a little about astrology, and in particular to explain its foundations. Unscientific astrologers have sometimes claimed that stars and planets compel our every action. As a scientific astrologer my first task is to correct that impression. The stars do not compel, they *impel*. Because of the difference between compulsion and impulsion, astrology is not committed to a rigid determination.

The main support for the truth of astrology came upon the scene only recently, several thousands of years after men began to practice our ancient of sciences. Modern physics teaches us that the stars are huge bodies, some of them thousands of time the size and mass of our own sun. Now these great bodies are constantly giving off vibrations. Ask any astronomer and he will tell you, for example, that the stars are the source of cosmic rays whose power in terms of electron-volts is fantastically high. Nor is there reason to think that cosmic rays are the only kinds of vibrations stars emit, or even the strongest kind. Astrologers have known for centuries how great is the force of stellar vibrations. The rediscovery of such forces in modern times is a corroboration of our views.

What about these vibrations? Consider how sensitive an instrument the human body is. Nothing is more intricate and responsive than the mind and nervous system of a creature that is capable of being affected powerfully by a new idea or by the words of someone long dead.

Put these two facts together: first, the incalculable forces involved in the vibrations given off by the stars, and second, the exceedingly frail and responsive machinery of the human mind. It seems exceedingly plausible that so delicate an instrument should respond to so gigantic an influence in a predictable kind of way. After all, the seismograph at Fordham registers earthquakes in the Sea of Japan and yet this seismograph is not nearly as sensitive as the mind, nor are the greatest quakes comparable in force to the vibrations given off by a star.

Thus, the colossal vibrations of the stars and planets under which we are born must set their stamp somehow on each human being. What astrology maintains is this: over the centuries these character-shaping influences have gradually been measured and have become incorporated into astrology as a science. Every newborn child's personality, plastic at birth, is given its character and permanent cast by the peculiar configuration of heavenly bodies at the time of the child's birth. (This, incidentally, is something that the modern behavioral sciences also maintain.) This truth has the further merit of explaining why no two persons are exactly alike. Even with so-called identical twins, the moment of birth is not the same, and with the passing of a few moments the star-picture can change significantly. We know, then, that astrology is a science like any other science.

People who dislike astrology will sometimes reply that the stars' distances are sufficient to make the effects of their vibrations negligible. The following refutes this. First, the space between the earth and the stars is nearly a vacuum, empty enough to let vibrations pass almost undiminished in vigor. Second, a ray of light from the star Arcturus hitting a photoelectric cell opened the Chicago Exposition in 1935. If a star as remote as Arcturus can open a fair in Illinois, it is very strange to argue that the stars are too far away to affect our lives.

—adapted from J. Oesterle, *Logic*

Bayesian Confirmation

The various criteria advanced above for the a priori evaluation of a proposed hypothesis naturally suggests similarities between this account and that of an earlier one--Bayesian confirmation theory. In fact, these various criteria can be interpreted to be criteria for the *plausibility* of a hypothesis: in determining how plausible an initial hypothesis is, these various criteria seem to be routinely utilized by scientists. To determine the initial plausibility of a hypothesis is to determine how probable it is prior to testing it, and this naturally suggests the notion of the prior probability of a hypothesis, an obvious Bayesian concept.

Although Bayesian confirmation theory is not, strictly speaking, an instance of the hypothetico-deductive model, there are obvious similarities between them, in fact, a sufficient number of similarities to warrant the notion that they belong to the same family. This is especially true if one includes, as a part of the hypothetico-deductive model, the important step of a priori evaluation, for it this step that is crucial to the Bayesian account. If one believes it makes senses to ascribe a numerical value to the a priori evaluation, one would have one component of Bayes' theorem: P(H). On the standard interpretation of the hypothetico-deductive model, one can calculate the likelihood of the evidence (P(E/H)), which will be 1.0 if H deductively entails E. In short, there are obvious similarities between the two. This is often denied by some individuals, especially Bayesians; so the reader should be forewarned that some individuals would contest the claim that they belong to the same family and hence that the above claim is somewhat controversial.

Hempel's Paradoxes of Confirmation

Carl Hempel published a seminal article in 1934, in which he layed out certain conditions for an adequate account of confirmation, when an item of evidence confirmed a hypothesis. He proceeded to show that given these very plausible conditions, paradoxes occurred, the most famous one being the *paradox of the ravens*.

Hempel's conditions of confirmation included the following, most of which seem to be necessary conditions for the employment of the hypothetico-deductive method. (In the following, *o* stands for observation, and *h* stands for hypothesis).

1. Entailment Condition: *if o implies h, then o confirms h*, i.e., any sentence (e.g., "this is black or gray") logically entailed by an observation report (e.g., "this is black") is confirmed by it.

2. Converse Entailment: *if h implies o, then o confirms h.*

3. Consequence Condition: *if an observation report o confirms every one of a class k of hypotheses, then it also confirms any hypothesis that is a logical consequence of this class k.*

4. Special Consequence Condition. *If an observation report o confirms h, then it also confirms every consequence of h*, e.g., if o confirms Newton's law of gravitation, then it also confirms Kepler's law of ellipses, since the latter is entailed by the former.

5. Converse Consequence Condition: *whatever confirms a given hypotheis h, also confirms any hypothesis that entails h*, e.g., if o confirms Galileo's law of free fall, then o also confirms Newton's law of gravitation (since the latter entails the former).

6. Consistencey Condition: *every logically consistent o report is logically compatible with the class of hypotheses that o confirms.* If o confirms h, and o confirms h', then o confirms (h and h').

6. Equivalence Condition: *if o confirms h, and h is logically equivalent to another hypothesis h', then o confirms h'.*

Hempel was trying to set out a set of conditions, or (what we might call) axioms of confirmation. Most of these seem to be intuitively plausible, and to be tacitly employed by scientists in their work. But

Hempel was able to show that they produced a paradox—the paradox of the ravens.

Consider the hypothesis that *all ravens are black*, which can be symbolized as: for anything that is, if it is a raven, then it is black. We might suppose there are various observations we could make of individual black ravens that would confirm this hypothesis. But this proposition is logically equivalent (via transposition: see Chapter 4) to: all non-black things are not-ravens (for everything that is, if it is not-black, then it is a non-raven). We might also imagine an observation o that would confirm this, e.g.., if we observed a shoe that was red, this would confirm it, since it is not-black and it is a non-raven. But then, by the principle of equivalence (above), together with certain of the other conditions, this observation report o would confirm the hypothesis that all ravens are black. But it is counter-intuive to think that observing a red shoe would confirm the hypothesis that all ravens are black. This is the paradox, and various strategies have been employed to solve it, e.g., some people have abandoned the Special Consequence Condition, some have abandoned the Consequence Condition, some have suggested it is correct and out intuitions are wrong, etc.

Like Goodman's new riddle of induction, Hempel's paradox of the ravens, which is just one example of set of paradoxes he was able to demonstrate, has been extensively discussed by philosophers of science; and like Goodman's new riddle, it remains a philosophical puzzle with no widely accepted solution.

For Further Reading

Most textbooks on logic and scientific method include a discussion of the hypothetico-deductive method. A particularly influential one is that of Hempel (1966, ch. 2). Popper (1935/1965) advocates a similar (but more advanced) account but does not accept the notion of confirmation or induction. My particular version owes a considerable amount to Dewey's version (Dewey, 1938).

There are three alternative accounts of the nature of confirmation besides the standard hypothetico-deductive model: (1) Hempel's (1945) instance account of confirmation (see above), (2) Bayesian confirmation, and (3) Glymour's (1980) bootstraps approach. Bayesian confirmation is presented in detail in Howson & Urbach (1993).

An excellent discussion of the "art" of scientific investigation can be found in Beveridge (1957).For more on the Semmelweis case, see Sinclair (1909).

On the Duhem-Quine thesis, see the anthology of collected articles (Harding, 1976). Duhem's argument can be found in Duhem (1914/1962), which is accessible to the beginning student. Unfortunately Quine's original presentation (Quine, 1951/1961) is not so accessible. Most discussions of ad-hoc hypotheses and crucial experiments begin at an intermediate level of difficulty.

Simon's views on the question of scientific creativity can be found in Langley, et al., (1987). The individual that reopened the issue of a logic of discovery was Hanson (1961/1970). For an excellent anthology, see Nickles (1980), including Nickles Introduction.

Notes

[1]The following account is taken from Howard W. Haggard (1946, pp. 80-93).

13

The Experimental Method

Introduction

The hypothetico-deductive method is an essential feature of the scientific method. In particular, it is necessary when the hypothesis in question concerns a theoretical entity of some kind. The existence and presence of this theoretical entity is not ascertained by a direct test, e.g., by observing it, but indirectly, by deducing from H a test-implication I. I is a proposition characterizing what would be true if H were true. If we can determine the truth or falsity of I, we can determine the likelihood of H.

But what is the nature of I? In some cases, I is an observable state of affairs, e.g., the prediction that a lunar eclipse will occur at a particular time. In these cases, one can ascertain the truth status of I simply by making the corresponding observation. Usually, however, these predicted observations are tacitly hypothetical; i.e., from H one does not merely predict that an observable state of affairs will occur, but one predicts that it will occur *under certain conditions:* if one waits until a certain date and looks at the moon under certain conditions, one will see an eclipse; if one waits a certain period of time and barring certain economic changes, one will record a rise in the Dow Jones Industrial Average. Hence, I itself has a hypothetical form: if certain test conditions T are created, then certain results R will be observed, i.e., $T \supset R$.

A classical example of this can be seen in the case of *atmospheric pressure*. From an initial hypothesis H concerning a presumed sea of air above us, the implication I—air pressure should be different at different altitudes—was deduced. From I, in turn, one deduced the *prediction* that if one performed a certain kind of test T, one should get a particular result R. T consisted of measuring the pressure at the top of the mountain and measuring the pressure at the bottom of the mountain, and R was the actual measurement at the top of the mountain.

There are two ways, therefore, a test hypothetical ($T \supset R$), is related to the test implication I: (1) either I is identical to $T \supset R$, in which case we have: $H \supset (T_1 \supset R_1)$; or (2) I is not logically equivalent to $T \supset R$ but entails it as a logical consequence, in which case $H \supset (I \supset (T \supset R)$. Since $T \supset R$ is one consequence of I and since there may be several others, we have the following:

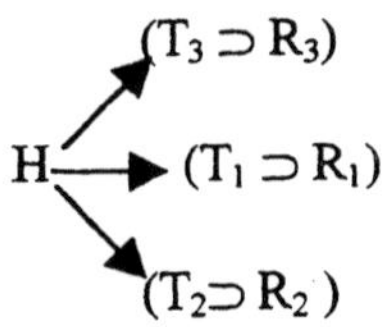

Sometimes, one can actually test the truth of I by making an observation, by merely waiting for the appropriate time, place, and circumstance (as specified in T) to occur and then making the relevant observation, which consists of R, e.g., the predicted rise of the stock market during the first quarter of the next year. Sometimes, an observation is the only way of testing I. This is true of many of the predictions in astronomy and in geology. Here, we are engaging in *naturalistic observation*: one is making an observation in nature, one is not performing an experiment in the laboratory.

Sometimes, however, it may not be possible to test H by simply waiting for T to occur naturally and then observing whether R is true or not. Often, nature simply will not provide a test of our hypothesis by creating the condition T; in such cases we will have to produce it by creating the test conditions T ourselves, which will then allow us to make the relevant observation R. This creation of test conditions is a hallmark of the experimental method and it is to this topic to which we now turn. To illustrate what is characteristic about this method, we will briefly describe a famous case in the history of science--the case of spontaneous generation.

Case Study: Spontaneous Generation

In an earlier chapter, the issue of spontaneous generation was used as an exercise involving the hypothetico-deductive method. It can also be used as an illustration of what an experiment is. The reader will recall that this issue concerned the question of the origin of life: Does life come only from the living or does inanimate matter "spontaneously generate" the living? **Louis Pasteur** was an advocate of what has since been called the germ theory, the view that life only comes from other kinds of living organisms, in this case, tiny germs. To test whether this hypothesis was correct or whether the hypothesis of spontaneous generation was correct, Pasteur performed a crucial experiment. Sugared yeast water was put into several flasks and boiled, presumably sterilizing the flasks and hence killing any germs. The flasks were then sealed so that no atmospheric air could get in. This yeast water was thus "fermentable"—as he showed later—but contained no living organisms. Left to itself, no life would arise from it. Then he collected dust from the outside air on a wad of cotton—by sucking air through this wad of cotton—and introduced this wad into one of the sterile flasks containing the sugared yeast water. In introducing the wad of cotton, Pasteur used air that had been sufficiently heated and hence could not be a source of life. In time, the flask that contained the cotton wad showed signs of fermentation—life—whereas the other flasks not containing such cotton wads showed no such signs.

In this example we have many of the features characteristic of an experiment. To see what these features are, compare it to the following observation of the embryonic stages in the development of the chick made by Aristotle in the fourth century BC:

> Generation from the egg proceeds in an identical manner with all birds, but the full periods from conception to birth differ. . .With the common hen after three days and three nights there is the first indication of the embryo; with larger birds the interval being longer, with smaller birds shorter. Meanwhile the yolk comes into being, rising towards the sharp end, where the primal element of the egg is situated, and where the egg gets hatched; and the heart appears, like a speck of blood, in the white of the egg. This point beats and moves as though endowed with life, and from it two vein-ducts with blood in them tend in a convoluted course [as the egg-substance goes on growing, towards each of the two circumjacent integuments]; and a membrane carrying blooding fibres now envelops the yolk, leading off from the vein-ducts. A little afterwards the body is differentiated, at first very small and white. The head is clearly distinguished, and in it the eyes, swollen out to a great extent. This condition of the eyes lasts on for a good while, as it is only by degrees that they diminish in size and collapse. At the outset the under portion of the body appears insignificant in comparison with the upper portion. Of the two ducts that lead from the heart, the one proceeds towards the circumjacent integument, and the other, like a navel-string, towards the yolk. The life-element of the chick is in the white of the

egg, and the nutriment comes through the navel-string out of the yolk. (*Historia Animalium*, Book 6, 56la3-562a20)

Although the scientific value of this passage is beyond question, it differs in important ways from Pasteur's research. Aristotle was engaged in scientific *observation*, whereas Pasteur performed an *experiment*. What is the difference between these two types of scientific research?

Observation versus Experiment

Both experiments and observations involve the noting of certain characteristics by observing natural phenomena through the senses. But there is an important difference between them. In an observation, one is noting phenomena as presented by nature, whereas in an experiment one is noting phenomena created by the experimenter. In particular,

> an **experiment** involves an experimenter (1) varying certain conditions, (2) holding everything else constant, and (3) noting any effect this variation produces on what is being observed.

Several features distinguish a simple experiment from an observation.

Active Intervention vs. Passive Observation

Observation involves noting and recording phenomena spontaneously produced by nature, whereas experimentation involves creating phenomena by means of an active intervention in nature. In observation, a scientist does not interfere with nature but waits for nature to produce this phenomenon. Since an observer must wait for nature to produce the phenomenon of interest before she can observe it, an observer is at the mercy of nature and is in this sense passive: she cannot produce the phenomenon at will, but rather must wait for such a phenomenon to be produced by nature. In an experiment, however, the scientist does not passively wait for nature to produce the phenomenon; instead, the scientist creates it by actively intervening in nature. She manipulates certain natural processes and changes the natural course of events, in order to observe what will happen.

To say that observation is passive whereas an experiment is active is not to say, however, that the process of observation is a totally passive affair; on the contrary, as we have suggested already, observation is always impregnated with active interpretation. Rather, the question rather is whether *the phenomenon is under the control of the experimenter or not.* Pasteur could vary the experimental conditions at will and produce what he considered to be the relevant cause into the flask. But in the case of Aristotle's observation of the chick embryo, the factor causing the development was not in his control. Consequently, the effect being studied could not be produced at will. Aristotle had to find such a phenomenon already existing in nature. In short, in order to experiment one must have control over nature.

Natural versus Artificial

In observing nature, one observes what nature has produced--spontaneously as it were. One must thus wait for this phenomenon. Hopefully, in time, nature will produce the phenomenon of interest as part of the normal course of events, just as hens regularly produce chicks, which in turn regularly develop in predictable ways. Thus, if one wishes to observe how chicken embryos develop, one must wait for nature to produce such an embryo. If Aristotle had been able to produce the development of the chick embryo at will, he would have had to intervene in the natural course of events and produce certain kinds of changes. He would have needed control over nature.

In an experiment, therefore, one disturbs the typical (natural) course of events: one actively produces some phenomenon which otherwise would not have occurred. In this sense it was "unnatural." A good example of this is a physiological study investigating the function of a bodily organ: an experimenter will perform an ablation study by removing or destroying some organ (e.g., the ventro-medial part of the hypothalamus), and then observe what effect the absence of this organ part has on the functioning of the animal, say, on its eating behavior. If one were to wait for nature to produce an animal whose ventro-medial hypothalamus was absent, one would wait perhaps indefinitely. As opposed to naturally occurring events, experimentally produced phenomena are artificial.

Manual Intervention versus Being a Spectator

In performing an experiment, one does not act like a spectator of nature; instead, one intervenes in nature by virtue of one's body and interacts with other natural bodies. In an observation, on the other hand, one's ideal is to produce a picture of what is being observed as exactly as possible; what one wants is to photograph nature without distortion or bias. This is precisely the goal of giving a naturalistic description. In an experiment, however, one is not a spectator of nature, but an intervener in nature. Hence, to be an experimenter, one must have a physical body that produces certain kinds of changes in other physical bodies. To do this, one must have power and control over nature.

Control

To have control over nature is to be able to actively intervene in nature and to produce a phenomenon at will. It is to have power over nature. To do this, one must know *what leads to what;* that is to say, one must know which causal factors to produce in order to produce certain effects. This is often put by saying that the experimenter *manipulates* certain factors and *varies* them. For example, Pasteur intentionally varied the condition of "presence of germs" versus "absence of germs." In this sense, he controlled whether germs entered into the natural course of events or not. This is often put by saying that *the experimenter has control over the independent variable.*

In a true or strict experiment, the independent variable must be under the control of the experimenter. Some variables are under experimental control, whereas other variables are not. A scientist can control the amount of a particular drug given to a group of subjects, just as he can control the amount of dust particles in the air. But a scientist cannot (typically) control such variables as atmospheric pressure, sex, I.Q., socioeconomic class, intelligence, etc. These variables have different values *in nature,* but scientists ordinarily cannot vary their value at will. Imagine, for example, how an experimenter could manipulate the values of the variable, sex. The experimenter cannot determine which value of this variable will be present in a particular individual since this decision has already been made by nature. Likewise, a scientist cannot assign individuals to different socioeconomic classes any more than she can assign different eye color to humans. If this is correct, then certain kinds of variables will not be variables that can figure in a true experiment. Here, one cannot conduct true experiments but rather *quasi-experiments.*

Not only does a scientist control whether and how many germs enter the picture, but there is another sense in which things were under his control: *he controlled all other factors except this one.* That is to say, not only did he intervene and produce a change in certain factors, but he also kept all other (known) factors constant; not only did Pasteur introduce a cotton wad into certain flasks, but he did not introduce a cotton wad into other flasks which were identical (in relevant ways) to these flasks. In keeping the flasks identical in all ways except for the presence versus the absence of the cotton wad, he was keeping all other factors the same. *He had control over extraneous variables.*

The Logic of Control Groups

Keeping all other factors constant except the one being studied is the basic idea of a **controlled experiment.** The underlying logic of a controlled experiment involves Mill's method of difference:

	Factors Controlled (constant)	Experimental Factor	Observed Effect
Experimental Group	A B C D	E	e
Control Group	A B C D	-	-

The logic behind a **control group**, therefore, is the following: if the **experimental group** and the control group are identical in all relevant respects except for the presence versus the absence of a particular factor, then any difference between the two group's results must be due to the presence of this factor. Correlatively, any other difference between the two groups other than the presence versus the absence of the experimental factor might be an explanation for the effect. Suppose, for example, that factors A, B and C are constant in both groups but that, in addition to the experimental factor E present in the experimental group and absent in the control group, another factor D is also present in the experimental group but absent in the control group:

	Factors Controlled (constant)	Experimental Factor	Observed Effect
Experimental Group	A B C D	E	e
Control Group	A B C -	-	-

In this situation, one can not know whether e is due to E or whether e is due to D. In such a case D, an **extraneous variable**, is **confounded** with E.

Fundamentally, therefore, the reason one must have control over the experimental arrangement--both control over the independent variable and the extraneous variables--is epistemological: one wants to know (to the greatest extent possible and with as much certainty as possible) what is responsible for what. To be able to do this, one must isolate this (causal) factor from all other factors.[1]

Epistemologically speaking, therefore, an experiment gives one knowledge only if there is some *comparison* between two conditions or groups and hence only if there is some *difference*.[2] For this reason, an experiment typically involves at least two groups: an experimental group and a comparison (or control) group. This is not always the case, but when it is not, other kinds of comparison are being made. In short, in an experiment one must have a comparison in order to have knowledge. But in order for the comparison to be revealing, the two things being compared must be equivalent in all relevant ways except for the presence versus the absence of the experimental factor.

Mill's method of difference, therefore, is a hallmark of the experimental method. One kind of differential comparison involves a control group and an experimental group in which all factors but one are kept the same. But this is not the only kind of differential comparison. One can also, for example, find such differential comparison in a pre-test--post-test (before and after study) involving a single group. Here, one notes a difference between the test results before and after the introduction of the experimental treatment condition. There is a comparison here but no separate control group.

The Four W's

In summary, we can say that for experimental control to be present, the experimenter must have control over "the four W's": when, where, what, and whom. The experimenter must have control over *what* experimental factor is being studied, and what amount of it is present. The experimenter must also have control over *when* this factor will be present. Furthermore, the experimenter must be able to decide *where* this factor will be studied--for example, in the laboratory or in a field setting. Finally, the experimenter must have control over "to *whom*" this factor will be presented, i.e., what individual subjects will be studied. The what, where, and when conditions relate to the experimental variable, whereas the whom relates to the subjects being studied. As we will see below, the experimenter may have control over some of these W's but not others.

Experimentation versus Observation

Although we have written as if there were a sharp distinction between an experiment and an observation, such a difference is really a matter of degree rather than of kind. We have already pointed out, for example, that observations are inherently active processes, since all observation involves theoretical conceptualization and interpretation. Furthermore, one can perform an observation in order to test a hypothesis, just as one can perform an observation with no prior hypothesis in mind. Indeed, the employment of the hypothetico-deductive method does not require the use of experimentation, since it is only necessary that a test implication I lead to a predicted observation.

Although observations can be undertaken in order to test a hypothesis, it is also the case that observations can be undertaken with no particular hypothesis in mind. The famous Swiss scientist Jean Piaget, for example, was sitting in a park one day and happened to notice an albino sparrow several feet away. Piaget proceeded to observe this sparrow and to record his observations with considerable care, enough care to be able to publish his observations in a professional journal. Piaget was not intending to test some hypothesis; he just happened to encounter this interesting sparrow. There are many cases in the history of science where chance observations have led to important scientific discoveries, just as there are numerous examples of **experiments in order to see**. In the latter case, an experiment is performed in order to see what will happen; the scientist here is reasoning: what would happen if. . . ? In such cases, there is no hypothesis a scientist has in mind. Furthermore, there are controlled observations, which are intermediate between experiments and observations. In a controlled observation, one controls the conditions under which an observation is made, e.g., experimental preparation of a microscopic object so that it can be observed, growing a culture, building a new instrument of observation, etc.

Exercise 13.1

In the following passages, indicate whether an experiment or an observation was performed and why.

1. Since Venus rotates so slowly, we might be tempted to conclude that Venus, like Mercury, keeps one face always toward the Sun. If this hypothesis were correct we could expect that the dark side would be exceedingly cold. Pettit and Nicholson have measured the temperature of the dark side of Venus. They find that the temperature is not low, its value being only -9°F, much warmer than our stratosphere in broad daylight. It is unlikely that atmospheric currents from the bright side of Venus could perpetually heat the dark side. The planet must rotate fairly often to keep the dark side from cooling excessively.

Fred L. Whipple, *Earth, Moon and Planets*

2. The wife of Epicrates, who was lodged at the house of Archigetes, being near the term of delivery, was seized with a violent rigor, and, as was said, she did not become heated; next day the same. On the third, she was delivered of a

daughter, and everything went on properly. On the day following her delivery she was seized with acute fever, pain in the cardiac region of the stomach, and in the genital parts. Having had a suppository, was in so far relieved; pain in the head, neck, loins; no sleep; alvine discharges scanty, bilious, thin, and unmixed; urine thin and blackish. Toward the night of the sixth day from the time she was seized with the fever, became delirious. On the seventh, all the symptoms exacerbated; insomnolency, delirium, thirst; stools bilious and highcolored. On the eighth, had a rigor; slept more. On the ninth, the same. On the tenth, her limbs painfully affected; pain again of the cardiac region of the stomach; heaviness of the head; no delirium; slept more; bowels constipated. On the eleventh, passed urine of a better color and, having an abundant sediment, felt lighter. . . .

Hippocrates

3. Two other theories assume that the Caribbean is basically oceanic. In one view, the sea opened when North and South America drifted apart; in the other, the Caribbean basin is a segment of Pacific crust that became wedged between the two large continents as they drifted westward. . . .

It was to test these hypotheses that scientists on Leg 15 of the Deep Sea Drilling Project went into the Caribbean for a two-month period ending Jan. 26. . . .

The findings were surprising. Each core, says Dr. Terence Edgar, co-chief scientist for the cruise and now chief scientist for the entire project, is accurately dated by microorganisms entombed in the sediments. The oldest sediments turned out to be only 75 million to 85 million years old. In contrast, the oldest part of the Atlantic is about 180 million years old, dating from the time when the continents first split and the present phase of sea-floor spreading began. If the Caribbean were formed when the continents broke up, Dr. Edgard and Saunders point out, the Caribbean crust would be about the same age as the Atlantic. If it were a piece of Pacific crust, they say, it would be even older. The youth of the Caribbean basins, therefore, essentially eliminates these two hypotheses, the scientists conclude.

Louise Purrett, "The Birth of the Caribbean" (*Science News*, 1971).

4. We have recently obtained conclusive experimental evidence that there can be no tooth decay without bacteria and a food supply for them. In germfree laboratories at the University of Notre Dame and the University of Chicago, animals innocent of oral micro-organisms do not develop cavities. Whereas animals in normal circumstances average more than four cavities each, the germ-free rats show no signs of caries. At the Harvard School of Dental Medicine we have demonstrated the other side of the coin: that food debris also must be present. Rats that have plenty of bacteria in their mouths but are fed by tube directly to the stomach do not develop cavities. In a pair of rats joined by surgery so that they share a common blood circulation, the one fed by mouth develops tooth decay, the one fed by tube does not.

Reider F. Sognnaes, "Tooth Decay" (*Scientific American*, 1957)

5. In the south of England there are many large shallow ponds, on hilltops known as "dew-ponds" because it is believed, locally, that they are replenished by dew. This local hypothesis is supported by the belief that a heavy layer of straw, placed under the basin of these ponds when they are being constructed,

acts as a non-conductor of heat, thus keeping the basin cooler than the surrounding earth, thus causing dew to be deposited, since dew is only deposited on something which is cooler than the dew itself. In 1911 the Royal Geographical Society sent one of their Fellows to investigate these peculiar ponds. His analysis of the case is interesting and instructive logically, whatever it may be geographically.

There were, he reasoned, only three sources from which the water could come, ignoring springs: dew, rain, or some hitherto unascertained source. The dew hypothesis he eliminated on three counts. The straw rots during the first winter and becomes useless as a non-conductor. The least likely place for a dew deposit would be the surface of the water in the pond, because in water the hottest part always rises to the top and forms the surface; and on a warm summer's night there is not sufficient time for this tepid surface water to fall below the dew point and it could, therefore, receive no dew. The investigator carefully checked the night temperatures of ground, air, and water, and found that the water surface remained the warmest. The maximum rate of dew-fall for all England is one and one-half inches per annum. That quantity would be required every night to compensate for what cattle drink from the pond each day.

The rain hypothesis was eliminated by the fact that 1911 was a dry year and practically no rain had fallen. In spite of these facts, the ponds did not dry up. Upon what remaining hypothesis could this fact be accounted for? "It is not until one stays all night by one of these ponds that the explanation is found, namely, that the ponds are replenished by the early morning sea mists which roll up over the hills and saturate their tops." This hypothesis explains "why these self-filing ponds are not to be found in valleys, but only on hilltops," since the mists roll up the hills, never stopping in the valleys. If this hypothesis is true, he proceeded to argue, then the pond water should contain a slight amount of salt, since the sea mists are evaporated sea water and by mechanical action a slight amount of salt would be taken up with the moisture. He analyzed the pond water and found that it did, as a matter of fact, contain a high amount of salt.

The Logic of Experiments

We can now address more explicitly this question: *what is the nature of an experiment,* and *how is it different from an observation*? In order to answer this question, it will be necessary to introduce several terms.

Variables

The concept of a **variable** is a crucial one in understanding what is unique about an experiment. In particular, it will be essential to distinguish several types of variables: independent variables, dependent variables, and extraneous variables. To begin with, we can say that

A *variable* is a property (factor, quality) that can take on different *values*.

More precisely put, let the variable be a *set* (i.e., a collection), consisting of certain members. These members will be the values, and the variable will be the entire set. For example, consider the set consisting of males and females. This set has two members or values--male and female--with the variable (sex) consisting of this set.

Sex = {male, female}

Likewise, if we suppose there is some factor known as intelligence (or I.Q.) of which there can be various numerical values, then we could represent it this way:

I.Q. = { . . ., 100, . . .120, . . .}

Once again, the variable would be I.Q. and its values would be all the numerical values such as 100, 120, etc. Finally, in the case of spontaneous generation, the variable mentioned was that of a germ, whose values consisted of the number of germs present.

Continuous versus Discrete Variables. There are two kinds of variables it is important to distinguish: discontinuous (or categorical) and continuous. (Sometimes discontinuous variables are called qualitative variables and continuous variables are called quantitative variables.)

If we consider the variable, gender, which has two values—male and female—we can see that this is a discontinuous variable: there are two and only two possible values. Someone is either a male or a female but not something in between. In short, gender is bivalent. Likewise, if we consider the variable color, we can see that the various values of this variable--blue, red, yellow, etc.—are discontinuous and qualitative.

Other variables are continuous in the sense that there is an indefinite number of values in between any two values. For example, if we take weight as our variable we can see that between, say, 100 pounds and 101 pounds there are other values, e.g., 100.3, 100.5, etc. and so forth.

Independent Variable. We now can introduce the notions of an independent variable and a dependent variable.

The **independent variable** is that variable under the control of the experimenter.

It is that variable whose particular value(s) the researcher creates at will. If Pasteur had been able to introduce at will differing quantities of germs into the flasks, this independent variable would have had different amounts. Similarly, if an educator were to believe he had hit upon a smart pill"--a pill that increases ones learning ability--the effect of this pill on the learning of students could be investigated by administering different amounts of this pill, say, 100 grams, 200 grams, etc., to a select group of students. The differing amounts of this pill would be different values of the independent variable.

As these examples indicate, the independent variable is typically thought to be the *cause* of some phenomenon. It is often described as the *experimental variable* or as the *treatment condition.*

Dependent Variable. If the independent variable is that causal factor one hypothesizes to be responsible for some phenomenon, the effect of this factor is the dependent variable.

> The **dependent variable** is that variable whose value changes as a function of the corresponding change in the independent variable.

In our example of the smart pill, for example, the dependent variable was the change in the amount or time of learning in students, just as the dependent variable in the case of spontaneous generation was: appearance of life.

Often, because the dependent variable requires measurement, the dependent variable will have different ways of being measured, and these various types of measurement are also referred to as the dependent variable. For example, a variable such as learning can be measured in various ways: number of errors, frequency of correct response, amount learned, etc. There are various ways of measuring the presumed effect, and the term *dependent variable* is often applied to what is measured and the particular measure of it.

"Independent" vs. "Dependent." The terms "independent" and "dependent" indicate a crucial difference between these two types of variables: the dependent variable "depends" upon the independent variable in the sense that when the independent variable takes on a particular value, the dependent variable also takes on a particular value.

There are several ways of understanding the meaning of *independent* versus *dependent.* We will distinguish four: (1) manipulated versus oberved, (2) cause versus effect, (3) "predicted from" vs. "predicted to" and (4) the mathematical sense.

We have already briefly mentioned the first way of distinguishing independent and dependent variables: the independent variable is that variable purposefully manipulated by the experimenter, whereas the dependent variable is what is observed. For there to be an experiment in the strictest sense of that term, the scientist must interrupt the natural course of events and physically manipulate some variable. The dependent variable thus "depends" upon one's intervention in nature and one; manipulation of these variables.

A second sense of independent versus dependent variable is a causal sense; the independent variable is the cause, whereas the dependent variable is the effect. Causes cause effects; thus, the cause is what is primary and the effect is what "depends" upon the cause. The arrow of causality proceeds from cause to effect (and not the other way around); hence the effect depends upon the cause. Consequently, introducing a change in the cause will be followed by a corresponding change in the effect. If we let "x" represent the independent variable and "y" represent the dependent variable, then we can represent this situation in the following way:

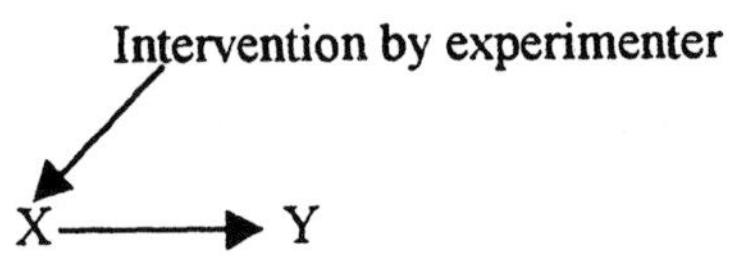

It is because of this presumed causal relation that one is said to have control over nature.

Thirdly, if there is an arrow going from the independent variable to the dependent variable, then one can change the dependent variable by changing the independent variable. Consequently, one will be able to predict certain changes in the dependent variable, which result from a change in the independent variable. One can say, therefore, that one "predicts from" the independent variable and one "predicts to" the dependent variable. Knowing the particular value of the independent variable allows one to predict the value of the dependent variable.

Before discussing the fourth sense of "independent" versus "dependent," namely, the mathematical sense, we will need to briefly discuss some preliminary mathematical concepts, especially that of *function*.

Functional Relation. As we have characterized the difference between the independent and the dependent variable, a change in the dependent variable results from a change in the independent variable; it changes "as a function of" a change in the independent variable. What does this concept "is a function of" mean?

The concept "functional relation" is the crucial relation, and it has two closely related senses, a causal sense (as we have seen) and a mathematical sense. In the *causal sense*, to say y is a function of x is to say that y is the effect of x, which is the cause. Thus, y depends on x as an effect depends upon a cause. Hence, to say that there is a functional relation between two variables is often to imply that there is some kind of causal connection between them.

In the *mathematical sense*, a dependent variable (y) is also said to be a function of the independent variable (x) but in a somewhat different sense. To begin with, suppose we have two variables, each of which is a set. The first set consists of intelligence test scores and the second consists of a group of individuals:

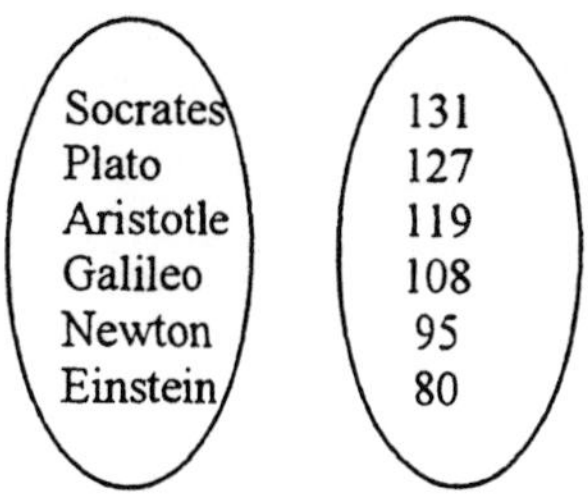

GreatThinkers I.Q. Scores

Suppose now we assign an I.Q. score to each of these individuals (or suppose we construct some test to guess what their I.Q.s are) in the following way:

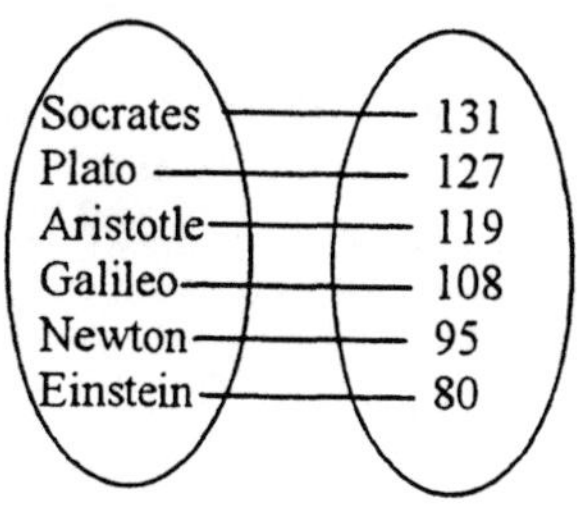

Great Thinkers I.Q. Scores

Clearly there is a relation between these two sets. In particular, there is a set of *ordered pairs:*

{Socrates, 131}
{Plato, 127}
{Aristotle, 119}

A **relation** is a set of ordered pairs.

In our example, the relation is [{Socrates, 131}, {Plato, 127}, {Aristotle, 119}, . . .]. The intuitive sense is that there is a relation between "Great Thinkers" and "Intelligence"; mathematically, this sense of *relation* is expressed as a set of ordered pairs. A *function* is a particular kind of relation:

A **function** is a relation in which, for each value of x, there is one and only value of y.

If one inspects the relation between Great Thinkers and Intelligence, one will see there is a function, since for every Great Thinker there is one and only one I.Q. score. Likewise, there is a function expressed in the following diagram:

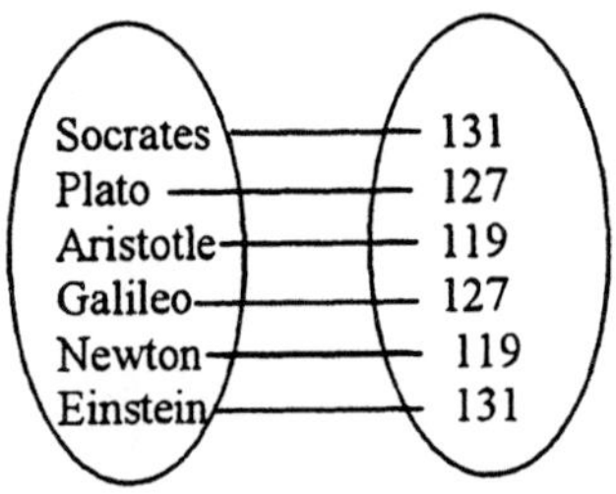

There is a function here because for each value of Great Thinker there is one and only one value of Intelligence; this is true even though some Great Thinkers have the same Intelligence score. In the following situation, however, we have a relation but not a function:

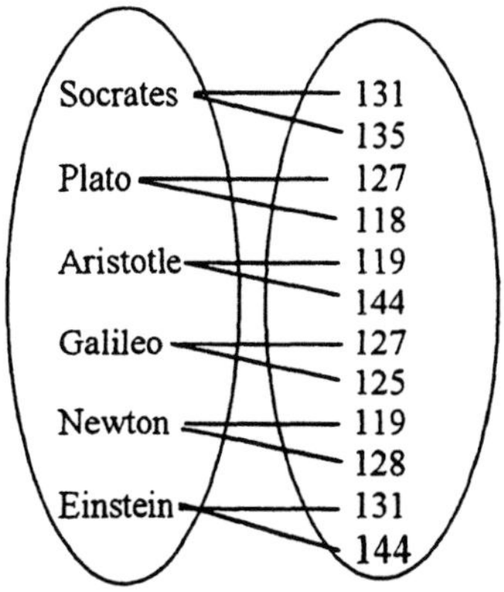

In the case of an experiment involving an independent variable X and a dependent variable Y, Y is a function of X (y = f(x)). In the mathematical sense, one would be saying that there is a mathematical function between x and y such that for every value of x there is one and only one value of y. In this mathematical sense, therefore, one can say that the dependent variable "depends" upon the independent variable: for every value of x, there is a unique value of y attached to it. In short, we can have a function which is a many-one relation between the dependent variable and the independent variable but one cannot have a function between x and y which is a one-many relation between them.

We have seen that there are two senses of "functional relation" expressed by the independent-dependent variable terminology: a causal sense and a mathematical sense. Actually, these two senses of "functional relation" seem to be connected since one often assumes that a particular effect can have several causes—a many-one relation—but a cause can not have several discrete, simultaneous effects. In causality, we usually believe there can be a one-one relation between cause and effect as well as a many-one relation; however, we usually do not believe there can be a one-many relation.[3]

Multiple Causes (many-one) Multiple Effects (one-many)

Extraneous Variables. In addition to the notion of the independent variable and the dependent variable, a third kind of variable is an **extraneous variable**. Extraneous variables are those variables the experimenter is not interested in but which may be causally relevant to the dependent variable. For example, suppose an experimenter is interested in the learning effects of our smart pill. Our experimenter is interested in studying the effects of this smart pill on rate of learning. But in addition to these two kinds of variables, there are extraneous variables that may also influence rate of learning, e.g., sex differences, year in college, motivation, etc. Hence, we can say:

> An *extraneous variable* is a variable in a study other than the independent variable that may (causally) influence the dependent variable.

If one wishes to investigate the effect of the experimental variable on the dependent variable, then one must do something about extraneous variables: one must make sure these extraneous variables are not influencing the dependent variable. To make sure these extraneous variables are not influencing the dependent variable, one must control their potentially causal effects.

Control of Extraneous Variables

How can we exercise experimental control over extraneous variables? There are at least five ways.

(1) The first way is simply by *eliminating* all extraneous variables. If a dry atmosphere is a possible extraneous variable in an experiment in chemistry, one can simply eliminate the dry atmosphere and substitute different laboratory conditions.

There are problems with this solution, however: often one doesn't *know* what the extraneous variables are, and it is not always possible to eliminate them. For example, if sex differences were an extraneous variable, how would one go about eliminating them? Other extraneous variables that cannot be eliminated include intelligence, age, socioeconomic status, etc.

(2) A second way to control extraneous variables is by holding them *constant*, that is, by making them the same in the experimental group and in the control group. If motivation were an extraneous variable, we might try to produce the same level of motivation in both groups of subjects. Similarly, with the other extraneous variables, resulting in the following situation:

Experimental Group	Females, 40 years of age, high school education, moderate motivation
Control Group	Females, 40 years of age, high school education, moderate motivation

The assumption here, of course, is that one knows what the extraneous variables are.

(3) A third way of controlling for known extraneous variables is by *balancing*. Suppose sex is an extraneous variable. We can control for it by making sure that both experimental and control groups have the same proportion of men and women in each group. The same would apply to the other extraneous variables. This would result in the following situation:

Experimental Group	18 male, 12 females 10 18-year old, 10 19-year old 10 20-year old 15, with I.Q. of 120, 15 with I.Q. of 140
Control Group	18 male, 12 females 10 18-year old, 10 19-year old 10 20-year old 15, with I.Q. of 120, 15 with I.Q. of 140

(4) *Counterbalancing* is a fourth way of controlling for extraneous variables. The need for counterbalancing occurs when subjects are tested and retested. Suppose, for example, one is interested in determining the effectiveness of two different teaching techniques—A and B—on rate of learning in students using a pre-and post-test. Using repeated measures on the subjects, one could determine which method was more effective. But the problem here is that there are **order effects,** he effectiveness of the second teaching method may be due not to the second method, but to the fact that one had practice taking the same test twice.

One solution, therefore, is to counter balance: First, present teaching method A, followed by B to one group, and present B, followed by A to a second group. Theoretically, since both groups have had a chance to practice, any difference between them should be due to the effectiveness of these respective teaching methods rather than to order effects. (The assumption here, of course, is the practice effects of A followed by B are the same as the practice effects of B followed by A.)

Before discussing the fifth method—randomization—several other concepts must be introduced.

Control Group. What is a control group? Recall that the experimenter is manipulating the value(s) of the independent variable. This in turn has some effect on a group of individuals. Let us call the application of a value of the experimental variable the **experimental treatment**, e.g., giving 100 grams of a drug to a group of college students. The group of individuals receiving the experimental treatment is the **experimental group**. The experimental group receives the treatment condition or experimental variable (which we may abbreviate as E). The effect of E we will symbolize as e. In addition to E, there are (we may suppose) numerous extraneous variables present: A, B, C, and D, which also may be influencing our experimental group.

A **control group** is a group of individuals that receive no treatment condition E, but have the same extraneous variables operating as does the experimental group. Once again, using Mill's method of difference, we can represent this in the following way:

Group	Extraneous Variables	Independent Variable	Dependent Variable
Experimental Group	A B C D	E	e
Control Group	A B C D	-	-

The logic here is just the logic of Mill's method of difference: If e is present when E is present and absent when E is absent, then one may feel confident in attributing the cause of e to E. If A, B, C and D are possessed by both groups, they presumably have the same effect on both groups; hence, one can ignore any effect they may have. These extraneous variables are acting as a constant and they can be factored out and ignored.[4] (If they are not, then they will be confounding variables.) The method of difference indicates that e cannot be the effect of any extraneous variable but may be the effect of the experimental variable E. All extraneous variables present in the experimental group are also present in the control group, but the experimental variable is absent.

Suppose, however, the extraneous variables are not "matched." In

Group	Extraneous Variables	Independent Variable	Dependent Variable
Experimental Group	A B C D	E	e
Control Group	A B - -	-	-

this case, one does not know whether the effect e is a function of E or D or C.

The use of a control group by the method of difference allows one to exclude C and D as possible causes. In a strict (true) experiment, the experimenter must not only *control the independent variable*, the experimenter must also use a *control group*. In the case of Pasteur we mentioned above, he employed a control group to investigate the effect of introducing a wad of cotton exposed tò germs in the air: the control flasks were like the experimental flasks in all relevant ways except for the fact that in the case of the experimental flasks wads of cotton had been introduced, whereas this was not true of the control flasks. What Pasteur observed was an increase in life in the experimental flasks and no increase in life in the control flasks.

Randomization

So far, we have said that the experimenter controls the independent variable and also establishes a control group. But we have said nothing about how this control group is created. Likewise, we have said nothing concerning how the experimental group is formed.

Random Assignment. In order to have an experiment, randomization must occur, i.e., the experimenter should assign individuals to each treatment condition in a random fashion. This means (roughly speaking) that every individual has an equal chance of being selected as a member of group. In the simplest case, one could flip a coin to see who goes into which group. In this way, individuals would be randomly assigned to the treatment versus control conditions. (A better way would be to employ a table of random numbers.)

The crucial point is that selection of membership in the group is under the control of the experimenter. What is being ruled out is the idea that the individuals should *themselves* decide to which group they will belong. Likewise, random assignment rules out the possibility that they already come to the experiment pre-selected.

To illustrate this idea of random assignment, suppose a scientist were interested in sex as an independent variable. Is sex under experimental control? Obviously not, since the experimenter cannot assign individuals to *maleness* versus *femaleness*. The sex of the subjects has already been pre-selected. Hence, membership in these different treatment conditions is not up to the experimenter. Consider, by contrast, the case in which a scientist randomly assigns individuals to two different amounts of a drug. Individuals don't already come to the experiment being in one of these two treatment conditions; hence, they can be randomly assigned.

Random Selection. Random assignment of subjects to treatment conditions is, for many thinkers, the sine qua non of an experiment. In discussions of randomization, a topic closely related is **random selection** of subjects. This concept of random selection is different from that of random assignment: using random selection, subjects are chosen from a population in such a way that each member in the population has an equal chance of being selected as a subject.

Suppose all the individual subjects belong to the same population P and that we select a sample from this population. If all our subjects belong to this population, we may assume that all variables will be uniformly distributed across members of this population. An extraneous variable V, therefore, will also be uniformly distributed. Now if we randomly sample from our population and form experimental and control groups by random assignment, any extraneous variable V present in P will have an *equal frequency in both experimental and control groups*, i.e., the probability distribution will be roughly the same in both groups. By the same logic, we can then suppose that *all* extraneous variables will be present to the same degree in both groups.

We do not need to *know* what these extraneous variables are since we can assume that all extraneous variables will be equally distributed in both groups. Since this is so, they will be constant across both groups and they can be ignored; we can "factor" them out and merely look at the *difference* between the presence of the experimental treatment vs its absence.

But suppose our groups are not randomly selected, i.e., suppose that the probability of being chosen as a member of the experimental group is not the same as the probability of being chosen as a member of the control group. Then we cannot assume that the distribution of extraneous variables will be the same across all individuals. Some individuals constituting the experimental group may possess characteristics not possessed by the other group; that is to say, some individuals may possess an extraneous variable not possessed by the other individuals.

In this case, one cannot factor out all extraneous variables because they will not be constant across both groups. Random selection of subjects is thus a more demanding characteristic than random assignment.

The most ideal situation would be one in which both random selection and random assignment occurs:

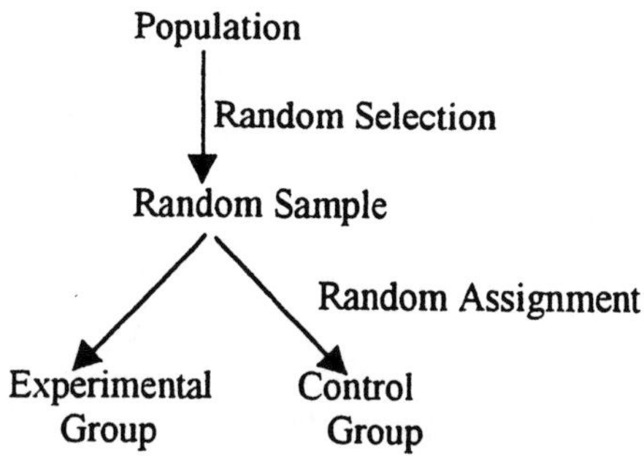

Although having both random selection and random assignment best captures the ideal of randomization, it is also possible to have random selection with non-random assignment to treatment conditions as well as nonrandom selection with random assignment to treatment conditions. The point, however, is that these are two different aspects of randomization.

Replication

Finally, something briefly must be said about replication and how it relates to the logic of experimentation. Throughout the earlier chapters of this book, it has been repeatedly stressed that the objectivity of science is partially ensured by the property that an experiment performed by one scientist can be replicated by another scientist. Science is a social enterprise and scientific objectivity results (in part) because of this feature of cooperation.

In order to have inter-subjective scientific knowledge, it must be possible to replicate an experiment and to obtain the same results as the first experiment did. If it is not possible to reproduce the empirical test performed by another scientist and to obtain the same results, this casts suspicion on the authenticity of the initial results. Conversely, replication can increase one's confidence in an experimental result. By reproducing the results of others, we corroborate their findings. It is partly because of this need for replication that the design of one's experiment must be explicitly specified, so that others can repeat it.

Because of this ever-present need to replicate the results of others, experiments are preferable to observations. Replication presupposes being able to control the independent variable and to produce it at will. Hence, replicability presupposes experimental control. In an observation, however, one cannot replicate at will. Suppose, for example, one wanted to corroborate the observations reported by an astronomer concerning Halley's Comet. One would have to wait years for Halley's Comet to return. In an experiment, however, the scientist is in control of the independent variable and can produce it whenever she desires.

Exercise 13.2

In the following passages, identify the experimental group, the control group, the independent variable, and the dependent variable. If there are any obvious extraneous variables not controlled for, indicate these. Numbers are provided in some examples, indicating either an observation or experiment was performed.

1. In the second decade of the twentieth century a disease called "pellagra" had become epidemic in the southern states. The disease was also endemic (i.e., its occurrence was confined to some localities and did not occur in others). One village might be affected; a neighboring one not. Pellagra is characterized by gastric disturbances, skin eruptions, and nervous derangement. It was the opinion of the world of medicine that the disease was caused by an unknown microbe. The U. S. Public Health Service decided to send Dr. Joseph Goldberger to Mississippi to try to find a way of eliminating the disease.

In tackling the problem of the cause of pellagra, [1] Goldberg *observed* the fact that the sufferers were always in contact with other victims. Since this is the consequence that would be expected if the disease was caused by microbes, he adopted the tentative hypothesis that the disease was caused in this manner. He then reasoned that if microbes were the cause, persons in close contact with the victims should catch the disease. But in the first hospital he visited, [2] he found that the orderlies, nurses, and doctors who were in close contact with the victims never caught the disease. This led him to discard the hypothesis that the disease was caused by microbes. He now needed another hypothesis, so he continued his *observations*. [3] He found that only the poor suffered from the disease. [4] He also noted that the victims' diet consisted of corn-meal mush, hominy grits, and similar foods, but practically no milk or fresh meat. [5] On one occasion, while visiting an orphanage, he found that only the children in the six to twelve age group had the disease. In investigating this further, he found that the children under six were given milk and the children over twelve were fed meat. The orphanage's funds were limited, so they skimped on the food of the six to twelve year old group, who were too young to work (and so did not "deserve" meat) and who were beyond the "baby" stage, the stage when milk was regarded as essential. This fact suggested to Goldberger the hypothesis that a diet lacking in milk and fresh meat was the cause of the disease. He then reasoned out the consequences of this hypothesis, namely, that remedying these deficiencies in diet would cure the disease. [6] He secured a sufficient amount of milk and fresh meat for the children at the orphanage. The result: All cases of pellagra disappeared. The cause of pellagra had been found.

2. Benjamin Franklin, it will be remembered, became convinced that lightning is nothing more than electricity passing from one cloud to another. This hypothesis was suggested to him by a consideration of the many respects in which lightning resembled electricity. Lightning sets things on fire; so does electricity. Lightning melts metals, kills people, causes blindness; so does electricity. Lightning is attracted to pointed bodies, e.g., spires, trees, mountaintops; so is electricity. His famous *experiment* with the kite was designed to confirm a prediction based on the hypothesis suggested by these many analogies. If lightning is electricity it ought to be possible to get a spark from it and to charge a Leyden jar. So he sent a kite up in a storm. To the end of the kite string he tied a key. To the key he tied a silk ribbon by means of which he controlled the kite. When the string had been dampened by the rain, sparks were drawn from the key, and a Leyden jar was charged by being attached to the key. Franklin thus believed his hypothesis to be confirmed.

3. Except from Pasteur's *Memoirs*:

> "We have now to test one of the principal foundations of the doctrine of spontaneous generation as formulated in this country. With this view, I place before my friend and co-inquirer two liquids which have been kept for six months in one of our sealed chambers, exposed to optically pure air. The one is a mineral solution containing in proper proportions all the substances which enter into the composition of bacteria, the other is an infusion of turnip--it might be any one of a hundred other infusions, animal or vegetable. Both liquids are as clear as distilled water, and there is no trace of life in either of them. They are, in fact, completely sterilized. A mutton-chop, over which a little water has been poured to keep its juices from drying up, has lain for three days upon a plate in our warm room. It smells offensively. Placing a drop of the fetid mutton-juice under a microscope, it is found swarming with the bacteria of putrefaction. With a speck of the swarming liquid I inoculate the clear mineral solution and the clear turnip infusion, as a surgeon might inoculate an infant with vaccine lymph. In four-and-twenty hours the transparent liquids have become turbid throughout, and instead of being barren as at first, are teeming with life. The experiment may be repeated a thousand times with the same invariable result. To the naked eye the liquids at the beginning were alike, being both equally muddy. Instead of putrid mutton-juice, we might take as a source of infection any one of a hundred other putrid liquids, animal or vegetable. So long as the liquid contains living bacteria, a speck of it communicated either to the clear mineral solution, or to the clear turnip infusion, produces twenty-four hours the effect here described. . .
>
> We now vary the experiment thus:--Opening the back-door of another closed chamber which has contained for months the pure mineral solution and the pure turnip infusion side by side, I drop

into each of them a small pinch of laboratory dust. The effect here is tardier than when the speck of putrid liquid was employed. In three days, however, after its infection with the dust, the turnip infusion is muddy, and swarming as before with bacteria. But what about the mineral solution which, in our first experiment, behaved in a manner indistinguishable from the turnip juice? At the end of three days there is not a bacterium to be found in it. At the end of three weeks it is equally innocent of bacterial life. We may repeat the experiment with the solution and the infusion a hundred times with the same invariable result. Always in the case of the latter the sowing of the atmospheric dust yields a crop of bacteria--never in the former does the dry germinal matter kindle into active life. What is the inference which the reflecting mind must draw from this experiment? Is it not clear as day that while both liquids are able to feed the bacteria and to enable them to increase and multiple, *after they have been once fully developed*, only one of the liquids is able to develop into active bacteria the germinal dust of the air?

The inference remains, that while the one liquid is able to force the passage from the inactive to the active state, the other is not.

4. Excerpt from Pasteur's *Memoirs*:

". . .the partisans of the doctrine of spontaneous generation hasten on to a quite reasonable deduction. If the smallest quantity of ordinary air suffices for the production of organisms in any medium whatsoever, and if these organisms do not originate spontaneously, then it follows necessarily that in this indefinitely small portion of common air there must be the terms of a multitude of different organisms. Finally, if things are thus, then ordinary air must, to use M. Pouchet's expressions, be clogged with organic matter, forming a dense fog.

But, here then is a serious and apparently well-founded difficult. . .Is it true, as has been supposed, that there is in the earth's atmosphere a continuity of cause of the generations said to be spontaneous? *Is it certain that the smallest quantity of ordinary air is sufficient to produce organized life in all infusions?*

The following experiments answer all these questions.

In a series of flasks of 250-cubic-centimeter capacity I place the same putrescible liquid (yeast water; the same with sugar, urine, etc.), the [fermentable] liquid occupying about a third of the total volume. I taper the necks of the flasks with a glassblowers' lamp [into the form of an S tube--hence the name "swan-neck flask"], I boil the liquid, and I seal off the tapered ends while the boiling is going on. A vacuum is thus formed in the flasks [since the steam has driven out all the air]. Then I break off the sealed tips of the flasks in a specified location. Ordinary air rushes in violently, carrying with it all the dusts that it holds in suspension and all the known or unknown principles associated with it. Then, barely touching them with a flame, I immediately reseal the flasks, and I place them in an incubator of 25° or 30°--that is to say, at the most favorable temperature for the development of animalcules and Mucors.

Here are the results of these experiments. . .

Generally the liquid changes after a very few days, and a great variety of organisms can be seen growing in the flasks..."

5. From Ivan Pavlov's *Conditioned Reflexes*:

"The dog used in the following experiment has been operated upon as described previously. It can be seen that as long as no special [external] stimulus is applied the salivary glands remain quite inactive [and do not secrete]. . .

Food is shown to the animal. The salivary secretion begins. . .[Food is followed by the sounds of a beating metronome. Then no food is presented.] [W]hen the sounds from a beating metronome are allowed to fall upon the ear, a salivary secretion begins after 9 seconds, and in the course of 45 seconds eleven drops have been secreted. The activity of the salivary gland has thus been called into by play by impulses of sound. . . .

--Ivan Pavlov, *Conditioned Reflexes: an Investigation of the Physiological Activity of the Cerebral Cortex.*

6. Avogadro (and some of his contemporaries) was impressed by the quantitative relationships that had been found to hold when *gaseous* elements combine. Here, we are considering not weight but volumes of gases, let it be carefully noted. One illustration will suffice. If a mixture of hydrogen gas and oxygen gas is exploded by a spark, the following relationship is found to hold: 1 volume of oxygen + 2 volumes of hydrogen = 2 volumes of water vapor. The relation between volumes is very simple: 1 to 2 to 2. Other gaseous elements were likewise found to combine in a volume relationship expressed by small whole numbers.

7. An experiment by Greenspoon on the reinforcement of a response (1950) provides another clear example of direct, automatic, or, in other words, unconscious learning. He had his subjects sit facing away from him so that they could not see him. He asked them to say all the words they could think if, pronouncing them individually without using any sentences or phrases, and he recorded their responses on a tape recorder. The response he was reinforcing was that of saying plural nouns; he did they by saying "Mmmm-hmmm" immediately after the subject said a plural noun. In this case the response was a highly generalized part of language habits and the reinforcing value of the stimulus "Mmmm-hmmm" must have been acquired as a part of social learning.

Greenspoon found that during the "training" period the experimental group, to whom he said "Mmmm-hmmm" after each plural noun, greatly increased the percentage of nouns spoken, while the control group, to whom nothing was said after plural nouns, showed no such increase. Furthermore this happened with subjects who on subsequent questioning showed that they had no idea what the purpose of the "Mmmm-hmmm" was and were completely unaware of the fact that they were increasing their percentage of plural nouns. This clearly demonstrates that the effect of a reinforcement can be entirely unconscious and automatic.

—John Dollard & Neal Miller, *Personality and Psychotherapy*

8. We shall begin our consideration of this material with a fairly detailed examination of one of the more striking studies, that by Heilig and Hoff. This concerns the induction of herpetic blisters (cold-sores) by means of suggested hallucinations. Three psychopathic women were employed as subjects. The typical experimental procedure was as follows: The experimenter recalled to the subject's mind when in deep hypnotic state an extremely unpleasant emotional experience connected with her particular neurosis. Thereupon the patient showed signs of great excitement, such as flushing, tossing about, and groaning in fear. At this juncture the psychiatrist stroked the patient's lower lip and suggested a feeling of itchiness such as the patient had often felt when a cold-sore was starting. Calming suggestions were then given, after which the patient was dehypnotized. During the following 24 hours or so the patient reported an itching on the lower lip, at the end of which time there appeared a slight swelling on the lip in question. At the end of 48 hours numerous small herpetic blisters appeared, which gradually merged into a large blister. After a few days this dried up forming a scab, and finally healed. The contents of the flesh blister, when transmitted to the cornea of a rabbit, caused a herpes in the rabbit on the third day, thus providing the technical genuineness of the cold-sore. This picture, with unimportant variations, holds for all three patients.

—Clark L. Hull, *Hypnosis and Suggestibility*

Types of Experiments

After laying out the basic framework, What makes an experiment distinctive from other forms of empirical inquiry?, we are now in a position to describe several of the most important types of experiments. In order to understand the logical and epistemological advantages and disadvantages of the various types of experimental design and to be able to evaluate them accordingly, one must be aware of the crucial differences between the various types. It must be pointed out, of course, that we are merely sampling several of the most frequently used experimental designs; we certainly are not attempting a definitive list. In fact, each of the sciences will employ designs appropriate to their own discipline. Many of these designs will not be found in (and will not be appropriate for) other sciences. What is being attempted, however, is an overview of the most important designs to be found in the sciences.

In the following discussion, we will symbolically represent these various designs by use of the following symbolic abbreviations: *O* stands for an observation (in the broadest sense of that term); O_a, O_b, etc., stand for various observations over time; *X* stands for the introduction of the experimental variable; X_1 stands for one value of this variable, X_2 for another value, etc. X with a circle around it indicates it is an "independent variable" not under experimental manipulation and control but a supposed causal factor. *Y* represents a second independent variable. *R* represents randomization, i.e., random assignment of subjects to treatment conditions; a *dotted line* represents there is no random assignment of subjects; *M* stands for matching subjects. The following table is a schematic summary of these various designs:

Table 13.2. Experimental Designs[5]

A. One-Group

1. *One-Shot Case Study*:

   X O

2. *One Group, Before and After*:

   O_a X O_b

3. *Time-Series (No Control Group)*:

   O_a O_b X O_c O_d

4. *Equivalent Time-Samples*:

   X_1O_a X_2O_b X_1O_c X_2O_d

B. Involving Two Groups

5. *Non-Equivalent Control Group*:

   X O

   O

6. *Experimental Group - Control Group*:

   a. *Randomized Subjects*:

      R X O

      R - O

   b. *Matched Subjects*:

      M X O

      M - O

7. *Non-Equivalent Control Group* (Before and After):

   O_a X O_b

   O_a - O_b

8. *Before and After Control Group*:

a. *Randomized*:

$R\ O_a\ X\ O_b$

$R\ O_a\ -\ O_b$

b. *Matched*:

$M\ O_a\ X\ O_b$

$M\ O_a\ -\ O_b$

9. *Counter balanced*:

$R\ X_1\ O_a\ X_2 O_b$

$R\ X_2\ O_a\ X_1 O_b$

10. *Factorial*:

$X_1\ X_2\ X_3$

Y_1

Y_2

Y_3

11. *Multiple Time Series*:

Oa Ob X Oc Od

Oa Ob - Oc Od

R Oa Ob X Oc Od

R Oa Ob - Oc Od

M Oa Ob X Oc Od

M Oa Ob - Oc Od

Designs Involving One Group

1. One-Shot Case Study. We will begin our discussion with the simplest examples of experiments[6]--those involving no control groups at all. In the one-shot case study design, an independent variable is introduced and one then observes its effect on some dependent variable. This is called a case-study approach because it involves only *one* subject, e.g., a patient in a hospital, one sample of ore, etc. A doctor, for example, will give a particular drug to a patient and then observe the effects of this drug on the patient's subsequent symptoms. If the drug improves the patient's condition, the drug would probably be judged effective.

The following appears to be a good example of this design:

> The primitive brain, as we saw it in the planaria, served chiefly as a sensory relay--a center for receiving stimuli from the sense organs and then sending impulses down the nerve cord. This is also true of the nereis, for, if the brain is removed, the animal can still move in a co-ordinated way--and, in fact, it moves about more than usual. If it meets some obstacle, it does not withdraw and go off in a new direction but persists in its unsuccessful forward movements. This very unadaptive kind of behavior shows that in the normal nereis the brain has an important function which it did not have in flatworms--that of *inhibition* of movement in response to certain stimuli.
>
> --Ralph Buichsbaum, *Animals Without Backbones*

Here the brain is removed (the independent variable), which is followed by observation of the animal's behavior (the dependent variable).

2. One Group, Before and After. In its pure form, the single-shot case study is perhaps difficult to find in science. This is because almost invariably there occurs an observation prior to the introduction of the independent variable. If there is, then we have a somewhat different design--the one group, before and after study. In this design, an initial observation is followed by the introduction of the independent variable, which is then followed by observation of the dependent variable. This is an example of a pre-test, post-test (or before and after design). The effectiveness of the independent variable is determined by comparing the difference between the "before" and the "after." The effectiveness of a drug, for example, is determined by noting whether it removes the patient's symptoms or not.

Sometimes the initial observation (the pre-test) will be explicitly recorded by the scientist, whereas at other times it will not. For example, in the quotation mentioned above, there appears to have been an implicit observation concerning the nereis, for the scientist already knew what kinds of behavior occurred when the brain was present. Hence, many examples of what appear to be single-shot case studies can also be classified as a one group, before and after study since it seems reasonable to assume that an initial pre-test observation occurred. The situation is even more complicated since scientists have access to observations of many things encountered by the average person during the normal course of life but which are not explicitly reported in their research.

Consider the following passage:

> In 1861 Pasteur at last carried general conviction against spontaneous generation. He boiled meat broth in a flask with a very long thin neck until no bacteria were left. This was shown by the fact that he could now keep the broth in the flask for an indefinite period without change

setting in, the narrow neck admitting nothing. Then he broke off the neck and in a few hours the liquid showed micro-organisms, and the meat was in full decay.

--H. T. Pledge, *Science Since 1500*

Here we have a clear-cut case of a before and after design: the initial observation consisted of no bacteria present (after boiling). This was followed by the introduction of the independent variable (breaking off the neck), which in turn was followed by a subsequent observation of the dependent variable (appearance of micro-organisms).

The problem with this design is the problem confronting all "before and after" studies, namely **order effects**. There are, as we have seen, ways to circumvent these **practice effects** through counter-balancing and using control groups. In addition, however, there is a particular problem here that cannot be so easily resolved. When one has only *one* group—the experimental group—there is no comparison group. But, of course, there is a comparison to be made between the pre-test and the post-test. In order to be confident about one's conclusion, however, there must be no extraneous variable operating simultaneously with the independent variable that might affect the dependent variable. But between O_a and O_b several things could occur that are outside of the control of the experimenter, e.g., events may be occurring in the environment or in the body of the organism. Suppose students take a test for prejudice (O_a) followed by the introduction of a film on the contribution of Hispanics to our intellectual heritage, followed by another test for prejudice (O_b). Suppose, however, that between these two testings, the subjects read about gang warfare in East L.A., see the movie "Colors," or are robbed by a group of Hispanics. Any of these historical variables may have a decisive causal influence on the subsequent test for racial prejudice and yet the experimenter has no control over any of these events. For this reason, a control group is often advisable.

3. Time-Series (No Control Group). In a time-series study, one makes a series of observations prior to the introduction of the independent variable and then one makes another series of observations afterwards. If one imagines a one-shot case study that involved several observations, one would have a time-series. For example, a doctor may give a patient medicine and then observe the effects of this medicine a week, two weeks, three weeks, etc., later.

The crucial question underlying a choice between a one-shot case study (or one group before and after) and a time series study is this question: is the phenomenon in question likely to have a *course of development*? This will occur most often in the biological, behavioral and social sciences, where there are phenomena such as maturation, development, historical progress, etc. In the physical sciences, such phenomena are not ordinarily present; once a change has occurred as a result of introducing the independent variable, the assumption typically made is that there will be no subsequent changes in the dependent variable. But when scientists introduce a drug or a chemical into an organism, it will probably be wise not merely to make a single observation of the dependent variable but several observations over time.

Consider, for example, the following passage:

> Diet soda junkies--and bottlers--might breathe a sigh of relief on the basis of a University of Minnesota study showing that even grotesque amounts of the artificial sweetener aspartame--marketed as NutraSweet—can be consumed safely.
>
> The study by [Arthur] Leon and co-director Dr. Donald Hunninghake . . . involved 108 healthy persons aged 18 to 62. Half took daily aspartame capsules with each meal and half took a placebo, or dummy pill. Each participated for 24 weeks in the study, which was conducted from 1984 until last summer [1987].
>
> Each person in the experimental group took an amount of NutraSweet proportional to his or her weight. . . .
>
> Each participant came to the university for evaluation every three weeks. Measurements were taken of resting heart rate, blood pressure and body weight. In addition cholesterol levels were taken, as well as tests of kidney, liver and blood cell functions. Records were also kept of any symptoms they reported.
>
> ---*The Coloradoan* (May 11, 1988)

In this study, observations were taken every three weeks for six months (or eight times). Why not take just one measurement of body weight, heart rate, blood pressure, blood cholesterol, etc.? Because effects of aspartame might not show up until the second or third observation. To safeguard against this, one might use a time-series (no control group). Of course, it would be better to have—as the present study did—a control group in addition to the experimental group. In this case, one would have a time-series (with control group), also known as a Multiple Time Series

4. Equivalent Time-Samples. In our final example of an experiment involving no control group, we will discuss a design often employed in medicine, physiology, and psychology. In an equivalent time-samples design, the experimenter does not use a different group of subjects as a control group; instead, the subjects themselves serve as their own comparison group. Ordinarily in such a design, one initially observes the dependent variable in the absence of the independent variable in order to determine a kind of natural *baseline* (normal frequency of occurrence) of this dependent variable. Then one introduces the independent variable, and its effect on the dependent variable is noted. Next, the independent variable is removed, and it is determined whether its absence results in the dependent variable returning to its original form--the baseline.

In operant conditioning experiments, pigeons are tested in a soundproof chamber. These pigeons can peck a key, which will deliver food reward. There are three phases in this experiment. First, a base rate is determined, that is to say, the feeder is disconnected and no food follows a key peck. It is then determined what rate of key-pecking occurs when no reward follows. In the second phase, food follows key-pecking. Over many weeks, the frequency or rate of key-pecking is recorded. Then in the third phase, the feeder is again disconnected and the rate of key-pecking is determined. In the first period, there is virtually no key-pecking; in the second period, there is a dramatic increase in the frequency of key-pecking; and in the third phase, there is a decline in the frequency.

Designs Involving Two Groups

5. Non-Equivalent Control Group. It often happens that researchers conduct an experiment in which there is an experimental group and a control group but these groups are not randomly assigned. Instead, the experimenter takes the groups as they are--already formed. For example, in an educational study one might use two different school systems, say, one from Loveland and one from Ft. Collins. Or in a study of the influence of diet on running speed, one might use a group of rats raised in Boston as the experimental group and a group of rats raised in San Francisco as the control group. Whenever one uses "intact" groups—groups constituted not by the experimenter but by nature—one is using non-equivalent control groups.

In a non-equivalent control group design, we find for the first time a control group. But the experimental group and the control group are not randomized, i.e., subjects are not randomly assigned to two treatment conditions. Instead, the two groups are taken "as is." The experimental group might be a second grade classroom from one school district, whereas the control group might be a third grade classroom from a different school district. One group functions as a control group, because it does not receive the treatment condition, whereas the other group is the experimental group. Because we have such groups, we can compare the observations made on the two groups and note any differences between them—presumably due to the introduction of the independent variable.

Consider the following passage:

> A Swiss investigator, Dr. G. Ritzel, reported the results of a double-blind study.[7] He studied 279 skiers, half of whom received 1 gram of ascorbic acid [vitamin C] per day and the other half an identical inert placebo. He reported a reduction of 61 percent in the number of days of illness from upper respiratory infections in the vitamin C group as compared with the placebo group.
>
> --R. Giere, *Understanding Scientific Reasoning*

In this study, two groups were used, one of which received the treatment condition and one of which was a control group. The key question is whether the two groups of skiers were "naturally constituted" groups or whether they were formed by random assignment. The text does not tell us which. Suppose, for example, Dr. Ritzel used two groups that were already intact; e.g., a group of skiers from Aspen, Colorado were given Vitamin C and a group of skiers from Killington, Vermont were given the placebo. These two groups would be *non-equivalent* because they were not randomly assigned to treatment and control conditions.

Might there be an extraneous variable present in the one group but not in the other that would confound the effect of Vitamin C on "colds"? Perhaps the difference in the dryness of the air in Colorado compared to Vermont makes a difference to the incidence of colds, with skiers from Vermont getting more colds than skiers from Colorado. Of course, one does not know this, but it is a possibility that needs to be considered. We do not know if the groups are *initially equivalent* on all extraneous variables or not. Hence, the certainty of our conclusion will be less than optimal.

6a. Experimental Group - Control Group. (Randomized Subjects) In this design, we have an experiment in its strictest sense. Here, we have two groups--the experimental group E and the control group C--members of which are randomly assigned. E is then given the treatment condition and C receives no treatment. Then both groups are tested to determine the effect of the independent variable on the dependent variable.

In our example of Dr. Ritzel and Vitamin C, we would have an Experimental Group - Control Group design with randomized subjects if Dr. Ritzel randomly assigned the 279 skiers to the two groups.

6b.Experimental Group - Control Group. (Matched Subjects). Although the randomized group design is one that many scientific researchers prefer--since it is an experiment that is epistemologically purest--for many other situations other designs are preferable. One of these widely used designs is the matched group design. In this design, individuals are not randomly assigned to different treatment conditions. It is desirable, nonetheless, to make them "equal." How might we do this? By matching.

The logic behind the matched groups design is the following: suppose we have two "intact" groups, one of which will be the experimental group and the other the control group. Since these are naturally constituted groups, there may be an extraneous variable present in, say, the experimental group and not present in the control group. A difference between these groups on the dependent variable, therefore, may not be due to the introduction of the experimental variable but rather to this extraneous variable. We would have a case of confounded variables, and any conclusion we would draw about the causal efficacy of our independent variable would be hazardous. But suppose there were some way to measure (or test) for the presence of such an extraneous variable before introducing the treatment condition. (It is important to note that this alleged extraneous variable should be correlated with the dependent variable.) Then suppose we "match" the two groups on the variable—sometimes called a "matching variable." Suppose, for example, after giving a test to all of our subjects, we find that two individuals A and B have scores of 100. Here is our first matched pair: A-B. Suppose we find next that C and D have identical scores of 120—our second matched pair. Suppose we continue with this process until we have a series of matched pairs:

A-B (matched pair #1)
C-D (matched pair #2)
E-F (matched pair #3)

Next, we randomly assign one member of each pair to the experimental treatment condition and the other twin to the control condition; e.g., A, D, E, etc, might be assigned to the experimental group and B, C, F, etc., to the control group. The experimental group then receives the treatment condition and the control group receives no treatment. After measuring any effect on the dependent variable, say, test A, one then determines if there is any difference. What we have, therefore, is the following:

	Pre-test	Treat-ment	Post-test
Control Group	B	--	A
Experimental Group	B	E	A
			Difference?

The assumption here is that the matched pairs are originally equal (as determined by the pre-test B). Any difference between them on the post-test A must be due to the effect of the treatment condition.

If there are several variables that might affect scores on the dependent variable, then subjects can be matched on them; e.g., one can match subjects on age, intelligence, socio-economic background, etc. For example, matched pair #1 might consist of two sophomore males, 20 years old, with an I.Q. of 120; matched pair #2 might be two senior females, 23 years old, with an I.Q. of 140, etc. Obviously, the situation grows more complex as we include more and more correlated variables, but the underlying rationale is the same: if a variable is thought to influence the dependent variable and hence to be correlated with it, one can control for it by matching subjects.

How would matching subjects work in the case of our Vitamin C example? Suppose Dr. Ritzel first gave his 279 subjects from Vermont and Colorado a preliminary test for proneness to colds. Such a test would conceivably consist of a blood sample and then a chemical analysis. Suppose such a test was available and that subjects were classified on a scale of 1 to 10 for their "proneness to colds." Dr. Ritzel would then take individuals with the same scores on this test and match them. Then he would randomly assign one "twin" to the experimental group and the other "twin" to the control group, introduce the independent variable, and then give them a post-test for "colds".

7. Non-Equivalent Control Group (Before and After)

Suppose we wanted to combine the above design involving experimental and control groups with a before and after type study. If one did this one would have a before and after study involving both an experimental group and a control group. However, if one did not randomly assign subjects (or attempt to match them), one would have a non-equivalent control group design.

In Dr. Ritzel's study, we would have an example of a non-equivalent control group (before and after) if he began with two intact groups--Aspen skiers and Killington skiers--and then observed whether they had a cold or not. Then the independent variable would be introduced and a second observation or test would be made.

8. Before and After Control Group. In the before and after control group design, one uses two groups, each of which receives a pre-test and a post-test. Subjects are either randomly assigned to the two groups or they are matched. If they are randomly assigned or matched, they cease to be non-equivalent.

The randomized version of the before and after study should be obvious. In the matched subjects version, everything is the same except that subjects are first tested on some variable thought to be correlated with the dependent variable and then, on the basis of this pre-test, matched and then randomly assigned to the two groups.

Consider the following passage:

> In a study done at the Common Cold Research Unit (in Salisbury, England), 91 human volunteers were divided into two groups. 47 subjects received 3 gm. of absorbic acid per day for 9 days and 44 subjects received a placebo. They were all inoculated with cold viruses on the third day. In each of the two groups, 18 developed colds. The incidence of colds observed in the subjects receiving ascorbic acid (18/47) was 6 percent less than in the control group (18/44).
>
> --Ron Giere, *Understanding Scientific Reasoning*

What was the design employed in this study? We have two groups, an experimental group and a control group. Were the two groups formed randomly? If they had been, we would have the following design--Experimental Group - Control Group (Randomized):

R X O
R -- O

Was there an initial pre-test performed? Although none is mentioned, the researchers must have determined if the subjects already had a cold or not before giving them the treatment condition. In that case, we would have

R Oa X Ob
R Oa - Ob

9. Counterbalanced Designs. Because of the presence of order effects, scientists often employ a counterbalanced design. In a counterbalanced design, the order of presentation of treatment conditions is varied so that the order of presentation will have no effect on the dependent variable. If, for example, we wish to investigate the teaching effectiveness of two methods of instruction--a straight lecture style (A) versus discussion method (B)--then one group would receive treatment A followed by treatment B, whereas the other group would receive B followed by A. If there were any practice effects--students improving as a result of any teaching method--then this would be present under *both* sequences B-A and A-B. As a result, these order effects would be a "constant" which we could ignore.

10. Factorial Designs. In the last chapter, we discussed John Stuart Mill's *method of concomitant variation* in which numerical degrees of the independent variable are related to numerical degrees of the dependent variable. Unlike other designs in which the variable is either present or absent, there are several values of the treatment condition. For example, instead of merely investigating the effect of a drug on the learning ability of students, one might use three levels or dosages, say, 100 grams, 200 grams, and 300 grams. In such a case, we would have the following:

Control Group	No Treatment	Test
Experimental Group	X Amount of Treatment	Test
	2X Amount of Treatment	Test
	3X Amount of Treatment	Test
		Difference?

One might naturally expect that the greater the amount of the smart pill, the higher the learning score. The following might then be the case:

	Test Score
No Drug	100
100 g. Drug	120
200 g. Drug	140
300 g. Drug	160

Here, we would have a nice and neat linear relation between the amount of the independent variable and the amount of the dependent variable. But suppose someone asked the question: How does level of motivation influence the efficacy of this new wonder drug? Does it make any difference on the influence of the drug that an individual is highly motivated to learn as opposed to not being motivated at all? Of course in a separate experiment, one could investigate the effect of motivation on learning, but what is wanted is an answer to the question: What are the joint effects of motivation *and* the drug? To answer this question one would have to investigate the effects of *two* independent variables *simultaneously*. Experiments that attempt to investigate more than one value of the independent variable (or several independent variables) at a time are called **factorial experiments**.

One might perform a factorial experiment involving, say, three quantities of the drug—100 gram, 200 gram, 300 gram—and three levels of motivation—high, medium and low. We could represent our factorial situation by means of the following matrix (or 2 dimensional cube):

Level of Motivation

		High	Middle	Low
Amount of Drug	100 g 200g 300g			

This is known as a 3 x 3 factorial design.

As our diagram indicates, there are nine different experimental groups: 100s gram with high motivation, 100 grams with middle motivation, etc. In a factorial design, we test each of our experimental groups, in this case nine, on the dependent variable, e.g., scores on a test. We could then represent the scores as falling inside each of the cells:

Level of Motivation

		High	Middle	Low
Amount of Drug	100 g	120	110	100
	200g	130	120	110
	300g	140	130	120

What conclusions should be drawn from this? To answer this question, inferential statistics (Chapter 14) involving the analysis of variance is required. Unfortunately, this topic is too advanced for our purposes and must be skipped here.

As the reader can guess, factorial designs can become extremely complex: not only are there 3 x 3 factorial designs, one can also have a 4 x 5, 3 x 5, etc. design. Moreover, one can also introduce a third independent variable, e.g., teaching technique, to obtain a three-way factorial design involving, say, three levels of motivation, three levels of drug dosage, and two levels of teaching technique (e.g., active versus passive learning). In this case, we would have a 3 x 3 x 2 design. In principle, it is possible to investigate any number of independent variables, although in practice most studies do not use more than three independent variables. The reason, of course, is that it is difficult to know how to interpret the meaning of the results when one has so many variables.

Interaction

Factorial designs allow for the possibility of investigating the effects of several values of several independent variables. In this respect, they are much truer to life than a simple experiment investigating the effect of the absence versus the presence of one independent variable.

But one issue that should at least be briefly mentioned concerns a problem peculiar to factorial designs--the problem of interaction. The effect of one independent variable may not be uniform in its operation. If one is investigating two independent variables, say, X_1 and X_2, not only might X_1 have an effect on the dependent variable Y, and not only may X_2 have an effect on Y, but one must also consider how X_1 and X_2 interact. A common example of the interaction of two variables is the case of taking medicine and drinking alcohol: the effect of doing both together is different than the effect of either singly.

Suppose, for example, we are investigating the effect of our smart pill on intelligence test scores. One might naturally expect that the greater the amount of the smart pill, the higher one's IQ scores. In such a case, we would have a linear relation:

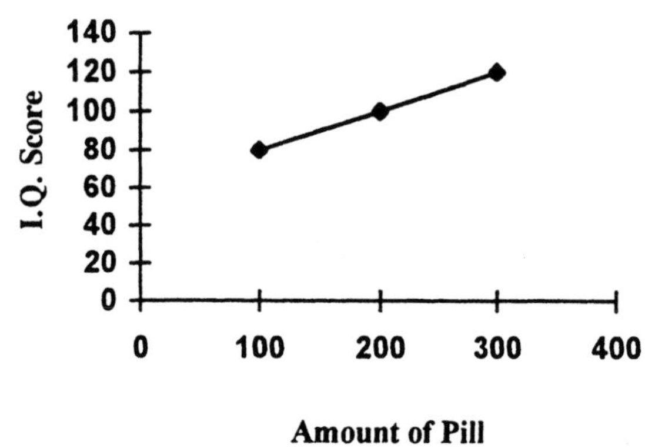

Suppose now we also investigate how the effect of this smart pill relates to the other independent variable—level of motivation. Suppose we have a 3 x 2 factorial design (3 levels of the drug and 2 levels of motivation).

		Level of Motivation	
		□High	◊Low
Amount of Drug	100 g	135	80
	200g	110	90
	300g	80	120

Here we have an interaction. We cannot say that the greater the amount of the drug, the greater the IQ score; rather, the effect of the drug depends upon the amount of motivation present. With 300 grams of the drug, the greatest effect is with low motivation not high motivation; with high motivation, the greatest effect is with only 100 grams, and so forth.

If one adds another variable X_3, then the possible interactions increase severalfold: X_1-X_2, X_2-X_3, X_1-X_3, X_1-X_2-X_3, etc. Conse-

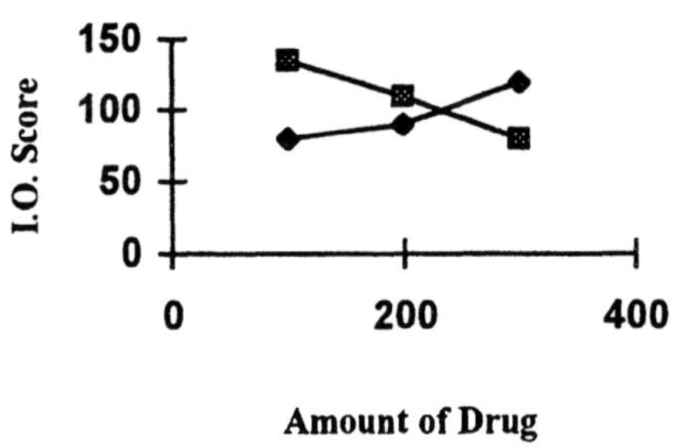

quently, if one investigates several independent variables, it becomes progressively more difficult to be certain one really knows what the effect of each is. This does not mean that one cannot engage in such research, known as *multivariate analysis*, in which several independent variables are manipulated and/or several dependent variables; it only means the scientist has a much greater responsibility to carefully control the various factors and to design his experiment with extreme care. In short, a multivariate experiment is much more difficult and the conclusions are often much more tentative than when one uses a univariate design.[8]

The above list of designs represents only a small sample of the possible ones used in science. Rather than attempt to be comprehensive, however, I prefer to list only those most widely used. Others are certainly possible, however. For example, one can combine a time-series design with a before and after study to get a *multiple time-series* design:

Oa Ob X Oc Od

Oa Ob - Oc Od

Furthermore, one can randomly assign subjects to the above treatment conditions to get a *randomized multiple time-series:*

R Oa Ob X Oc Od

R Oa Ob - Oc Od

On the other hand, instead of randomizing subjects, one can match them to produce a *matched multiple time-series:*

M Oa Ob X Oc Od

M Oa Ob - Oc Od

The point being made here is that the number of possible designs is almost unlimited.

Prospective Designs

Up to now we have been discussing experimental designs, that is to say, designs involving experiments. It is characteristic of such studies that the experimenter have control over the independent variable and the extraneous variables. Often, however, the scientist will not have such control (or will have less than complete control). This is true in an observational study, which is not an experiment in the strict sense of that word. But between observational studies and real experiments there is another class termed **quasi-experiments**, and one very important kind of quasi-experiment is a retrospective design. Since this is an important kind of empirical study, especially in medicine and sociology, we will briefly discuss it. But before doing so, we must first discuss a class of studies termed **prospective studies.**

In a prospective study, one examines the future effects of what one thinks is a causal factor. For example, scientists may be interested in tracing the future course of the disease AIDS in a patient. AIDS is the cause of a variety of biological effects one may wish to study for several years. But clearly one will not have control over the variable. One will not, for example, randomly select a group of human beings and inject them with AIDS. One can do this with animals, of course, but in addition to ethical issues involved, one may have theoretical questions concerning how similar animals are to humans and how useful such information will be for human medicine.

Suppose, then, one wished to study the effects of AIDS. What would one do? One solution, a prospective study, is to find a group of individuals who already have AIDS and then to compare the future effects of this virus with that of a control group that does not have AIDS. Although AIDS would be the independent variable, the experimenter would not have control over it. This could be represented as:

(X) Oa Ob Oc

-- Oa Ob Oc

where (X) indicates that we do not have control over this variable.

It is clear that a prospective design is an *observation* and not a true experiment: one must wait for nature to produce a particular instance of the independent variable in question. One must, in other words, look around for a case of it instead of intervening in nature and actively producing one.

Retrospective (Ex Post Facto) Designs

In a prospective design, one begins with the presumed cause and looks for its effects in the future (*pro*-spective). But one can also begin with the effects of some cause and look *backwards* to its earlier cause. In this case one has a *retro*-spective design, also sometimes called an *ex post facto* (after the fact) design.

In a retrospective design, one begins with some phenomenon and one attempts to discern its antecedent cause. For example, a scientist may begin with an observation of cancer in a group of patients. One may then look backward to determine what might have caused this cancer. Not surprisingly, researchers are apt to suggest that these cancerous patients smoked cigarettes earlier and hence that cigarette smoking was the cause of this cancer.

Suppose it was discovered that the cancerous patients did indeed smoke cigarettes. All we would have so far would be a correlation between smoking cigarettes and cancer. However, several other factors or variables are also correlated with cancer: sex, age, exercise habits, diet, type of environment, etc. For example, males get cancer more often than females do, older individuals get cancer more frequently than younger people do, individuals who do not regularly exercise get cancer more frequently than individuals who do, etc. It might easily turn out, therefore, that in our "experimental group" of smokers who have cancer, they are predominantly older males who do not exercise, who eat certain kinds of food, who live in polluted cities, etc. How can these variables be ruled out as possible causes of cancer? Clearly we cannot perform an experiment to determine if smoking cigarettes is a cause. We might of course do a prospective study and observe the future course of patients that begin smoking. But this would take perhaps decades. There is an easier method to use--a retrospective study. But first we will need a comparison or control group. How can we find one?

A control group is a group that does not receive (or, in this case, does not already have) the independent variable--in the present case, a group that does not smoke. Suppose we look for a group of earlier non-smokers and then determine the frequency of subsequent cancer in this group compared with the frequency of cancer in our smoking patients. One group would have the independent variable and the other group would not. Hence, the essential feature of a control group would be met.

But another feature of a control group would be absent, namely, one would not be controlling for extraneous variables--all the other factors correlated with cancer. What can be done to control for these? One obvious answer is to *match* the control group and the experimental group on these extraneous variables. Thus, we should find a group of individuals who are as similar to our experimental group as possible except for the absence of smoking. Thus, we should find a control group of older males, who do not exercise, who have the same dietary patterns as our experimental group, who live in polluted environments, etc. Now, after we match both groups on these extraneous variables, we can determine whether our smokers still contract cancer more than do our non-smokers. If so, then we can feel more confident that smoking causes cancer.

The essence of a retrospective design, therefore, could be represented in the following way:

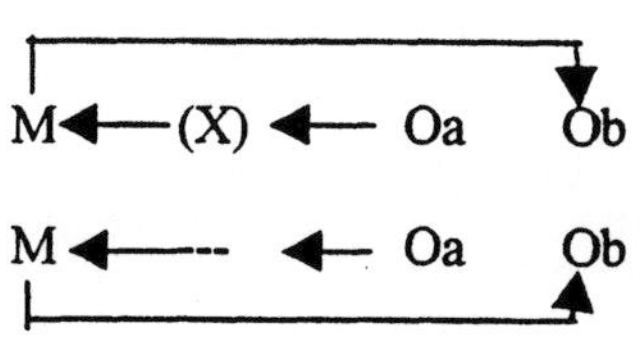

Exercise 13.3

In the following passages, identify the experimental design in question. Point out how the example in question illustrates the particular experimental design.

1. In the second decade of the twentieth century a disease called "pellagra" had become epidemic in the southern states. The disease was also endemic (i.e., its occurrence was confined to some localities and did not occur in others). One village might be affected; a neighboring one not. Pellagra is characterized by gastric disturbances, skin eruptions, and nervous derangement. It was the opinion of the world of medicine that the disease was caused by an unknown microbe. The U. S. Public Health Service decided to send Dr. Joseph Goldberger to Mississippi to try to find a way of eliminating the disease.

In tackling the problem of the cause of pellagra, [1] Goldberg *observed* the fact that the sufferers were always in contact with other victims. Since this is the consequence that would be expected if the disease was caused by microbes, he adopted the tentative hypothesis that the disease was caused in this manner. He then reasoned that if microbes were the cause, persons in close contact with the victims should catch the disease. But in the first hospital he visited, [2] he found that the orderlies, nurses, and doctors who were in close contact with the victims never caught the disease. This led him to discard the hypothesis that the disease was caused by microbes. He now needed another hypothesis, so he continued his *observations*. [3] He found that only the poor suffered from the disease. [4] He also noted that the victims' diet consisted of corn-meal mush, hominy grits, and similar foods, but practically no milk or fresh meat. [5] On one occasion, while visiting an orphanage, he found that only the children in the six to twelve age group had the disease. In investigating this further, he found that the children under six were given milk and the children over twelve were fed meat. The orphanage's funds were limited, so they skimped on the food of the six to twelve year old group, who were too young to work (and so did not "deserve" meat) and who were beyond the "baby" stage, the stage when milk was regarded as essential. This fact suggested to Goldberger the hypothesis that a diet lacking in milk and fresh meat was the cause of disease. He then reasoned out the consequences of this hypothesis, namely, that remedying these deficiencies in diet would cure the disease. [6] He secured a sufficient amount of milk and fresh meat for the children at the orphanage. The result: All cases of pellagra disappeared. The cause of pellagra had been found.

2. Benjamin Franklin, it will be remembered, became convinced that lightning is nothing more than electricity passing from one cloud to another. This hypothesis was suggested to him by a consideration of the many respects in which lightning resembled electricity. Lightning sets things on fire; so does electricity. Lightning melts metals, kills people, causes blindness; so does electricity. Lightning is attracted to pointed bodies, e.g., spires, trees, mountaintops; so is electricity. His famous *experiment* with the kite was designed to confirm a prediction based on the hypothesis suggested by these many analogies. If lightning is electricity it ought to be possible to get a spark from it and to charge a Leyden jar. So he sent a kite up in a storm. To the end of the kite string he tied a key. To the key he tied a silk ribbon by means of which he controlled the kite. When the string had been dampened by the rain, sparks were drawn from the key, and a Leyden jar was charged by being attached to the key. Franklin thus believed his hypothesis to be confirmed.

3. In 1772 Antoine Lavoisier wrote this note and sent it to the French Academy: "About eight days ago I discovered that sulfur in burning, far from losing weight, on the contrary gains it;. . . it is the same with phosphorus; this increase of weight arises from a prodigious quantity of air that is fixed during the combustion and combines with the vapours.

"This discovery, which I have established by experiments that I regard as decisive, has led me to think that what is observed in the combustion of sulfur and phosphorus may well take place in the case of all substances that gain in weight by combustion and calcination; . . . . "

Lavoisier's explanation is that something from the air, or the air itself, is absorbed in combustion.

His classic experiment to support his theory is:

"Mercury heated in common air produces a red material [an oxide, we would say, but a 'calx' to the chemists of the eighteenth century]. The red material weighs more than the metal from which it was formed. Therefore something has disappeared from the air and combined with the metal. The red material, the oxide or calx, is next strongly heated in an enclosed space with the sun's rays brought to a focus by a large lens or 'burning glass,' a gas is evolved, and the metal regenerated. The new gas is the 'something' which disappeared from the original air, for the amount is the same, and the calx has lost weight in the right amount. The new gas (oxygen) mixed with the residue from the first experiment yields a mixture which is identical with common air."

--from James Conant, *Science and Common Sense*

4. Avogardro (and some of his contemporaries) was impressed by the quantitative relationships that had been found to hold when *gaseous* elements combine. Here, we are considering not weight but volumes of

gases, let it be carefully noted. One illustration will suffice. If a mixture of hydrogen gas and oxygen gas is exploded by a spark, the following relationship is found to hold: 1 volume of oxygen + 2 volumes of hydrogen = 2 volumes of water vapor. The relation between volumes is very simple: 1 to 2 to 2. Other gaseous elements were likewise found to combine in a volume relationship expressed by small whole numbers.

5. For several years there had been". . . a great to-do about a cure for anthrax, invented by the horse doctor, Louvrier, in the Jura mountains in the east of France. Louvrier had cured hundreds of cows who were at death's door, said the influential men of the district; it was time that this treatment received scientific approval. . . .

Pasteur arrived there, escorted by his young assistants, and found that this miraculous cure consisted first in having several farm hands rub the sick cow violently to make her as hot as possible; then long gashes were cut in the poor beast's skin and into these cuts Louvrier poured turpentine; finally the now bellowing and deplorably maltreated cow was covered--excepting her face!—with an inch thick layer of unmentionable stuff soaked in hot vinegar. This ointment was kept on the animal--who now doubtless wished she were dead--by a cloth that covered her entire body.

Pasteur said to Louvrier: "Let us make an experiment. All cows attacked by anthrax do not die, some of them just get better by themselves; there is only one way to find out, Doctor Louvrier, whether or not it is your treatment that saves them."

So four good healthy cows were brought, and Pasteur in the presence of Louvrier and a solemn commission of farmers, shot a powerful dose of virulent anthrax microbes into the shoulder of each of these beasts; this stuff would have surely killed a sheep; it was enough to do to death a few dozen guinea-pigs. The next day Pasteur and the commission and Louvrier returned, and all the cows had large feverish swellings on their shoulders, their breath came in snorts—they were in a bad way, that was very evident.

"Now, Doctor," said Pasteur, "choose two of these sick cows—we'll call them A and B. Given the your new cure and we'll leave cows C and D without any treatment at all." So Louvrier assaulted poor A and B with his villainous treatment. The result was a terrible blow to the sincere would-be cure of cows, for one of the cows that Louvrier treated got better--but the other perished; and one of the creatures that had got no treatment at all died—but the other got better.

"Even this experiment might have tricked us, Doctor," said Pasteur. 'If you had given your treatment to cows A and D instead of A and B-

-we all would have thought you had really found a sovereign remedy for anthrax.' "

---Paul de Kruif, *Microbe Hunters*

6. "I explained before how I was once led to study the part played by sugar in nutrition and to investigate the mechanism by which this nutritive principle was destroyed in the organism. To solve this problem, I had to hunt for sugar in the blood and follow it into the intestinal vessels which absorbed it, until I could note the place where it disappeared. To carry out my experiment, I gave a dog sweetened milk soup; then I sacrificed the animal during digestion and found that the blood in the superhepatic vessels, which hold all the blood of the intestinal organs and the liver, contained sugar. It was quite natural and, as we say, logical to think that the sugar found in the superhepatic vessels was the same that I had given the animal in his soup. . . . However, I made a comparative experiment. . . .

So for comparison with the dog fed on sugary soup, I took another dog to which I gave meat to eat, being careful moreover to exclude all sugary or starchy material from its diet; then I sacrificed the animal during digestion and examined comparatively the blood in its superhepatic vessels. Great was my astonishment at finding that the blood of the animal which had not eaten any sugar also contained sugar.

7. An experiment by Greenspoon on the reinforcement of a response (1950) provides another clear example of direct, automatic, or, in other words, unconscious learning. He had his subjects sit facing away from him so that they could not see him. He asked them to say all the words they could think of, pronouncing them individually without using any sentences or phrases, and he recorded their responses on a tape recorder. The response he was reinforcing was that of saying plural nouns; he did they by saying "Mmmm-hmmm" immediately after the subject said a plural noun. In this case the response was a highly generalized part of language habits and the reinforcing value of the stimulus "Mmmm-hmmm" must have been acquired as a part of social learning.

Greenspoon found that during the "training" period the experimental group, to whom he said "Mmmm-hmmm" after each plural noun, greatly increased the percentage of nouns spoken, while the control group, to whom nothing was said after plural nouns, showed no such increase. Furthermore this happened with subjects who on subsequent questioning showed that they had no idea what the purpose of the "Mmmm-hmmm" was and were completely unaware of the fact that they were increasing their percentage of plural nouns. This clearly demonstrates that the effect of a reinforcement can be entirely unconscious and automatic.

---John Dollard & Neal Miller, *Personality and Psychotherapy*

8. A variety of experimental results show that alcohol produces a reduction in fear.

Conger. . . tested hungry rats in a simple approach-avoidance situation. He trained them to approach the distinctively lighted end of an alley to secure food and then threw them into an approach-avoidance conflict by giving them electric shocks at the goal. Five minutes after a control injection of water the rats would not approach the food-shock end of the alley; five minutes after one of alcohol (1.5 parts to 1,000 of body weight) they ran up to get the food.

Having determined that results could be secured in a simple situation, presumably not involving any complex habits, he performed a second experiment to determine whether the alcohol strengthened the approach habits based on hunger, weakened the approach habits based on fear, or modified both. He trained one group of rats to approach the lighted end of the alley for food and a *different* group to avoid the lighted end of the alley to escape shock. The strength of the tendencies to approach or avoid was measured by having the animal wear a little harness, temporarily restraining it, and measuring how hard it pulled. Half the animals in each group were tested when mildly drunk and half when sober. The alcohol produced little if any decrease in the pull of the hungry animals toward food and a marked decrease in the pull of the frightened animals away from the place where they had been shocked on previous trials. Since the same responses of running and pulling were involved in both groups, it seems reasonable to interpret the difference as indicating that the alcohol reduced the strength of fear without markedly affecting hunger.

--John Dollard and Neal Mill, *Personality and Psychotherapy*

9. There is a deep-red dye, well known to analytical chemists, which is used for the analysis of aluminum and a lesser extent for beryllium and which, on paper, seemed to meet our requirements. This dye is known by the trade name "aluminon," and by the chemical name aurintricarboxylic acid, or simply ATA. In the first test of ATA we injected mice with enough beryllium salt to kill them within a few days. We then injected half the animals with a small dose of ATA and left the other untreated. The results were dramatic: virtually every animal treated with ATA survived and lived on normally, while all of the untreated animals died. We have repeated this experiment with hundreds of animals of different species, with the same high degree of protection.

--Jack Schubert, "Beryllium and Berylliosis" (*Scientific American*, 1958)

10. In Nebraska, in 1976, a new state law went into effect to regulate involuntary commitment to mental hospitals. Nebraska's previous

commitment law had been ruled unconstitutional for failing to protect adequately the rights of mental patients. The new law was stricter, requiring a series of evaluations and hearings, four steps in all, which served as safeguards against the continued commitment of any citizen who was not dangerous or unable to care for himself or herself.

One group expected that the number of involuntary admissions would be reduced whereas another group claimed it would now be more difficult to provide treatment to those who really needed it. Luckey and Berman (1979) tried to determine which of these claims were true and, in general, what the effect of the new law was.

Records were available for the two years prior to the implementation of this law and for the two years subsequent to the law. These records were divided into three-week periods, each period representing one unit. There were thirty of these before and a similar number after the introduction of the law.

Luckey and Berman found that introducing the new law did precede changes in several measures, e.g., the number of involuntary admissions did drop immediately after the law took effect, but it began to rise after a short time, reaching the old level again in about 18 months.

--from Paul D. Cherulnik, *Behavioral Research*

11. In Connecticut, in 1955, there was a crackdown on speeding. After a record high of traffic fatalities in 1955, a severe crackdown on speeding was initiated. A year after the crackdown, the number of fatalities decreased from 324 to 284 and continued to decrease thereafter. The following figure illustrates this decrease.

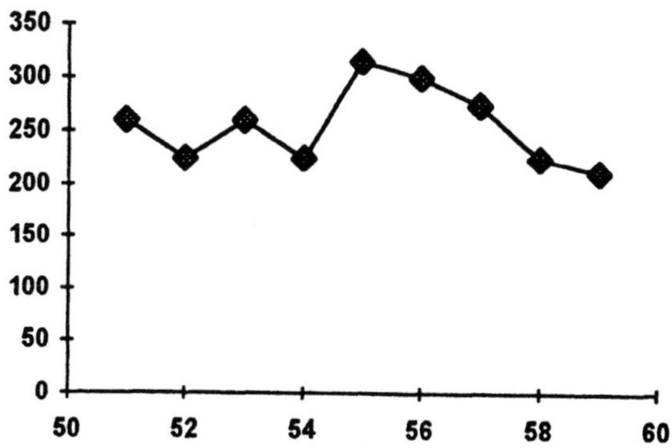

12. "One day, rabbits from the market were brought into my laboratory. They were put on the table where they urinated, and I happened to observe that their urine was clear and acid. This fact struck me, be-

cause rabbits, which are herbivore, generally have turbid and alkaline urine; while on the other hand carnivores, as we know, have clear and acid urine. This observation of acidity in the rabbits' urine gave me an idea that these animals must be in the nutritional condition of carnivore. I assumed that they had probably not eaten for a long time, and that they had been transformed by fasting, into veritable carnivorous animals, living on their own blood. Nothing was easier than to verify this preconceived idea or hypothesis by experiment. I gave the rabbits grass to eat; and a few hours later, their urine became turbid and alkaline. I then subjected them to fasting and after twenty-four hours or thirty-six hours at most, their urine again became clear and strongly acid; then after eating grass, their urine became alkaline again, etc. I then repeated this very simple experiment a great many times, and always with the same result. I then repeated it on a horse, an herbivorous animal which also had turbid and alkaline urine. I found that fasting, as in rabbits, produced prompt acidity of the urine, with such an increase in urea, that it spontaneously crystallizes at times in the cooled urine. As a result of my experiments, I thus reached the general proposition which then was still unknown, to wit, that all fasting animals feed on meat, so that herbivore then have urine like that of carnivore.

--Claude Bernard, *An Introduction to the Study of Experimental Medicine*

13. A study was conducted by Sanford and Hemphill to decide whether midshipmen who took a psychology course developed greater confidence in social situations. The second-year class was chosen to take the psychology course whereas the third-year class constituted the comparison group. The second-year class reliably increased confidence scores on a social situations questionnaire from 43.26 to 51.42, but the third-year class only increased their scores from 55.80 to 56.78.

14. A number of investigators have reported that hospitalized patients who showed high anxiety had larger secretions of hydrocortisone (one of the hormones secreted by the adrenal cortex) than matched normal subjects. In light of these reports, an experiment by Weiner, et al. (1963) was designed to see if injection of this hormone into volunteer medical students would produce increases in subjective feelings of anxiety.

Sixteen subjects were injected with one dose level of the hormone and immediately tested with the Rorschach test, as well as with tests designed to measure both immediate feelings of anxiety and persistent levels of anxiety. Three months later the same subjects returned to the laboratory, were injected with a placebo, and retested with the same tests.

Another 16 subjects were handled in the same way, except for the fact that they were injected with the placebo first and then with the hormone 3 months later.

15. A team of New York medical researchers said recently that a heavy upsurge of cigarette smoking among American women is causing a soaring increase in heart attacks. These warnings were based on a study of 194 Westchester County, New York women who died between 1967 and 1971. The researchers autopsied the women and questioned relatives about the women's cigarette smoking habits. The researchers said 41 percent of the women suffered sudden, fatal heart attacks and the other 153 women died from other causes.

Comparing the two groups, the researchers found 62% of the heart attack victims had been heavy cigarette smokers--20 or more cigarettes daily. In contrast, only 28% of the women who died from other causes had been heavy smokers.

16. Ever since Columbus took Tobacco from the New World back to the Old, people have been saying that it is an "evil weed" that causes all manner of illness and disease. But it wasn't until the 1950's that this was really shown. In one of the most important of these studies, a sample of almost 200,000 men, both smokers and nonsmokers, between the ages of 50 and 69, was used. The subjects, gathered from 394 counties in nine different states, were personally interviewed by a team of 22,000 trained volunteers. They recorded the history of each subject's smoking habits plus other information that was thought might be relevant. All subjects were in good health at the beginning of the study.

The subjects were followed for almost 4 years. During that period almost 12,000 deaths were reported, more than 2,000 from some form of cancer. It was discovered that the smokers died 2 ½ more than the nonsmokers did.

--from Ronald Giere, *Understanding Scientific Reasoning*

17. Since the mid-1960s, when the use of birth control pills became widespread, many hospitals and clinics have followed the medical histories of their patients to determine whether women who choose to use the pill experience effects different from those who choose other contraceptives or none at all. In the early 1970's it was discovered that women on the pill had a higher death rate from heart attacks than others. Typically, one found results like the following (death rate from heart attacks given in deaths per 100,000 women of ages 40-44):

women not on pill: 7
women on pill: 24

For Further Reading

Until recently, there were few philosophical discussions of the experimental method and experimental design. Recently, this has changed. Hacking (1983) was one of the first of this new genre; Gooding, Pinch & Schafer (1989) represent a good anthology on this subject.

Experimental design has been an area closely connected with statistics and statistical inference (see Chapter 14). On experimental design, see Campbell & Stanley (1963).

Notes

[1]This is the reason why many individuals manipulate only one factor at a time while keeping all others constant. More complicated experiments involve several variables--a **multi-variate** experiment.

[2]As we will see in the next chapter, inferential statistics is concerned with whether the difference in scores between two groups really is significant or not.

[3]The basis for this belief may be the principle of sufficient reason: for every thing there must be a reason that suffices to explain it completely. C cannot be a sufficient reason for E_1 if C is also a sufficient reason for E_2, for in that case given C, why should E_1 occur instead of E_2? There must be something that explains why E_1 occurs instead of E_2. But if so, this will represent a difference in the causal sources of E_1 and E_2. Hence, C cannot be the cause of both.

[4]The assumption here is that there is no **interaction** between an extraneous variable and the treatment variable. Suppose, however, there was. Then although E by itself might have a particular effect on e and D by itself might not, D and E together might have a still different effect than either D or E do singly.

[5]Many of these designs are borrowed from the classic work by Campbell and Stanley (1963), although I have modified various aspects of their discussion and introduced other designs.

[6]Several individuals prefer to label these as pre-experiments or quasi-experiments. However, since these are often found in sciences such as physics and chemistry, which are paradigm cases of experimental sciences, I have chosen to label them experiments.

[7] In studies involving drugs, scientists have found it necessary to give a **placebo**--an inert substance with no pharmacological properties--to control for the effectiveness of the drug. In a **blind study,** the subjects do not know whether they are receiving the drug or the placebo. In a **double blind study,** the re-

searchers also do not know which subjects are receiving which treatment condition, i.e., both subjects and researchers are "blind."

[8]Individuals often argue that one should investigate <u>all</u> the relevant factors at the same time. Although this advice is well worth considering, the practical problems of interpreting one's results become prohibitive in many cases. For example, suppose one is investigating the influence of six independent variables on success in school. If the effects of these variables were all linear we would not have insurmountable problems in interpreting our data. But suppose one has an extremely complex higher-order interaction involving all six variables. Then the ability to understand what interpretation to give to this interaction becomes extremely difficult. The modern statistical technique of multivariate analysis has greatly facilitated the interpretation of such data, but several problems involving the interpretation of data remain.

14

Statistical Inference

Introduction

In previous chapters we have noted several examples of scientists conducting observations, quasi-experiments, and experiments from which they drew conclusions regarding a hypothesis on the basis of evidence. For example, a typical experimental situation involves a comparison between the experimental group (E) and a control group (C), in which there is a difference between the numerical values of the dependent variable for these two groups--what we can call a *difference score*. For example, in our experimental group, we may find a 30% frequency of cancer compared to 12% frequency of cancer in our control group. The difference score of 30%-12% is attributable (we presume) to the presence versus the absence of our experimental variable (e.g., smoking), and in particular, to the amount of smoking. Once again, Mill's methods of difference and concomitant variation provide part of the underlying logical rationale for employing the experimental method.

But suppose, however, we had observed a difference of only 30% versus 29% incidence of lung cancer in our two groups, a difference score of 1%? We would have much less confidence in our conclusion. Why? According to Mill, if a variable is a causal variable, there should be a clear and obvious difference between its presence and its absence, and a difference score of 1% does not reflect the presence of the cause. After all, we know that there is always a certain amount of error present whenever we do an experiment: a certain amount of *measurement error* occurs whenever we measure something; *sampling error* occurs whenever we use a sample instead of the population from which the sample came; there is *systematic error*, etc. Hence, the 1% difference may be due to chance fluctuations in the scores of the two groups. The real difference score may be 0% but because of chance, the difference score may vary from this amount.

But what about intermediate difference scores, scores of 4% or 8%? Here things are not quite so clear. Whereas a 1% difference score is negligible (and perhaps due to chance) and a 15% difference score substantial (probably not due to chance), a difference score of 4% is equivocal: we do not know if it is really significant or just due to chance. To answer this question, the discipline of inferential statistics is of crucial importance.

In an earlier chapter (Chapter 9, which the reader may wish to review at this point), we distinguished descriptive statistics from inferential statistics. Descriptive statistics is concerned with providing numerical figures summarizing certain properties of a sample and population, e.g., mean, standard deviation, etc. By contrast, inferential statistics is concerned with the logic behind using a sample (and sample statistics) to infer certain characteristics of a population (population parameters). There are two main subdivisions of inferential statistics: statistical estimation and statistical hypothesis testing.

Given a sample of I.Q. scores with a mean of 107 and a standard deviation of 16, what may we reasonably conclude about the mean and standard deviation of the population of I.Q. scores from which our sample was drawn? In short, how can we estimate the probable population parameters? A second part of inferential statistics is concerned with the question of how one can determine whether a statistical hypothesis concerning the values of a population parameter (e.g., that the population mean of IQ scores is equal to 110) is correct or not. These two parts of inferential statistics will be the subject of this chapter. It should be pointed out that we are providing only the most basic and rudimentary introduction to inferential statistics and that the reader should consult the standard textbooks mentioned at the end of this chapter for a more complete introduction to inferential statistics.

Statistical Estimation

Ordinarily, we tend to think of sampling as involving the drawing of a single sample from a population, after which certain characteristics of the sample (e.g., a sample mean) are determined

(Figure 14.1). But we can also imagine *several* samples being drawn from a population as in figure 14.2.

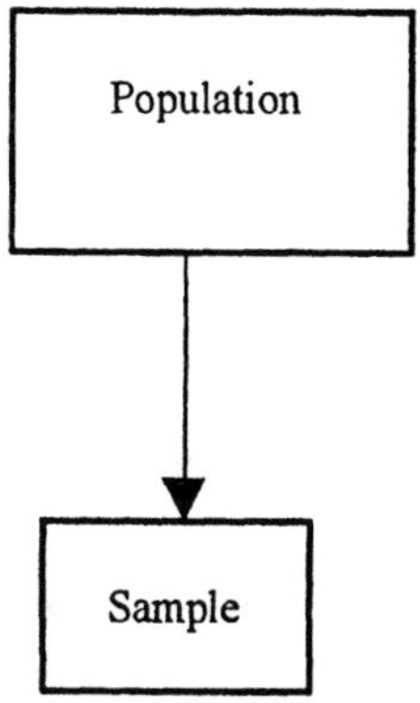

Figure 14.1 A Single Sample from a Population

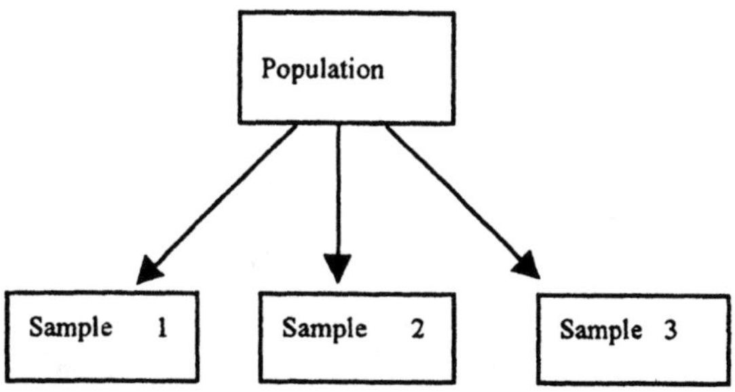

Figure 14.2. Several samples from a population

In this situation. we can imagine calculating the mean and standard deviation of each of these three samples. Suppose, for example, our population consists of just three items (the numerical observations 2, 4, 6), and suppose we draw (with replacement) three samples of the same size, say, two and record each of the three samples' mean and standard deviation. The situation might look like this (Figure 14.3). Sample 1 consists of the items 2,2, with a mean of 2, and a standard deviation of 0, etc. Notice that if we compare the three sample means (2, 3, 5) with the population mean of 4, no one of the sample means equals the population mean but the three sample means cluster around the population mean.

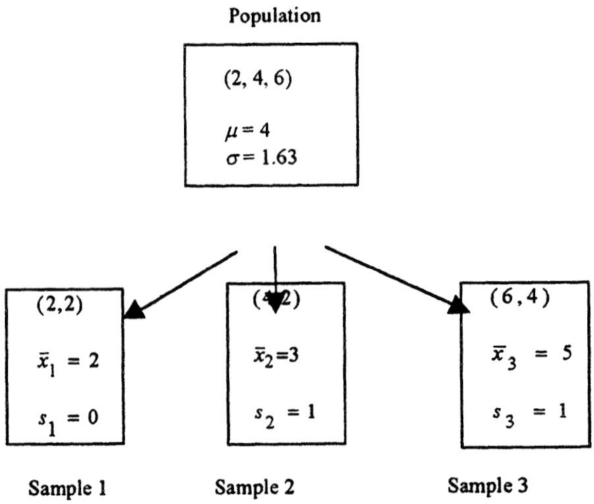

Figure 14.3. Three hypothetical samples

Although no one of the three sample means is a very good estimate of the population mean, suppose we take the average of these three means (*the mean of the sample means*, abbreviated $\mu_{\bar{x}}$):

$$\mu_{\bar{x}} = \frac{\bar{x}_1 + \bar{x}_2 + \bar{x}_3}{n} = \frac{2 + 3 + 5}{3} = 3.33$$

The average of the sample means is a better estimator of the population mean than is any one particular sample means.

The sample means also have a standard deviation (abbreviated $\sigma_{\bar{x}}$), *the standard deviation of the sample means* calculated in the ordinary way:

$$\sigma_{\bar{x}} = \sqrt{\frac{(\bar{x}_1 - \mu_{\bar{x}}) + (\bar{x}_2 - \mu_{\bar{x}}) + (\bar{x}_3 - \mu_{\bar{x}})}{3}}$$

$$= \sqrt{\frac{(2-3.33)+(3-3.33)+(5-3.33)}{3}} = 1.25$$

Here, we also notice that although the individual standard deviations of our samples are not good estimates of the population standard deviation, a better estimate is the standard deviation of the sample means.

Let us now introduce some additional concepts. Suppose we consider not a particular sample, nor a particular set of three samples of size n, but all possible samples of size n that can be drawn from the population. For example, what are all the possible samples of size two that can be drawn from our population of three scores? The answer depends on how one samples--with or without replacement--and whether one is ignoring the order in which a sample is drawn. Suppose we sample with replacement and do not ignore order. Then there are 9 possible samples of size 2 that can be drawn from our population.[1] These are listed below, together with the mean and standard deviation of each:

Sample	$\bar{x}$	s
2,2	2	0
2,4	3	1
2,6	4	2
4,2	3	1
4,4	4	0
4,6	5	1
6,2	4	2
6,4	5	1
6,6	6	0
	$\mu_{\bar{x}} = 4$	$\sigma_{\bar{x}}=1.1$

Figure 14.4 All possible samples of size 2

If we take our 9 sample means and calculate the average of these means ($\mu_{\bar{x}}$), it is the same as the population mean, 4. It turns out that this is true regardless of the sample size, as long as one takes all possible samples of that size:

> *the mean of the means of all possible samples of the same size drawn from the same population is equal to the mean of that population:* $\mu_{\bar{x}} = \mu$

At this point, it is useful to recall the concept of a random variable, which we discussed in Chapter 9: a variable is a random variable if the particular value of the variable is due to chance.

There are several important random variables in statistics, especially $\bar{x}$--the sample mean. $\bar{x}$ can take on different values: each of the possible sample means $\bar{x}_1$, $\bar{x}_2$, $\bar{x}_3$, etc. The particular value of this variable depends on chance, on which particular mean was chosen. Hence $\bar{x}$ is a random variable.

As we just saw, the mean of the random variable $\bar{x}$ (i.e., $\mu_{\bar{x}}$) equals the population mean μ. What about the standard deviation of $\bar{x}$, which is also a random variable? In each particular sample, there will be a standard deviation s. But since we are calculating the mean of these means, these means will also have a standard deviation ($\sigma_{\bar{x}}$). In Figure 14.4, $\sigma_{\bar{x}}$ was 1.1. Since σ (our population standard deviation) is 1.63, it is not the case (as it was for the relation between $\mu_{\bar{x}}$ and μ.), that $\sigma_{\bar{x}} = \sigma$. But if we divided σ by $\sqrt{n}$, they are equal:

$$\sigma_{\bar{x}} = \frac{\sigma}{\sqrt{n}} = \frac{1.63}{\sqrt{2}} = \frac{1.63}{1.4} = 1.15$$

Hence,

> *the standard deviation of the mean all possible means of size n drawn from a population is equal to the population standard deviation divided by the square root of n*: $\sigma_{\bar{x}} = \frac{\sigma}{\sqrt{n}}$.

$\sigma_{\bar{x}}$ is called the **standard error of the mean**, i.e., the standard deviation of $\sigma_{\bar{x}}$. $\sigma_{\bar{x}}$ is the amount of *sampling error* present when one estimates a population mean by a sample from the population since it represents the variance of $\bar{x}$.

Random variables such as $\bar{x}$ have a certain distribution of values. Calculating their frequency results in a probability distribution. (This probability distribution is also known as *the sampling distribution of the mean*). In the case of our random variable $\bar{x}$, whose values are shown in Figure 14.4, the probability distribution is that shown in Figure 14.5.

$\overline{x}$	Probability of $\overline{x}$
2	1/9= .11 (one of the samples has $\overline{x}$=2)
3	2/9 = .22 (two of the nine samples have $\overline{x}$=3)
4	3/9= .33 (three of the nine sample have $\overline{x}$=4)
5	2/9= .22 (two of the nine samples have $\overline{x}$=5)
6	1/9 = .11

= 1.0 (rounding off from .99)

Figure 14.5 Probability Distribution of $\bar{x}$

If we were to use all possible samples of a different size, say, three, we would get a different probability distribution, namely, one with less variance—the probability distribution would be squashed together around the mean. It turns out that *as the sample size increases, the variance decreases,* clustering more and more around the mean. This means that there is less sampling error, since there is less variance. So, to decrease sampling error and provide a more accurate estimate of the population mean, one should use a larger sample.

When one does an experiment, one draws a sample from a population—a particular sample with a particular sample mean. Suppose we want to use this sample mean to estimate the value of a population mean. We know there is sampling error surrounding our sampling mean, so our estimate of the population mean will be off somewhat. How much it is off is determined by the sampling error (the sampling distribution of the mean). If you know the sampling distribution and the population mean, one can determine how good an estimate a particular sample is. But this is often unrealistic since you don't know either one. However, even if you don't know the actual sampling distribution—because it is too difficult to calculate—one can approximate it by means of an ideal model involving the concept of a normal distribution: if our random variable x were normally distributed, we would obviously know the sampling distribution. To have a normally distributed random variable $\bar{x}$, we would need to know the mean and standard deviation of the distribution. What about the distribution of our population? Usually, of course we do not know its distribution, whether for example it is normal or not. But an important theorem—the famous *central limit theorem*—shows that it doesn't matter what the distribution of the population is.

Central limit theorem: Whether the population is normally distributed or not, if one randomly samples from any population with a sufficiently large sample size (e.g., about 30 or more), the random variable $\bar{x}$ will be normally distributed. As the size of the sample increases, the distribution of the mean of all possible samples of the same size drawn from the same population becomes more and more like a normal distribution.

In short, we may assume our random variable $\bar{x}$ has a normal distribution. Its mean will be $\mu_{\bar{x}}$, and its standard deviation will be $\sigma_{\bar{x}}$. Furthermore, as we have shown

$$\mu_{\bar{x}} = \mu,$$

$$\sigma_{\bar{x}} = \frac{\sigma}{\sqrt{n}}$$

Because $\bar{x}$ will be normally distributed, it provides the basis for estimating the value of the population mean using a sample mean.

Suppose you draw a particular sample and calculate its mean, which we will abbreviate as $\bar{x}_i$ to indicate that we are referring to a particular value of our random variable $\bar{x}$, e.g., 3. How accurate is 3 as an estimate of our population mean? To answer this question, we have to know the sampling error of the mean—$\sigma_{\bar{x}}$. Suppose you know what the population standard deviation is—in this case 1.63—but you do not know the population mean, which you want to estimate. Suppose we ask: what is the probability that our sample mean $\bar{x}_i = 3 = \mu$? By virtue of being a normal distribution, $\bar{x}$ will have an area under the normal curve such that 95% of it will be within two standard deviations $\sigma_{\bar{x}}$ of its mean $\mu_{\bar{x}}$ (actually 1.96 standard deviations—see Chapter 9). We are interested in estimating the population mean with a probability of, say, 95%. So, the probability (that $\bar{x}_i$ is within 1.96 standard deviations of μ) = .95, which we can abbreviate: $P(\bar{x}_i = \mu \pm 1.96\sigma) = .95$. But since $\mu_{\bar{x}} = \mu$ and $\sigma_{\bar{x}} = \frac{\sigma}{\sqrt{n}}$, this becomes

$$P(\bar{x}_i = \mu_{\bar{x}} \pm 1.96\frac{\sigma}{\sqrt{n}}) = .95.$$

Now suppose you know σ=1.63 (later we'll deal with the case in which you don't know this) and you already know that n=2. Substituting, we get: $P(3 = \mu_{\bar{x}} \pm 1.96\frac{1.63}{\sqrt{2}} = .95)$, which in turn is: $P(3 = \mu_{\bar{x}} \pm 2.3 = .95)$. This says that the probability that our sample mean 3 equals the population mean plus or minus 2.3σ is 95%, or (in other words) the probability that our sample mean 3 falls within the interval from .7 to 5.3 (3 minus 2.3 to 3 plus 2.3) is .95. This interval is called a **confidence interval,** and it says that we can be 95% confident that our population mean falls within the above interval. So our original question, How can we estimate the population mean by taking a sample mean?, has been answered.

Example. Suppose a scientist wants to estimate the average age of individuals with diabetes. She takes a random sample of 35 individuals with diabetes and determines their ages. The mean $\bar{x}$ is 55.3. (Assume a population standard deviation of 21.2 years.) $\bar{x}$ is used as an estimate of μ. How confident can she be that $\mu =$ 55.3? Suppose the desired confidence level is 95%. Then the probability that μ will be within 1.96 standard deviations $\sigma_{\bar{x}}$ of $\bar{x}$=95%:

$$P(\mu = \bar{x} \pm 1.96 \frac{\sigma_{\bar{x}}}{\sqrt{n}})=.95$$

or

$$P(\bar{x} - 1.96 \frac{\sigma_{\bar{x}}}{\sqrt{n}} < \mu < \bar{x} + \frac{\sigma_{\bar{x}}}{\sqrt{n}})=.95.$$

Substituting we get:

$$P(55.3 - 1.96 \frac{21.2}{\sqrt{35}} < \mu < 55.3 + \frac{21.2}{\sqrt{35}})=.95$$

$$= P(55.3 - 1.19 < \mu < 55.3 + 1.19) = .95$$

$$= P(54.2 < \mu < 56.6) = .95$$

Hence, we can be 95% confident that μ is between 54.2 and 56.6.

In the above example, 1.19 is known as *the margin of error*—half the confidence interval. Survey reports often report a margin of error. For example, one might read that 47% of the voters sampled approve of the job the President is doing with a margin of error of 4% (.47 ± .04); this means that in 95% of such polls, one can be confident that the percentage of voters approving the job the President is doing is within the range 43% to 51%.

In summary, it is important to remember that we have three different kinds of means and standard deviations that must be distinguished: those of a particular sample ($\bar{x}, s$), those of a population (μ, σ) and those of the set of samples ($\mu_{\bar{x}}, \sigma_{\bar{x}}$). From a particular sample, one makes an inference about a set of samples, and from this one makes an inference about a population.

Exercise 14.1

1. The heights of the five starting basketball players on the CSU men's basketball team are listed below (in inches).

76	78	79	81	86

(1) Draw all possible samples of size two (with replacement and not ignoring order) from the population of basketball player heights. Find the mean of each sample.
(2) Make a frequency table (probability distribution) for the sample means.
(3) Draw a histogram for the population and for the distribution of the sample means. Compare their shapes. What does this comparison show?
(4) Find μ, $\mu_{\bar{X}}$. What is the relation between them? What is the relation between $\bar{x}$ and μ ?
(5) Find σ, $\sigma_{\bar{X}}$. What is their relation?

Exercise 14.2

1. An educational psychologist at a large university wants to estimate the mean IQ of the students in attendance. A random sample of 15 students gives the following data on IQs.

113	120	103	118	100
88	110	126	112	120
98	110	126	101	115

(1) (a). Use the data to obtain a point estimate for the population mean IQ of all students attending the university (The sum of the data is 1664). (b) Is it likely that your point estimate is exactly equal to μ? Explain.

(2) (a). Assume the population standard deviation is 10. Determine σ at 95% confidence interval for μ. (b) Does the population mean IQ lie in the confidence interval you obtained? Explain.

2. To estimate the gestation period of domestic dogs, 15 randomly selected dogs are observed during pregnancy. Their gestation periods in days are recorded and a mean is determined. Obtain a 95% confidence interval for the mean gestation period μ of the domestic dog. (Assume $\sigma = 10$).

3. An earlier problem involved a pediatrician who tested for cholesterol levels of young patients. She found that many had levels over 100g. per 200mL. (She sampled 20 individuals). Assume $\bar{x} = 210, \sigma = 5.9$.

(1). Use the data to obtain a point estimate for the population mean. Is it likely that your point estimate is exactly equal to μ? Explain.

(2). Determine a 95% confidence interval for μ. Does the population mean lie in the confidence interval you obtained? Explain.

Statistical Hypothesis Testing

Statistical inference involves the making of inferences about characteristics of the population on the basis of evidence contained in one's sample. Often a scientist is interested in determining whether a hypothesis about a population parameter is correct or not, e.g., whether a population mean is a particular numerical value, whether there is a difference between two population means, etc. To do this, one must engage in statistical hypothesis testing, a topic of central importance to scientific methodology.

The theory of statistical hypothesis testing was largely initiated by Sir **Ronald Fisher** in the early twentieth century and later developed by **Jerzy Newman** and **Egon Pearson**. Although there are several things about which both schools agree, there are also fundamental disagreements between these two schools of thought. (The Bayesian approach to statistical inference departs from both of these accounts.) The final version of the standard (classical) account of statistical hypothesis testing taught in standard statistical textbooks testing is thus a synthesis of Fisher and Newman-Pearson, with the Newman-Pearson approach receiving pride of place. Although there are fundamental objections that individuals continue to make against this standard view and although there are several more recent developments that it would be important to discuss (e.g., the inclusion of p values and confidence intervals), it is this traditional account that we will present.[2]

On the standard account of statistical hypothesis testing, one begins with a *statistical hypothesis*—a statement about the probability (distribution) of a random variable—called the **null hypothesis** (symbolized H_o). The null hypothesis is the hypothesis one wishes to "nullify" or cast doubt upon, thereby leading to its rejection and to an increase in one's confidence in the alternative or experimental hypothesis (abbreviated H_1). This is the statistical hypothesis one wishes to accept, certify or "validate." H_o is often thought of as the hypothesis that there is a null difference between the experimental and control groups; that is to say, the effect of one's independent variable is null and that any difference between the two groups is due to chance (sampling error). H_1, on the other hand, is the hypothesis that there is a systematic difference between the two groups, one due to the independent variable and one not due to chance. Ordinarily, H_o hypothesizes that a population parameter, e.g. μ, has a particular value μ_o ($\mu = 100$) or that the means of the two populations are the same ($\mu_1 = \mu_2$). H_1 is then the denial of H_o (either its contrary or contradictory): $H_1 = {\sim}H_o$. How does one "nullify" H_o? H_o predicts there should be a particular distribution (typically a normal distribution) of the alleged random variable. The initial premise therefore is: if H_o is true, then there should be a particular normal distribution of a variable. One then casts doubt upon H_o by showing that the experimental results obtained in one's sample are very unlikely (improbable) given H_o; i.e., the results are assigned a very low probability by H_o, but are much more probable on H_1. Since one should accept that hypothesis that makes the results the more likely (probable), one should accept H_1 and not H_o. This argument is similar, therefore, to the Modus Tollens argument form, only cast in a probabilistic dress:

$H_o \supset E$.
But the experimental results E' are improbable (unlikely) on H_o
and more probable (likely) on H_1.
Thus, E is probably false
Thus, H_o is improbable and disconfirmed
Thus, H_1 is more probable and confirmed.

Such negative results fall into one of the extreme tail regions of the normalized (standardized) distribution, the area that is unlikely on H_o. These extreme regions are known as the *rejection regions*. Assume, for example, that H_o claims $\mu = \mu_o$. Then H_1 can deny this claim in one of three ways, by claiming (1) $\mu \neq \mu_o$, (2) $\mu < \mu_o$, or (3) $\mu > \mu_o$. In the first case, H_1 is the contradictory of H_o and the negative results for H_o would fall in either end of a two-tailed test. In the second case, H_1 is the contrary of H_o, and the negative results for H_o would fall into the left tail. In the third case, H_1 is the contrary of H_o, and the negative results for H_o would fall into the right tail region. All of this is represented below and graphically displayed in Figure 14.6.

	Two-tailed test	Left-tailed test	Right-tailed test
Sign in H_1	$\neq$	<	>
Rejection region	Both sides	Left side	Right side

Example: Calcium is abundant in one's body and works to build and maintain bones. The recommended daily allowance (RDA) for adults is 800 milligrams (mg). A nutritionist thinks the average person with an income below poverty level does not get the RDA of 800 mg. She wants to perform a hypothesis test to answer this question. Suppose she takes a sample of 24 individuals and determines that the sample mean $\bar{x}$ is 747 mg. (Assume a population standard deviation of 188 mg). Suppose μ is the mean calcium intake of all people with income below poverty level. Then $H_0{:}\mu = 800mg$ and $H_1{:}\mu \neq 800mg$. The relevant test here is a two-tailed test.

Exercise 14.3

In the following problems, determine H_o, H_1 and classify the hypothesis test as two-tailed, left-tailed or right-tailed.

1. The Food and Nutrition Board of the National Academy of Sciences states that the RDA of iron for adult females under the age of 51 is 18 mg. A hypothesis test is to be performed to decide whether adult females under the age of 51 are, on the average, getting less than the RDA of 18mg. of iron2. As reported in the U.S. Office of Juvenile Justice and Delinquency Prevention in *Children in Custody*, the mean age of all juveniles held in public custody in 1989 was 16 years. The age of a random sample of juveniles currently being held in public custody are to be used to decide whether this year's mean age of all juveniles being held in public custody is less than the 1989 mean of 16 years.

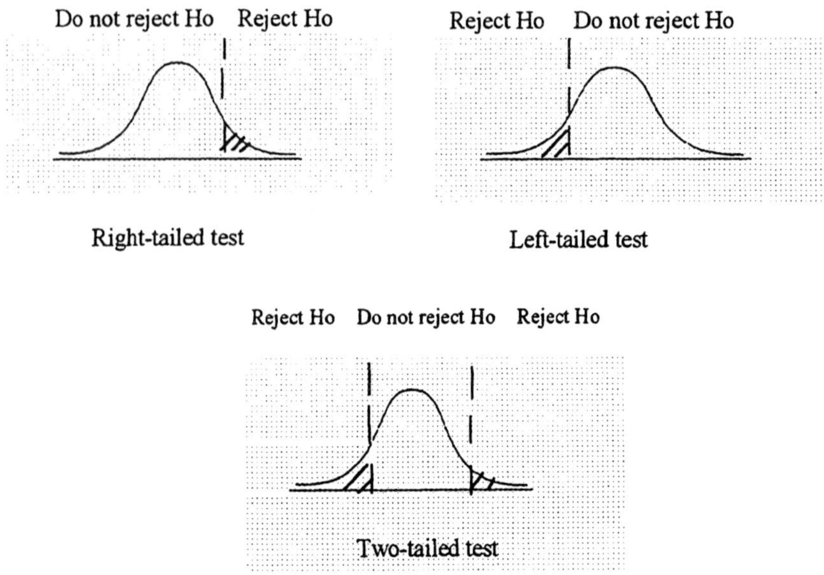

Figure 14.6. Graphical Displays of Different "tail" Tests

3. The World Almanac reports that the mean travel time to work in 1980 for all South Dakota residents was 13 minutes. A transportation official wants to use a sample of this year's travel time for South Dakota residents to decide whether the mean travel time to work for all South Dakota residents has changed from the 1980 figure of 13 minutes.

4. A study was concerned with the tendency of alcoholic middle-aged en to relapse after having undergone a certain new course of treatment. In the past, under an old treatment, 39% of such men relapse into alcoholism within six months. It was felt that the 6-month relapse rate might be different after the new treatment. A random sample of 150 male alcoholics was given the new treatment. After its conclusion, the number that relapsed within 6 months was noted. A second scientist became interested in the possibility that the new treatment was better, as there is little practical value in a treatment that increases relapse rate.

5. There was concern in a public school system that the method of teaching reading then in use might be inferior to other methods. A standardized test was available, giving national norms on reading achievement. For fifth-graders, this test showed a national average of 172, with a standard deviation of 16. A random sample was taken consisting of

250 fifth-graders taught by the method in question, who were each given the test.

The Logic of Hypothesis Testing

In our *Example*, H_o claimed our sample came from a population whose mean was 800 mg. Hence, $\bar{x}$ should also be 800 mg. However, we do not expect $\bar{x}$ to be exactly equal to μ because of sampling error. Hence, it is natural to expect a difference between $\bar{x}$ and μ. So the question is whether the difference between the predicted mean and the actually obtained mean is due to error (chance) or not. This in turn is the question of whether the difference score is sufficiently large to warrant the conclusion that $\mu \neq 800mg$. How likely would such a difference be if H_o was actually correct and the difference score was due to chance? This can be symbolized: P(e/h), the probability of the evidence (data) given the hypothesis. Because H_o claims $\bar{x}$ is a random variable whose value depends on chance, $\bar{x}$ should be normally distributed with $\mu_{\bar{x}} = \mu$ and $\sigma_{\bar{x}} = \frac{\sigma}{\sqrt{n}}$. Therefore, according to H_o, the probability that $\bar{x}$ will be within two standard deviations of $\mu_{\bar{x}}$ is (roughly) 95% and the probability that $\bar{x}$ will be greater than two standard deviations away from $\mu_{\bar{x}}$ is 5%. (See Figure 14.7). Hence, if the actual values of $\bar{x}$ equals 800mg ± 2 standard deviations or less, then the difference between the predicted and the actual values of $\bar{x}$ would be due to sampling error and we should accept H_o. The probability of this is 95% (on H_o). The probability that the actual value of $\bar{x}$ is more than 2 standard deviations away from the predicted value is 5% (on H_o.) So, if we obtained results falling into these extreme regions, the probability of it happening is .05. Since it did happen, this renders H_o unlikely (since H_o led us to expect it only to degree .05). If the probability is small on H_o, it is much greater on H_1 because these results are more likely if it is true. The precise probability of the results given H_1 is not specified since H_1 is not an exact or precise hypothesis. Nevertheless *the likelihood of H_1 seems greater than the likelihood of Ho*.

In Figure 14.7, the shaded areas under the two tails is the rejection region, the area under the curve such that results obtained there (any value of the test statistics greater than 2) lead to the rejection of H_o. The non-shaded area is the acceptance region—

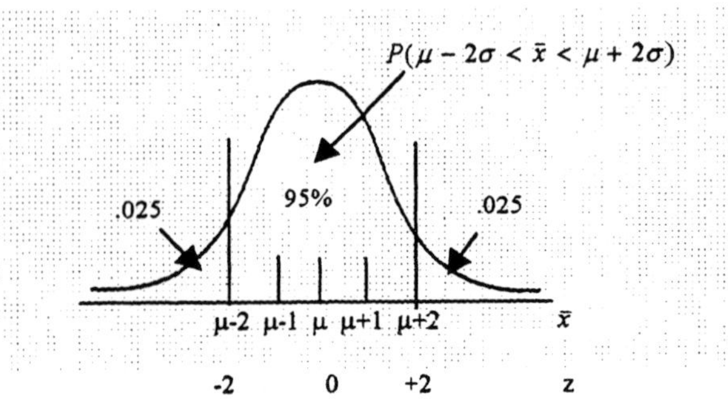

Figure 14.7. Probability that x will be within two standard deviations of μ

(actually the *non-rejection region*)--the area from -2 standard deviations to +2 standard deviations. Results in this area lead to the non-rejection of H_o. The value separating these two areas is the *critical value*. When one chooses acceptance and rejection regions, one is setting up a *decision rule*, a rule indicating under what conditions one will reject H_o and under what conditions one will not. See Figure 14.8

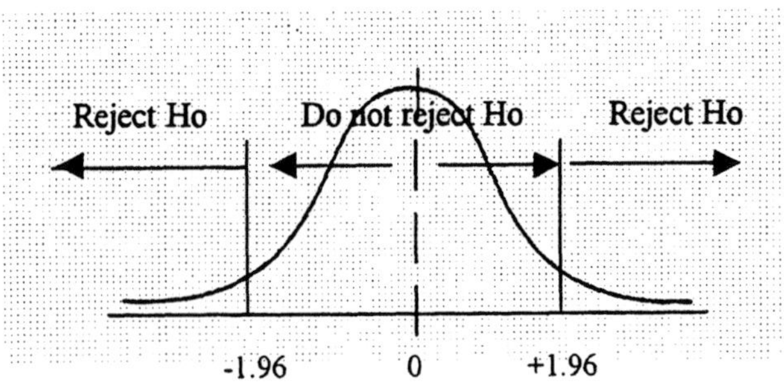

Figure 14.8 Possible Rejection Regions in a Normal Sampling Distribution

In testing a statistical hypothesis, one can make a correct decision or an incorrect decision about H_o, and a correct decision or an incorrect decision about H_1. There are, therefore, four possibilities depending upon whether H_o is true or false and whether one accepts it or rejects it. (See Table 14.1). When H_o is true, we are in a Type I situation; when H_o is false, we are in a Type II situation. When we reject H_o when it is false, we have made a correct decision; when we don't reject H_o when it is true, we have made a correct decision. When H_o is true and we reject it, we have made a **Type I Error**; when we don't reject H_o when it is false, we have made a **Type II Error.**

Table 14.1. Four Possible Outcomes of a Hypothesis Test

	True	False
H_o is not rejected	Correct Decision (1-α)	Type II Error (ß)
H_o is rejected	Type I Error (α)	Correct Decision (1-ß)

What are the respective probabilities of making these various errors? The *probability of making a Type I error is the probability of rejecting H_o when it is true*. This can occur only when our results fall in the rejection region. In the above discussion, the rejection region is the shaded area under the two tails of the curve, which constitutes 5% of the distribution. So the probability of rejecting H_o when it is true is only 5%. This percentage is called the **significance level** of the test (also called the *α-level*. It follows that the probability of a Type I correct decision is 1-α. The probability of committing a Type II error is known as the β-level; and the probability of a Type II correct decision is 1-β. Ordinarily, we do not know the β level (see below).

Although we have been using a significance level of 5%, this is just a convention; we could have used other significance levels, e.g., 10%, 1%, .01%, etc. The convention is that one ordinarily uses a 5% or a 1% level of significance.[3] To each of these levels (and depending on whether we are using a one-tailed or a two-tailed test), we would have different critical values and rejection regions. The following table (Table 14.2) lists the most commonly used critical values. Figure 14.9 displays this graphically.

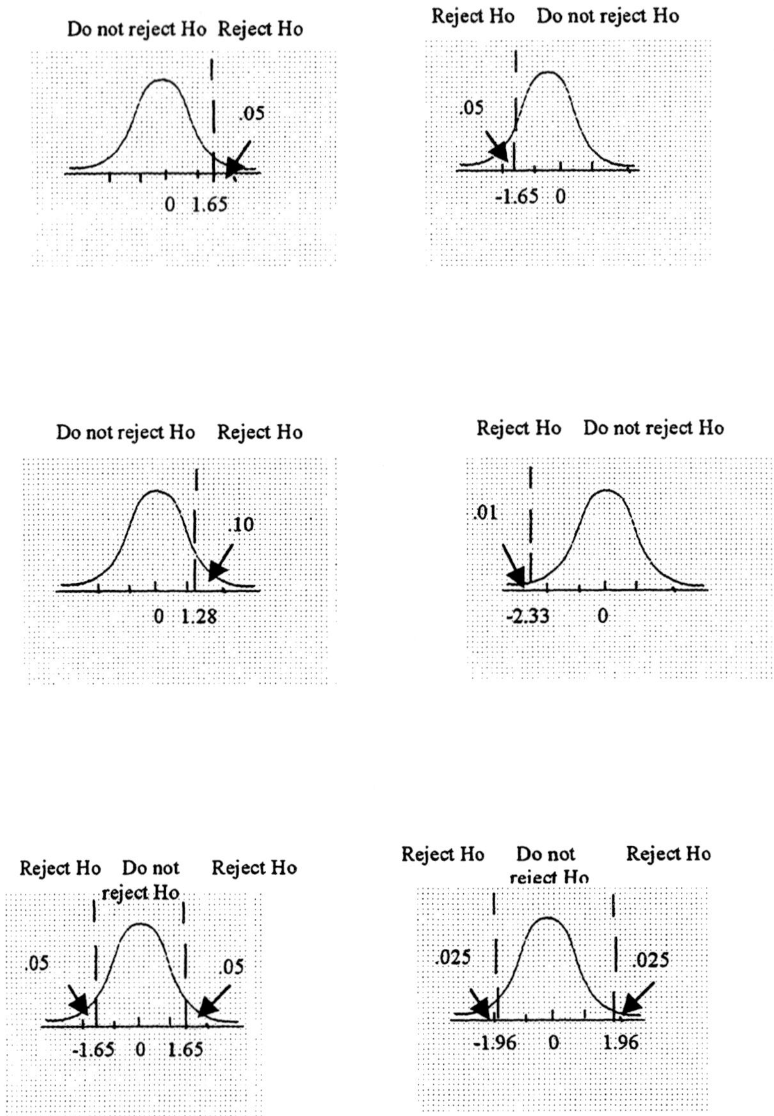

Figure 14.9 Graphical Displays of Decision Criteria

Table 14.2. Critical Values for commonly used Significance Levels

Level	.1	.05	.01	.005
One-tailed test	± 1.28	± 1.65	± 2.33	± 2.58
Two-tailed test	± 1.645	± 1.96	± 2.58	± 2.8

When one's sample results fall into the rejection region and H_o is rejected at a particular level of significance, the test results are said to be **statistically significant** at the particular α-level. It should be born in mind that the phrases "statistical significance" and "statistically significant result" mean nothing more than this: our sample results fell into the rejection region.

What then is the underlying "logic" of statistical hypothesis testing? This question has been the subject of a great deal of debate both among statisticians and logicians. It was debated, for example, by Fisher and Newman-Pearson in the 1930s and it has continued to be debated up until the present time. It is not possible, therefore, to give a standard answer to this question since there isn't one. The following is offered, however, as a reasonable approximation to what many individuals believe is the underlying logic of statistical hypothesis testing.

1. H_o specifies that a population parameter has a certain value (e.g., $\mu = \mu_o$) and H_1 specifies a contrary value (e.g., $\mu \neq \mu_o$).
2. If H_o is correct, then the sample mean $\overline{x}_i$ equals $\mu \pm$ sampling error.
3. $\overline{x}_i$ is (according to H_o) a normally distributed random variable such that $\mu_{\overline{x}} = \mu$ and $\sigma_{\overline{x}} = \sigma / \sqrt{n}$.
4. A significance level is chosen, e.g., $\alpha = .05$.
5. Thus, the probability that $\bar{x}_i = \mu \pm 1.96\sigma_{\overline{x}} = .95$.
6. A sample (x_i) of size n is randomly drawn from the population.
7. The mean of x_i ($\overline{x}_i$) is calculated.
8. $\overline{x}_i$ fell into the rejection region.
9. The probability of this occurring if Ho is true is .05; the probability of this occurring if H_1 is true is greater than .05.
10. Therefore, H_o is rejected and H_1 accepted.
11. Therefore, probably $\mu \neq \mu_o$.

This has a very complicated (and controversial) argument structure with both deductive and inductive components. As we indicated, a key step is the inference from the likelihood of the data given H_o (the P-value) to the probability of H_o. This was and is hotly debated by many statisticians, as is the question of specifying the precise likelihood of H_1 (which in general is unknown). Furthermore, in many cases Ho and H_1 are not contradictories but only contraries and this contributes to muddying the waters.

Additional conceptual questions that are debated include: if one fails to reject H_o, is this equivalent to accepting H_1? (Standard answers: no). If one rejects H_o, is this equivalent to (does this entail) accepting H_1? (Standard answer: yes). What does it mean to "accept/reject" a hypothesis? The standard answer is: it means acting as if it were true and this in turn means acting as if it were true for practical purposes. So, acceptance is not equivalent to believing H is correct. As Newman maintained, instead of inductive belief, there is just inductive behavior.

Tests of Significance

Up to now, our discussion has remained on a somewhat abstract and general level. In order to perform a test of a statistical hypothesis, we need a *test statistic,* a statistical test. The statistical hypothesis is a hypothesis about the probability distribution of a random variable (e.g., a population mean). To determine whether our sample results are consistent with H_o or not, we need to be able to compare them to the sampling distribution as predicted by H_o. The test statistic is a measure of the distance of our sample results from the sampling distribution as indicated by H_o. On the basis of this test statistic we can then determine whether the distance is due to chance or not.

There are many test statistics available and we can not discuss these in even a cursory way. Since our concern is with conceptual and theoretical questions and not with providing a manual on the various tests, we will illustrate the use and value of statistical tests by reference to just a couple of examples. For pedagogical purposes, we will begin with an extremely simple test: a *z-test statistic.*

Z-Test Statistic. As we saw earlier:

$$z = \frac{\bar{x} - \mu_{\bar{x}}}{\sigma_{\bar{x}}} = \frac{\bar{x} - \mu}{\frac{\sigma}{\sqrt{n}}}$$

We can use this formula to perform a statistical test.

Example: assume we are interested in testing the hypothesis that the average person with an income below poverty level does not get the RDA of 800mg of calcium per day. Hence $H_1 : \mu \neq 800mg$. By contrast, the null hypothesis must take on a specific value so one can generate a sampling distribution. Hence $H_0 : \mu = 800mg$. We must test H_o and attempt to "nullify" it, thereby increasing our confidence in H_1. In an earlier example involving this experiment, our sample mean was 747 mg, with a population standard deviation of 188mg and n was 24. Substituting in the above z-statistic, we get:

$$z = \frac{\bar{x} - \mu}{\frac{\sigma}{\sqrt{n}}} = \frac{747 - 800}{\frac{188}{\sqrt{24}}} = \frac{-53}{\frac{188}{4.9}} = \frac{-53}{38} = -1.39$$

Now, according to H_o, the population mean is 800 and the sample mean should also be (approximately) 800 (since it came from this population). Hence, the difference between the two should be 0. We know, however, that there will be some deviation from 0 due to chance (sampling error). We expect such deviations to be normally distributed, with 95% of the results falling within 1.96 standard deviations of the mean. Since the standard deviation $\sigma_{\bar{x}}$ is 38, this means that 95% of all the results should fall within 74 points on either side of the mean (1.96 x 38 = 74), i.e., within the interval 726-874. One can expect (at most) that 5% of the chance results will fall outside of this interval. Now if we transform all of this into a standardized distribution, then 95% of all the results should fall into the region 0 ± 1.96, i.e. -1.96 to +1.96 and at most 5% of the sample should fall outside of this region. Hence, if our Z score falls into the region -1.96 to +1.96, we will not reject H_o but if our Z score is either more than -1.96 or more than +1.96, we will reject it. Since our Z score was -1.39, it does *not* fall into the rejection region and so we do not reject H_o. (see Figure 14.10). Suppose, however, our sample mean were 721 mg. In that case, our z-statistic would have been:

$$z = \frac{\bar{x} - \mu}{\frac{\sigma}{\sqrt{n}}} = \frac{721 - 800}{\frac{188}{\sqrt{24}}} = \frac{-79}{\frac{188}{4.9}} = \frac{-79}{38} = -2.1$$

In this case, our z-statistic would have fallen into the rejection region, leading to the rejection of H_o; our results would have been statistically significant at the .05 level. (Figure 14.11)

It should be pointed out that all statistical tests are based upon certain assumptions. What is assumed varies from test to test, but common assumptions of statistical tests include the following: (1) it is assumed that the population is normally distributed, and (2) it is assumed that one has taken a random sample. Some statistical tests make further assumptions, e.g., that one knows the population standard deviation, that the standard deviations of the two populations are equal, etc. A discussion of this is beyond the scope of this elementary discussion. But what if these assumptions are false or are questionable? The situation is not so bleak and for two reasons: (I) there are other kinds of statistical tests that do not make assumptions about the parameters of the populations—called *non-parametric statistics*; and (2) under certain conditions, even the violation of these assumptions does not matter too much. (If this is true, the test is said to be *robust*, meaning that the violation of the assumption is not too damaging).

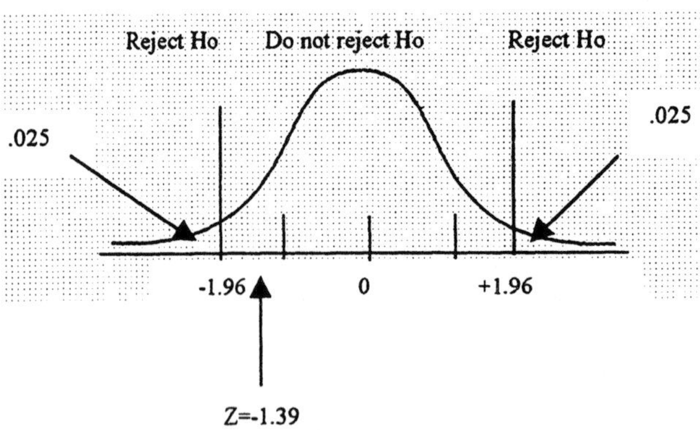

Figure 14.10. Criterion for Deciding to Reject Ho.

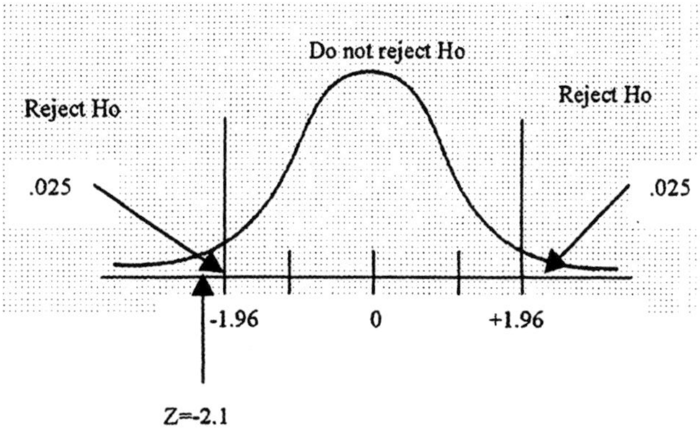

Figure 14.11. Criterion for Deciding to Reject Ho.

According to the standard account of statistical hypothesis testing, the following procedures are involved in statistical hypothesis testing:

Procedures for Statistical Hypothesis Testing

1. *State the Assumptions (e.g.,):*
   a. *Normal Population*
   b. *Random sample*
   c. *Population standard deviation is known*
2. *State the Null hypothesis:* $\mu = \mu_o$, *e.g.,* $\mu = 800$,
3. *State the Alternative hypothesis:* $\mu \neq \mu_o$, *e.g.,* $\mu \neq 800$
4. *Decide on the Level of significance, e.g., 5%*
5. *Determine the Critical Values: e.g.,* ± 1.96
6. *State the Decision Rule:*

if the Z-statistic is ± 1.96*, then reject* H_o*; If the Z-statistic is not* ± 1.96*, then do not reject* H_o

7. *Compute the Test-statistic. For example:*

$$z = \frac{\bar{x} - \mu}{\frac{\sigma}{\sqrt{n}}} = \frac{747 - 800}{\frac{188}{\sqrt{24}}} = \frac{-53}{\frac{188}{4.9}} = \frac{-53}{38} = -1.39$$

8. *State the Conclusion: since the z-statistic was not −1.96 or more, do not reject* H_o

Exercise 14.4

In the following problems, suppose a hypothesis test is to be performed for a population mean, with $H_o: \mu = \mu_o$, Suppose we use a z-test. For each exercise, follow the procedures set out above for the testing of statistical hypotheses.

1. The Food and Nutrition Board of the National Academy of Sciences states that the RDA of iron for adult females under the age of 51 is 18 mg. A hypothesis test is to be performed to decide whether adult females under the age of 51 are, on the average, getting less than the RDA of 18mg. of iron. For 45 randomly selected females under the age of 51, the sample mean obtained was 14.68 mg (assume the population standard is 4.2mg). At the 5% level of significance, do the data suggest that adult females under the age of 51 are, on the average, getting less than the RDA of 18 mg of iron?

2. As reported in the U.S. Office of Juvenile Justice and Delinquency Prevention in *Children in Custody*, the mean age of all juveniles held in public custody in 1989 was 16 years. The age of a random sample of juveniles currently being held in public custody is to be used to decide whether this year's mean age of all juveniles being held in public custody is less than the 1989 mean of 16 years. The mean age of 250 randomly selected juveniles currently being held in public custody is 15.86 years. Assuming σ = 1.01 years, does it appear that the mean age of all juveniles being held in public custody this year is less than the 1989 mean of 16 years? Perform the appropriate hypothesis test using α = .10.

3. *The World Almanac* reports that the mean travel time to work in 1980 for all South Dakota residents was 13 minutes. A transportation official obtained this year's travel time, in minutes, for a random sample of 35 South Dakota residents. At the 5% level of significance, do the data provide sufficient evidence to conclude that the mean travel time to work for all South Dakota residents has changed from the 1980 figure of 13 minutes? Assume σ = 11.6. The sum of the data is 491 minutes.

4. A study was concerned with the tendency of alcoholic middle-age men to relapse after having undergone a certain new course of treatment. In the past, under an old treatment, 39% of such men relapse into alcoholism within six months. It was felt that the 6-month relapse rate may be different after the new treatment. A random sample of 150 male alcoholics was given the new treatment. After its conclusion, the number that relapsed within 6 months was 40. Assuming σ = 4, test the hypothesis that there was no change in the percentage at the .05 level of significance.

5. There was concern in a public school system that the method of teaching then in use might be inferior to other methods. A standardized test was available, giving national norms on reading achievement. For fifth-graders, this test showed a national average of 172, with a standard deviation of 16. A random sample was taken consisting of 250 fifth-

graders taught by the method in question, who were each given the test. Assume the sample mean is 170. Perform a statistical test at the .01 level.

Z-test for Differences between Groups. A z-test can be used to test whether a population parameter has a certain value, e.g., whether the population mean is equal to a certain numerical value. But z-tests can be used as test-statistics in other ways. For example, one of the most important uses is to test whether there is a significant difference between two samples. As we have seen, an experiment typically consists of collecting a difference score between the experimental and the control group.

Suppose an experimenter believes that infants breast-fed milk have higher IQs than infants not breast-fed. An experiment is done on two groups of infants: the experimental group (E) is breast-fed on milk, whereas the control group (C) is not. Later, both groups are given IQ tests and their average scores are compared. Suppose there were 90 infants in each group and suppose the mean IQ of the experimental group was 103.7, as opposed to the mean IQ of the control group (which was 92.8). Is this difference score of 10.9 a statistically significant difference, one due to the difference between being breast-fed and not being breast-fed, or is the difference due to chance? The null hypothesis claims the latter is the case, i.e., there is no difference between the two populations from which these two samples came: $H_o : \mu_1 = \mu_2$ and hence that $\bar{x}_1 = \bar{x}_2$. The alternative hypothesis (H_1) claims that there is a real (causal) difference between the two samples, that they come from different populations, and that the mean of the one sample is greater than the mean of the other sample because the population from which the first sample came has a mean greater than the population from which the second sample. Hence $H_1 : \mu_1 > \mu_2$ and $\bar{x}_1 > \bar{x}_2$. Following the logic of statistical tests, there will be a sampling distribution of the differences between the sample means ($\bar{x}_1 - \bar{x}_2$), and the standard deviation of the differences between the two samples will be the sum of their respective standard deviations. Hence, there is a standard error of the difference, which is: $\sqrt{\frac{\sigma_1^2}{n_1} + \frac{\sigma_2^2}{n_2}}$ This is because $\bar{x}_1 - \bar{x}_2$ itself is a random variable (just as, in earlier discussions, $\bar{x}$ was a random variable). Hence,

$\mu_{\bar{x}_1-\bar{x}_2}=\mu_1-\mu_2$ and $\sigma_{\bar{x}_1-\bar{x}_2}=\sqrt{\frac{\sigma_1^2}{n_1}+\frac{\sigma_2^2}{n_2}}$. We can use a z-test statistic to test whether the difference scores between the two samples are statistically significant or not.

$$z=\frac{(\bar{x}_1-\bar{x}_2)-(\mu_1-\mu_2)}{\sqrt{\frac{\sigma_1^2}{n_1}+\frac{\sigma_2^2}{n_2}}}$$

Since according to null hypothesis $H_o:\mu_1=\mu_2$, their difference should be 0 and this can be ignored in the above z-test. Suppose our two populations variances are equal, say, 360. Then our z-test for a difference between sample means is:

$$z=\frac{(\bar{x}_1-\bar{x}_2)-0}{\sqrt{\frac{\sigma_1^2}{n_1}+\frac{\sigma_2^2}{n_2}}}=\frac{(103.7-92.8)-0}{\sqrt{\frac{360}{90}+\frac{360}{90}}}=\frac{10.9}{\sqrt{4+4}}=\frac{10.9}{2.8}=3.89$$

Assuming a 5% significance level and a right-tailed test, the critical value for a z-test is 1.645 and our z-statistic of 3.89 clearly exceeds this. Hence, it falls into the rejection region, and we reject the null hypothesis that the difference is due to chance. It appears as if being breast-fed does influence one's IQ score.

Exercise 14.5

1. The Northeastern University Placement Center conducts surveys on starting salaries for college graduates. Results of the surveys can be found the *The Northwestern Lindquist-Endicott Report*. The annual salaries obtained from independent random samples of 35 liberal arts graduates and 32 accounting graduates were obtained. (The sum of the liberal arts data is 892.9 and the sum of the accounting data is 887.6). At the 5% significance level, can we conclude that the mean starting salaries of liberal arts and accounting graduates differ? Assume the population standard deviations of starting salaries are 1.82 ($1820) for liberal arts graduates and 1.73 ($1730) for accounting graduates.

2. Independent samples of 17 sophomores and 13 juniors attending a large state university gave the statistics below for cumulative grade point average.

Sophomores	Juniors
$\bar{x}_1 = 2.54$	$\bar{x}_2 = 2.68$
$\sigma_1 = .52$	$\sigma_2 = .31$

Can you conclude from these data that the mean GPA's of sophomores and juniors at the university differ? Use $\alpha = .05$.

The above discussion of statistical tests of significance should have given the student a basic introduction to the concepts underlying its use. As we have pointed out, statistical hypothesis testing is a child of the 20th century. One should not, therefore, expect to find its employment in prior centuries. It is true, however, that our knowledge of these tests should allow us to make sense out of experimental results obtained in the past. And this is indeed the case. But the problem of applying statistical tests to earlier scientific studies is the fact that these investigators did not provide complete information needed to conduct such tests. For example, in an earlier chapter (Chapter 12) , we discussed the case of Semmelweis and childbed fever. He observed differences between the proportions (percentages) between different groups: e.g., between Division I and Division II (.099 vs. .033), between Division I before the experimental treatment and Division I after the treatment (.12 vs. .033). In retrospect we can ask: were these differences statistically significant or due to chance? (Their differences certainly appear to be substantial). There are statistical tests to determine this (e.g., a test for difference between proportions) but we would need to have additional information in order to perform the calculations (e.g., the total numbers in both groups) and this information is lacking. Hence we cannot use these tests.

Not only can we use a z-test to determine whether a difference score between two samples is statistically significant, we can also use a statistical tests to determine whether a difference score involved *matched subjects* (groups) and a difference score involving a *before and after* design are statistically significant (Chapter 13).

In actual cases of statistical hypotheses testing, one must estimate population parameters (e.g., the population standard deviation σ), on the basis of the sample standard deviation s). Because of sampling error, we do not expect our estimate to be exactly correct but only approximately so, with the degree of approximation depending on our sample size. We need, therefore, a correction but only approximately so, with the degree of approximation depending on our sample size. We need, therefore, a correction factor, and this results in a somewhat different sampling distribution; instead of a z-distribution, it is called the *t-distribution* and, hence, instead of a z-test, one would employ a t-test (also called a student's t-test). The t-test has the same form as a z-test except that in place of the population standard deviation, we use the sample standard deviation, and in place of a normalized (z) distribution, we use a t-distribution. As a result, the critical values needed to reject the null hypotheses are somewhat larger than in a z-test.

The basic logic behind the t-test, however, is the same as the logic behind a z-test and involves the same concepts: random variable, sampling distribution, etc. In fact, this same logic has produced several statistical tests of population parameters using estimates from the sample. In place of the z-tests discussed above, there are corresponding forms of t-tests (t-tests for the difference between sample means, t-tests for a difference score, etc). In addition, there are statistical tests concerning the population variance, several population variances, correlation and regression equation, confidence intervals, etc. None of these tests will be discussed here. Instead, we will simply refer the interested reader to standard texts in statistics. But one other statistical test will be discussed because of the special importance it has assumed with respect to the remarkable results of the founder of genetics—Gregor Mendel.

The χ^2 Test and Mendel's Results. In 1866 Gregor Mendel—an Austrian monk--published an article reporting the results of a study on the breeding of plants (sweat peas) over an eight-year period in an obscure Bohemian journal. Although he discovered two basic laws of inheritance, which are the foundations of the modern genetics, his study was basically unknown to the scientific world for 30 years. Then about 1900 three other individuals appear to have simultaneously made the same discovery.

Mendel bred the plants with various characteristics: height, color, seed texture, etc. He found that when red-flowered plants were crossed with white-flowered plants, the hybrid offspring were all red. But when two hybrid reds were crossed, the offspring were both red and white but in a certain frequency (proportion). To explain these facts, Mendel introduced the idea of a *factor* (later called a gene) that controlled hereditary traits of which one type was dominant and the other recessive. Let us symbolize the dominant trait by a capital letter and the recessive one by a lower case letter (e.g., R: red, w: white). Assume each parent has a pair of genes for color. A pea plant, for example, might contain two dominant genes for color, two recessive genes, one of each, etc. Furthermore, in passing on genes to offspring, a parent can produce genes of both types equally often. This is true of both males and females. Suppose now that one parent has Rw and the other one also has Rw. Then we have the following situation:

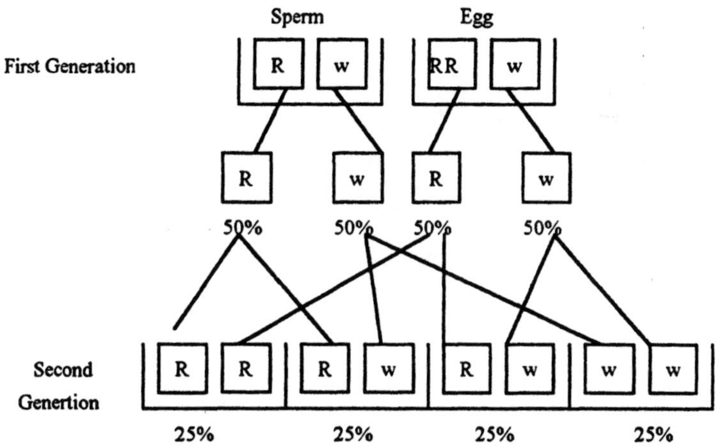

Assume that sperm from the male randomly combines with the egg of the female. Then we have four possibilities (a frequency distribution): RR, wR, Rw, ww, each with a probability (frequency) of 25%. Now if R is dominant and w is recessive, we would have three red plants and one white plant, i.e., a frequency distribution of 75% red and 25% white. This is just what Mendel found. (It is important to note that Mendel's hypothesis is based upon a chance model, i.e., which genes combine with which genes is due to chance.)

Mendel (1865/1966) reported several studies. Here are two:

> Experiment 1. Seed shape. From 253 hybrids 7324 seeds were obtained in a second experimental year. Of them, 5474 were round or roundish and 1850 angular wrinkled. This gives the ratio 2.96:1.
> Experiment 2. Albumen coloration. 258 plants yielded 8023 seeds, 6022 yellow and 2001 green; their ratio, therefore, is 3.01:1 (p. 11).

How closely do the obtained frequencies match the predicted frequencies? The predicted frequencies are 3:1 (75% to 25%) in hybrids. The observed results were 2.96:1 and 3.01:1. This is a remarkable agreement between theoretical prediction and obtained results. In fact, out of a total of 19,959 plants, Mendel found 14,949 were dominant, a frequency of 74.9% These are impressive results about which many textbooks are naturally impressed. But perhaps the results are too impressive--at least this is the claim made by a famous statistician-geneticist Ronald Fisher (1936/1966).

A χ^2-test can be used to test hypotheses concerning the difference between predicted and obtained frequencies. A χ^2 (Chi-squared) test is a statistical test for the equality of population variances based upon sample variances. It has a corresponding distribution (a χ^2-distribution) based upon sample size--one with an unusual shape. This distribution can be used to determine if the expected frequencies of data match the frequencies predicted from one's hypothesis. For example, one expects that if one has a fair coin and tosses it a certain (large) number of times, there will be a certain frequency of heads and tails. We can actually observe the frequency of heads and tails in a large number of tosses and see how closely the two match each other. One might do this, for example, if one believed there was an unbiased coin (or roulette wheel, etc.). Naturally one expects some sampling error, so that if one tossed a coin 1000 times, the observed frequency of heads and tails would not match the theoretical frequencies exactly. If one obtained a run of 1000 heads, this would make one alarmed even if such a frequency could be due to chance (since it might also be due to the fact that the coin was biased). On the other hand, suppose one got exactly 500 heads and 500 tails. This is also very unexpected because one would expect a certain amount of error. The amount of error expected can be calculated as sampling error. In 1000 coin tosses, one would not expect 1000 heads, nor would one expect 500 heads but some frequency around 500 plus or minus an error term. Whenever one is dealing with probability and chance, the situation is the same: one always expects some sampling error.

The χ^2-test allows one to decide if the comparison of observed versus expected frequencies is due to chance or not.[4] The null hypothesis claims that the observed and expected frequencies are equal and that any difference between them is due to chance.

Hence, their difference should be 0. The alternative hypothesis denies this and claims the difference is not due to chance, hence that it should not be 0.

In one of Mendel's experiments, he reported a total of 8023 hybrid plants. The predicted frequency was 3:1, which would be: 6017:2006. The actually obtained frequency was 6022:2001. Performing a χ^2-test gives:

$$\chi^2 = \frac{\Sigma\ (\text{observed frequency} - \text{expected frequency})^2}{\text{expected frequency}} = .012$$ [5]

This gives a P-value (the likelihood of H_o) of between .95 and .90.

Fisher pooled all of Mendel's results and got a P-value of .999; this is interpreted to mean that the probability of getting this result if H_o is true is almost certain. The results mean two things (1) the obtained frequencies were very close to the predicted frequencies, and (2) the tests showed one could not reject H_o, which (we will remember) claimed there was no difference. So, what's the problem here? It looks like H_o is correct that there is no significant difference between observed and expected frequencies, which is just what Mendel was claiming was correct. The problem is that the results are too good; the difference between the observed and expected frequencies is too small. Remember this is a chance model. We expect sampling error when one takes random samples. H_o claims that any difference between the observed and expected frequencies is due to chance. But here it looks like there is no chance, no sampling error to speak of and such an event, although possible, is extremely unlikely.

For example, in an actual coin tossing experiment, 10 coins are tossed over 500 times (presumably with an unbiased coin). A certain number of tails are observed. How many tails do we expect and how many do we actually get? One expects to get a frequency somewhere around .5 but not exactly; in fact, it turns out that you get about 50% plus or minus 15% (i.e., one will get between 35% and 65% tails). By contrast, Mendel's results show an error factor of less than 1% over thousands of trials. Suppose, by analogy, that in 500 tosses, one got exactly 250 tails? That is extremely unlikely (improbable). In fact, in (say) 8000 tosses of a coin, one expects the chance error to be about 45, i.e., 4000 tails plus or minus 45. But in Mendel's case, his sampling error was between 4 and 5, in effect, no sampling error at all.

How can we explain these puzzling facts? Did Mendel fudge his data? Did an assistant do it? Did Mendel unconsciously do it? Did he quit collecting data when the results seemed correct? Did he discard what he thought was "contaminated" data? All of these suggestions have been made with no consensus except that Mendel's results are too good to be (probably) true. What is significant (for our purposes) is the fact that Fisher used a χ^2 -test to show there was a problem in Mendel's data. This shows, once again, the crucial importance of understanding statistical analysis in scientific methodology since (arguably) this result would never have occurred prior to the rise of modern statistics.

For Further Reading

Among the several introductions to statistical inference, one that is especially good is Freedman, et al (1991). It is about as simple as one can get and yet conceptually and pedagogically well thought out. Another one I heartily recommend is Weiss (1996). There are numerous intermediate textbooks in statistics. Unfortunately, there are not numerous philosophical discussions of the underlying logic of statistical inference. One very good one is Giere (1979a), who also gives a nice overview of the field of philosophy of statistics in his (1979b). Baird (1992) is an excellent textbook covering a range of topics in probability and statistics including the neglected area of factor analysis.

I have yet to find a good discussion of the conceptual differences between the Fisher school and the Newman-Pearson school. All statistics textbooks gloss over the many important differences between these two schools, so that students go away thinking there are no disagreements in statistics. Likewise there are very few (if any) full-length discussions of the philosophical debates about statistics accessible to the beginning student. Most discussions are written at an intermediate or advanced level.

Fisher's important writings are difficult but important, as are those of Newman-Pearson. Fisher's major works have been reissued, in a one-volume paperback edition, as Fisher (1990). Newman and Pearson's many papers are not as accessible but also collected together.

In the last twenty years there has been considerable controversy and discussion surrounding the "significance test controversy". Although we did not include a discussion of it in this chapter, the reader can consult the classic anthology of Morrison & Henkel (1970).

Finally, it should be mentioned that, recently, Bayesian statistics has become more widely discussed as an alternative to classical statistical hypothesis testing. In addition to the Howson & Urbach (1993), a shorter introduction (for those interested in the more practical aspects) can be found in Edwards, et al (1963).

Notes

[1]This is calculated as N^n, where N=the number of possible observations in one's population and n=the size of the sample.

[2] Our account thus ignores the widely debated controversy about the use of significance tests, especially in the behavioral sciences. In recent decades, numerous individuals have criticized the use to which scientists have put significance tests. See, e.g., Morrison & Henkel (1970) to which the interested reader is referred.

[3] The reason for this convention, so the story goes, is that when setting up statistical tables in the early twentieth century, the column headings were labeled: 5%, 1%, etc. for the sake of convenience. In time, however, it became a matter of convention to use these. It should be pointed out that this convention has been challenged by many individual statisticians and scientists who favor reporting the significance of one's results at whatever level obtained, 4%, 12%, etc. This is one way of avoiding the dichotomy: rejecting H_o and failing to reject H_o. This is part of what has become known as the "significance test controversy."

[4] The formula for a χ^2 tests is: Σ(observed frequencies-expected frequencies)^2/expected frequencies.

[5]

O	E	(O-E)	$(O-E)^2$	$(O-E)^2/E$	
6017	6022	-5	25	25/6022	.00415
2001	2005	-4	16	16/2005	.00798

$\chi 2 = .012$

Answers to Selected Exercises

Chapter 3

Exercise 3.1

1. [Logic is a good subject to study,] for it constitutes the intellectual backbone of Western science.

3. Because logic is one of the principal means which assures intellectual discipline and integrity, [it can, if properly applied, only promote the achievement of desirable social ends.]
5. [The earth is spherical in shape.] For the night sky looks different in the northern and southern parts of the earth, and this would not be so if the earth were not spherical in shape.

Exercise 3.2

1. Since the secondary light [from the moon] does not inherently belong to the moon, and is not received from any star or from the sun, and since in the whole universe there is no other body left but the earth, what must we conclude? What is to be proposed? Surely we must assert that [the lunar body (or any other dark and sunless orb) is illuminated by the earth.]

3. [Women of the working class, especially wage workers, should not have more than two children at most.] because The average working man can support no more and the average working woman can take care of no more in decent fashion.

Exercise 3.3

1. Since (1) secondary light [from the moon] does not inherently belong to the moon, and is not (2) received from any star or from the sun, and since (3) in the whole universe there is no other body left but the earth
(4) we must assert that [the lunar body (or any other dark and sunless orb) is illuminated by the earth.]

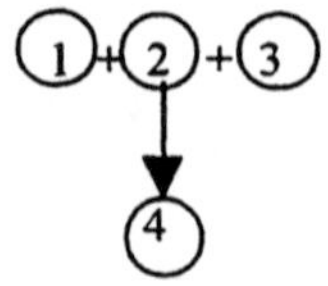

3. (1) [Women of the working class, especially wage workers, should not have more than two children at most].(2) The average working man can support no more and the average working woman can take care of no more in decent fashion.

Exercise 3.4

1. (1) If the fetus is a human life, then there are only a few circumstances in which cases of abortion are justified, and (2) if the fetus is just a mass of cells, then all cases of abortion are justified. (3) Either the fetus is a human life or just a mass of cells. (4) [either there are only a few circumstances where abortion is justified or all cases of abortion are justified.]

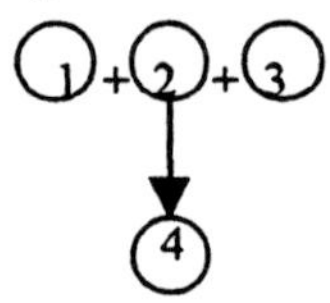

The argument is deductive.

3. (1) [A new drug, acebutol, already available in Great Britain and now under study here, looks excitingly good for dangerous heart rhythms.] (2) At Stanford University not long ago, doctors tried it for a man with a potentially lethal arrhythmia that resisted all the standard antiarrhythmia drugs such as quinidine and procainimide, and had required repeated electrical shocking of his heart.(3) In the eight months he has been on acebutol, he has had no recurrence of arrhythmia.

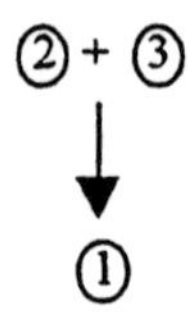

The argument is inductive and employs a sample. The key words are "looks excitingly good;" in addition, an experiment is reported.

5. (1) The governor's commission strongly recommends that use of marijuana continue to be illegal since marijuana leads to the use of heroin.] This is evidenced by the fact that (2) 94 percent of 1,000 heroin addicts interviewed reported that they had smoked marijuana before becoming addicted to heroin.

This is an inductive argument based upon a sample; the key phrase is "is evidenced by the fact that."

Exercise 3.5

1. False; false; false. Valid. Unsound because the argument has false premises.

3. False. False. True. Valid. Unsound because of false premises.

Chapter 4

Exercise 4.1

1. $\sim(p)$. This is false because p is true.
3. $(c \vee l)$. The disjunctive proposition is true because c is true.
5. $(p \,\&\, q)$. The compound proposition is true because p is true and q is true.

Exercise 4.2

1. $p \supset (q \,\&\, r)$.
   T T TTT

The compound proposition is true because the antecedent is true and the consequent is true.

3. $\sim p \supset (q \,\&\, \sim s)$
   F T TTT

The compound proposition is true because the antecedent is false and the consequent is true.

5. $p \supset (\sim s \supset r)$
   T T TTT

The compound proposition is true because the antecedent is true and the consequent is true.

Exercise 4.3

1. Birth is a necessary condition for death—depending on what is meant by "birth"—since, if something is not born, then it cannot die. (A possible counterexample would be uncreated life that died, e.g., God dies.) Birth is a sufficient condition for death, since, if something is born, then it will die (assuming there are no examples of everlasting life).

3. Being married is not a sufficient condition for pregnancy (since there are many married females that are not pregnant), nor is it necessary (since there are many unmarried females that are pregnant).

5. Being at least 35 years old is a necessary but not sufficient condition for being the President of the U.S.

7. A fever is a necessary condition for having the flu but it does not appear to be a sufficient condition.

9. Having a parent is a sufficient condition for being a child (since if you have a parent you must be that parent's child). Having a parent is also a necessary condition (since if you don't have a parent, then you can't be anything, much less a child). Once again, we have a necessary and sufficient condition for something that is not a definition.

Of course, you can take a different interpretation of what "having a parent" means, namely, one currently has a living parent. If this were the interpretation, then the answer would be: Having a parent is a sufficient condition for being a child but it is not necessary (since your parent may have died).

11. Being a male is not a sufficient condition for being an uncle, but it is a necessary condition (since only males can be uncles).

13. Education is not a sufficient condition for a B.A. (since one can be educated and not have a B.A.). It seems to be a necessary condition for a B.A. (since presumably if you are not educated, then you can't have passed the course work necessary to obtain a B.A.). Of course, part of the question concerns what one means by "education," and on a different interpretation, the answer might be different.

15. Oxygen is a necessary but not sufficient condition for a fire.

Exercise 4.4.

A. 1. Big cities voting democratic is a sufficient condition for a Democratic candidate winning.

3. Ross' helping Bill is a necessary condition for Bill's success.
5. Getting a band is a necessary condition for a dance.
7. Going home is a necessary condition for getting a drink.
9. A good library is a necessary condition for greatness in a University.
11. Being a doctor or hospital administrator is a necessary condition for being eligible.
13. Doing a heart transplant is a necessary condition for living (= not dying).

B.

1.

		If. . .then___	. . .only if___
(A) is a sufficient condition for (B)	entails	If A then B	= A only if B
(C) is a necessary condition for (D)	entails	If D then C	= D only if C

2. (a) "(X) is a necessary and sufficient condition for (Y)" implies:

Check each	(X) if X, then Y
correct	(X) X only if Y
answer	(X) If Y, then X
	(X) Y only if X

C.

1. The necessary conditions were: a supply of fuel (the leaves), an increase in the temperature of the fuel (the lighted match), and oxygen. The sufficient condition was the conjunction of these three factors.

D. Answers:

1. The bartender is claiming that an I.D. is a necessary condition for getting some booze, i.e., if no I.D. then no booze = $(\sim q \supset \sim p)$.

3. Here, the teacher is claiming that if there's no snow on Friday, then we will have class = $(\sim q \supset p)$

5. If I don't have to write a term paper, then I will go to Mesa Verde over Spring break: $(\sim q \supset p)$

Exercise 4.5

1. Invalid. Fallacy of affirming the consequent.
3. Valid. Hypothetical Syllogism.
5. Invalid. Fallacy of Affirming a Disjunct.

7. Valid. Modus Tollens.
9. Valid. Modus Tollens.

Exercise 4.6

1. It is false that (the U.S. did not pledge support to the Zulu.) = ~(~t). This is false because t is false.

3. It is false both that (the U.S. pledged support to the Zulu) and (Russia pledged support to the Afrikaners) = (~t & ~u) is true because t is false and u is false.

Exercise 4.7

1. This is Thursday and we are not in Belgium.

3. Either Sally is not a good candidate for president or Susan is not a good candidate for president.

5. Louise will win the Coors Classic and she can not [or did not] beat Karen once more; or Louise can [or did] beat Karen once more and Louise will not [or did not] win the Coors classic.

Exercise 4.8

1.

p q	~(p & q) ≡ (~p v ~q)
T T	F T T T T F F F
T F	T T F F T F T T
F T	T F F T T T T F
F F	T F F F T T T T

Exercise 4.9

1.

p q	[p ⊃ (p v q)]
T T	T T T T T
T F	T T T T F
F T	F T F T T
F F	F T F F F

3.

p q r	{[(p ⊃ q) & (q ⊃ r)] ⊃ (p ⊃ r)}
T T T	T T T T T T T T T T T
T F T	T F F F F T T T T T T
F T T	F F T F T T T T F T T
F T F	F T T T F T F T F T F
T F F	T F F F F T F T T F F
F F T	F T F T F T T T F T T

FTF FT T F TFF T FT F
FFF FT F T FTF T FT F

5. p q	{[~q & (p ⊃ q)]} ⊃ ~p}
T T	F F T T T T F
T F	T F T F F T F
F T	F F F T T T T
F F	T T F T F T T

Chapter 5

Exercise 5.1

1. Strength of Conclusion: the conclusion drawn is moderately strong ("looks excitingly good").

Scope of Conclusion: The conclusion appears to be generalized to a large population: all people with arrhythmia.

Amount of Evidence: weak, since only one subject was used.

Variety of Evidence: weak, since only one kind of evidence (lethal arrhythmia) is cited.

Overall Evaluation: this appears to be a weak argument (primarily because of the amount of evidence), but since the strength of the conclusion is itself weak, the argument is a relatively strong one.

3. Strength of Conclusion: the conclusion drawn is moderate one: "seem."

Scope of Conclusion: the conclusion concerns a general hypothesis, and raises questions about its truth; hence its scope is extensive.

Amount of Evidence: only a single subject was used, but since it is being used to disconfirm a hypothesis, one doesn't need a great deal of negative evidence.

Variety of Evidence: only one type of evidence is mentioned.

Overall evaluation: this is strong as an inductive argument because the evidence disconfirms the hypothesis. If the experiment had been one involving confirmation, it would have been much weaker.

5. Strength of Conclusion: the conclusion indicated is drawn in a strong way: marijuana use leads to heroin use.

Scope of Conclusion: extensive since the claim is that smoking marijuana leads to heroine.

Amount of Evidence: considerable evidence is cited (940 patients).

Variety of Evidence: not very varied, since they only used heroin addicts.

Overall Conclusion: a strong inductive argument because the amount of evidence seems more important than the variety. But as we will see later, "correlation does not entail causation." Hence, it will have to be weakened somewhat.

7. Strength of Conclusion: strong ("must").

Scope of Conclusion: narrow since it applies only to one case—Barbara. The presumption seems to be, however, that everyone with this skin color and loss of energy probably has hepatitis. Hence, the scope would be more extensive on this interpretation.

Amount of Evidence: not extensive since it mentions only skin color and loss of energy of one patient, but this is probably a sufficient amount.

Variety of Evidence: strong since two different kinds of evidence are offered.

Overall evaluation: a strong inductive argument, especially if the conclusion only includes Barbara.

Exercise 5.2

1. Inference to the best explanation:
   - Certain facts are puzzling: primitive man discovered fermentation, effects of alkaloids and toxins, arts of roasting, etc.
   - If a certain hypothesis were true (primitive man was a keen observer of accidents), then these phenomena would be explained
   - Thus this hypothesis is very plausible
3. Statistical Generalization:
   - 1/3 of sampled Americans over 12 have tried some kind of illicit substance
   - Thus, 1/3 of all Americans over 12 have tried some kind of illicit substance
5. Inference to the best explanation:
   - Puzzling facts are observed (medical researchers can identify no therapeutic powers in Laetrile; patients have recovered from cancer after taking Laetrile)
   - These facts would be explained if the placebo hypothesis were true
   - Thus, the placebo hypothesis is the best explanation
7. Statistical syllogism:
   - 3% of cigarette smokers die from lung cancer
   - I am a cigarette smoker.
   - Hence, I will die from lung cancer [probability is very low]
9. Hypothetico-deductive method involving falsification:
   - If H is true (sleep is needed to restore energy), then O should be true (runners should need more sleep than ordinary people and should fall asleep immediately after their race)

But O is false
Thus, H is false

Chapter 6

Exercise 6.1

1. a. 1/52 b 1/52 + 1/52 = 2/52. c. 4/52. d. 4/52 + 4/52 + 4/52 = 12/52. e. = 40/52. f. 4/52 + 13/52 - 1/52 = 16/52. g. 4/52 + 13/52 - 1/52 = 16/52. h. 4/52 + 39/52 - 3/52 = 40/52.

3. .2 + .15 = .35 .2 + .15 - .1 = .25.

5. Probability (College Graduate or Mechanic) = P(C) + P(M) - P(C & M) = .4 + .2 - .05 = .55.

7. In three tosses of a fair coin, there are eight possible combinations of heads and tails ($2^3 = 8$), only one of which is not "at least one tail," i.e., HHH. The probability of this is 1/8. Hence, the probability of at least one tail in 3 tosses = 1 - probability (not at least one tail in 3 tosses) = 1 - 1/8 = 7/8 = .875.

Exercise 6.2

1. 2/2v4v6 = 1/3. 2v4v6/1v2v3v5v6=3/5.

3. Since the two events are independent, P (Heads and Heads) = P(Head) x P(Head) = .5 x .5 = .25.

5. a. P(A&C)=P(C)P(A/C). b. P(C & B) = P(C)P(B/C). c P(A/C) = P(C & A)/ P(C).

Exercise 6.3

1. Probability (Coal Miner/Respiratory Disease) =

$$\frac{\text{P(Coal Miner) x P(Resp.Disease/Coal Miner)}}{\text{P(Resp.Disease)}} = \frac{.005 \times .4}{.01} = .2$$

3. P(Cancer/Work in Sun) =

$$\frac{\text{P(Cancer) x P(Work in Sun/Cancer)}}{\text{P(Work in Sun)}} = \frac{.01 \times .9}{.4} = .0225$$

Exercise 6.4

1. Use the second form of Bayes' Theorem. P(H) = .2 P(~H) = .8. P(E/H) = .0625, P(E/(~H) =1.0. Then

$$= \frac{.2 \times .0625}{[.2 \times .0625] + [.8 \times 1.0]} = \frac{.0125}{.8125} = .015$$

3. Use the third form of Bayes' Theorem. Then

$$P(H_1/E) = \frac{.7 \times .6}{[.7 \times .6] + [.2 \times .1] + [.2 \times .6]} = \frac{.42}{[.42] + [.02] + [.12]} = .75$$

Exercise 6.5

1. a 1/6. b 5/6. c. 1/2 d. ½.

3. 1/8. The Principle of Indifference is not appropriate here, because the eight competitors do not all have an equal chance of winning due to different abilities, diet, training, etc.

Exercise 6.6

1. .51.
3. .27.

Exercise 6.7

1. First calculate all possible state descriptions (of which there are 8). In the c+ function, each of these receives an equal probability value: $1/8^{th}$. This results in:

$$c+(Pc/Pb) = \frac{m+(Pc/Pb)}{m+(Pb)} = \frac{(Pc \,\&\, Pb)}{Pb} =$$

$$\frac{\text{state description } \#1\&\#4}{\text{state description } \#1,\#2,\#4\&\#6} = \frac{1/8+1/8}{1/8+1/8+1/8+1/8} = .5$$

Then recalculate the probability assignments, assigning equal probabilities to structure descriptions (of which there are 4). Each of these structure descriptions receives a value of $1/4^{th}$. Because there are three state descriptions in one of these structure descriptions, each of these three receives a value of $1/12^{th}$. The same applies to the other structure descriptions. This results in:

$$c^*(Pc/Pb) = \frac{m^*(Pc/Pb)}{m^*+(Pb)} = \frac{(Pc \,\&\, Pb)}{Pb} =$$

$$\frac{\text{state description \#1\&\#4}}{\text{state description \#1,\#2,\#4\&\#6}} = \frac{1/4+1/12}{1/4+1/12+1/12+1/12} = .67$$

2. Our hypothesis is everything: Ba & Bb & Wa & Wb. Our evidence is: Ba & Bb. With a c+ function, each state description

$$\frac{c+(\text{Ba \& Bb \& Wa \& Wb})}{\text{Ba \& Bb}} = \frac{\text{state description \#1}}{\text{state description \#1,\#2,\#3\&\#6}} =$$

$$\frac{1/16}{1/16+1/16+1/16+1/16} = 1/4$$

But with a c* function, we assign each probability to structure descriptions of which there are five: no negatives, 1 negative, 2 negatives, 3 negatives and all negatives.

(1)	Ba	Bb	Wa	Wb	1/5
(2)	Ba	Bb	Wa	~Wb	1/20
(3)	Ba	Bb	~Wa	Wb	1/20
(4)	Ba	~Bb	Wa	Wb	1/20
(5)	~Ba	Bb	Wa	Wb	1/20
(6)	Ba	Bb	~Wa	~Wb	1/30
(7)	Ba	~Bb	Wa	~Wb	1/30
(8)	~Ba	Bb	Wa	~Wb	1/30
(9)	Ba	~Bb	~Wa	Wb	1/30
(10)	~Ba	Bb	~Wa	Wb	1/30
(11)	~Ba	~Bb	Wa	Wb	1/30
(12)	~Ba	~Bb	Wa	~Wb	1/20
(13)	Ba	~Bb	~Wa	~Wb	1/20
(14)	~Ba	Bb	~Wa	~Wb	1/20
(15)	~Ba	~Bb	~Wa	Wb	1/20
(16)	~Ba	~Ba	~Wa	~Wb	1/5

With this assignment, we have:

$$\frac{c+(\text{Ba \& Bb \& Wa \& Wb})}{\text{Ba \& Bb}} = \frac{\text{state description \#1}}{\text{state description \#1,\#2,\#3\&\#6}} =$$

$$\frac{1/5}{1/5+1/20+1/20+1/30} = .6$$

Exercise 6.8

With odds of 2;3, his subjective probability is 2/2 +3 = 2/5 = .4. With odds of 6:7, it is 6/13 = .46; with odds of 7:7 it is 7/14 = .5; with odds of 8:2, it is 8/10 = .8

Exercise 6.9

1.

$$P'(H) = P(H/E) = \frac{P(H)P(E/H)}{P(E)} = \frac{.3x1.0}{.4} = .75$$

You should update your revised assessment of H from .3 to .75. This difference shows the effect of observing the new evidence E.

3. Originally, P(H) was .56 (5/9), and later it was .65, i.e., P'(H).65. Hence P(H/E) = .65. By Bayes' Theorem,

$$P(H/E) = \frac{P(H)P(E/H)}{P(E)} = \frac{.56xP(E/H)}{.3} = .65.$$

Solving for P(E/H) gives a value of .35.

Chapter 8

Exercise 8.3

1. A. Animal, vertebrate, mammal, feline, lynx, wildcat.
2. A. Connotation, denotation, connotation, denotation, connotation, extension, connotation

C. Extension, connotation, denotation, connotation.

Exercise 8.5.

Definiendum	*Definiens*	
	Genus	*Differentia*
Bachelor	Male	Unmarried
Boy	Man	Young
Brother	Sibling	Male
Child	Offspring	Young
Giant	Man	Very Large
Midget	Man	Very Small
Snack	Meal	Very Small

Exercise 8.6.

1. a. If Joan is placed on a set of scales, the pointer reads '120.'

c. If litmus paper is placed in this liquid, then the litmus paper turns read.

e. If a body is placed near this object, then this body will be attracted by this object.

2. a. Operational.
   c. Ostensive
   e. Synonymous and connotative.
   g. Genus and differentia
   i. Operational
   k. Connotative

Exercise 8.7.

1. Theoretical and reportive. Reportive is better since "sublimation" has an established usage in chemistry and the definition does not appear to be very theoretical.
3. Reportive.
5. Stipulative
7. Precising, and theoretical (to a small degree).
9. Precising, stipulative. Precising seems better.

Exercise 8.8.

1. This does not state the essential characteristics.
3. Too narrow.
5. Circular.
7. A. Too broad. B. Too narrow: it excludes mathematics and those parts of science that are not concerned with empirical testing.
9. Circular.

Chapter 9

Exercise 9.1

1. Age of Degrees

Age	Frequency	Relative Freq.	Cum. Frequency	Cum. Rel.Freq.
20-24	2	2/12=.17	2	.17
25-29	3	3/12=.25	5	.42
30-34	1	1/12=.08	6	.5
35-39	3	3/12=.25	9	.75
40-44	3	3/12=.25	12	1.0
	12	1.0		

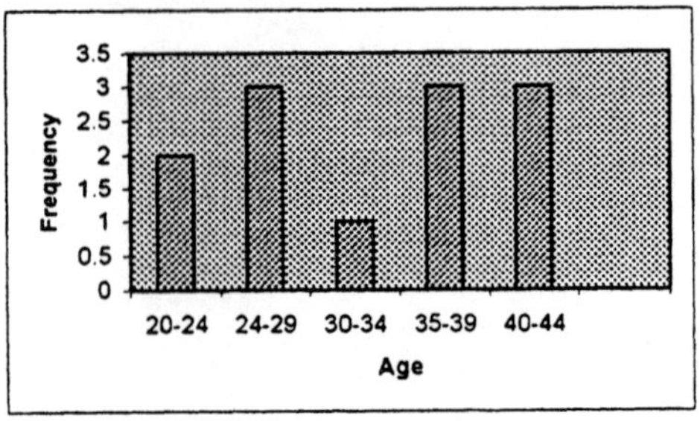

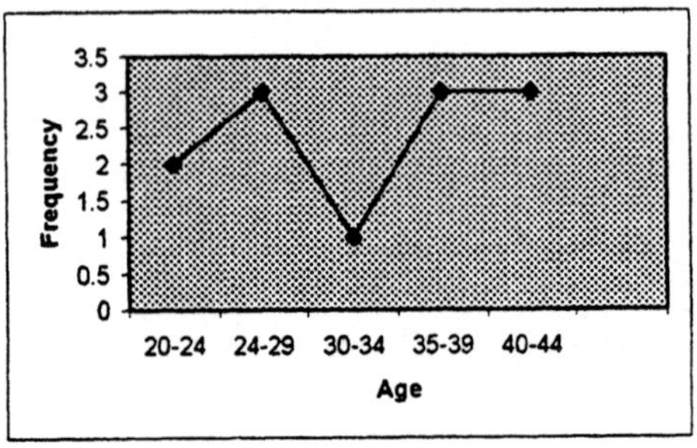

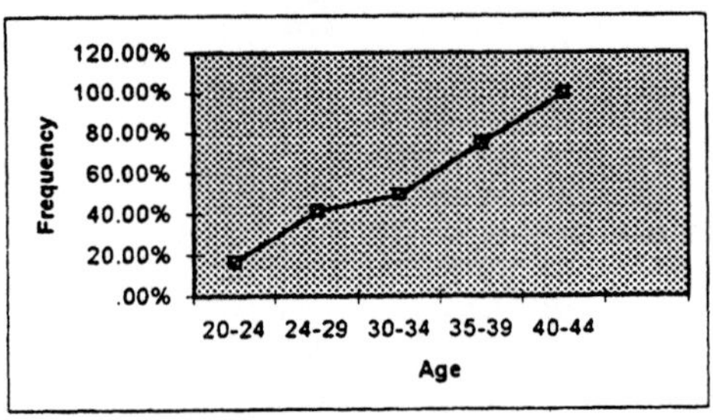

The distribution is multi-modal

3. Cholesterol:

Cholesterol Level	Frequency	Relative Freq.	Cum Freq	Cum. Rel.Freq.
195-199	1	1/20=.05	1	.05
200-204	3	3/20=.15	4	.2
205-209	4	4/20=.2	8	.4
210-214	7	7/20=.35	15	.75

215-219	4	4/20=.2	19	.95
220-224	1	1/20=.05	20	1.0

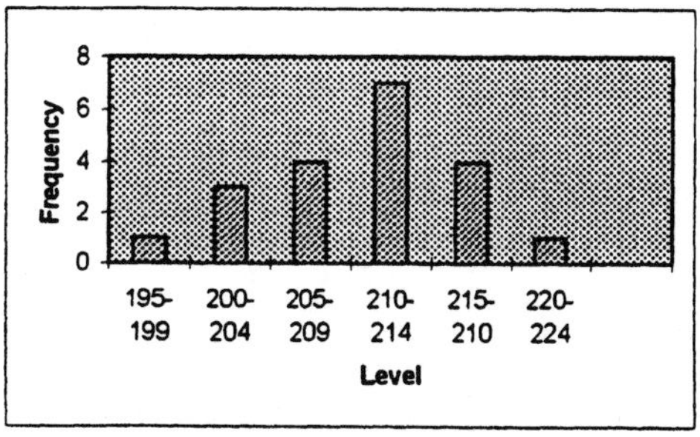

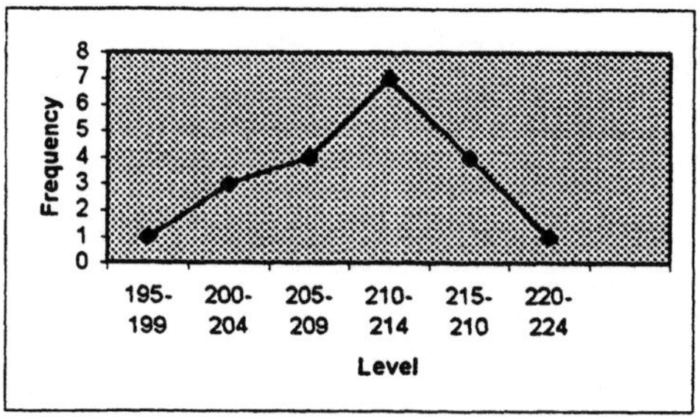

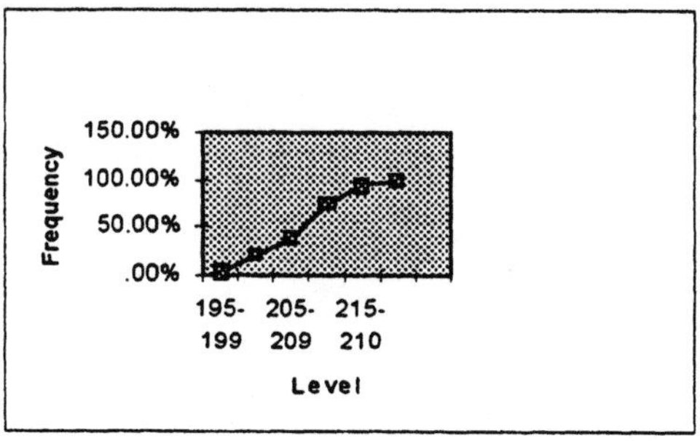

The distribution is slightly left-skewed

Exercise 9.2

1. Mean: 33.5. Median: 34.5. Mode: 37, 28, and 24. Variance: 47.6. Standard Deviation: 6.9

3. Mean: 210. Median: 210. Mode: 210. Variance: 35. Standard Deviation: 5.9

Exercise 9.3

1. The z-scores for 32 and 36 are:

$$z = \frac{32 - 33.5}{6.9} = -.2 \qquad z = \frac{36 - 33.5}{6.9} = +.36$$

3. The z-scores for 204 and 210 are:

$$z = \frac{204 - 210}{5.9} = -1.0 \qquad z = \frac{210 - 210}{5.9} = 0$$

Exercise 9.4

1. Heart Rate:

(1) Scatter Diagram:

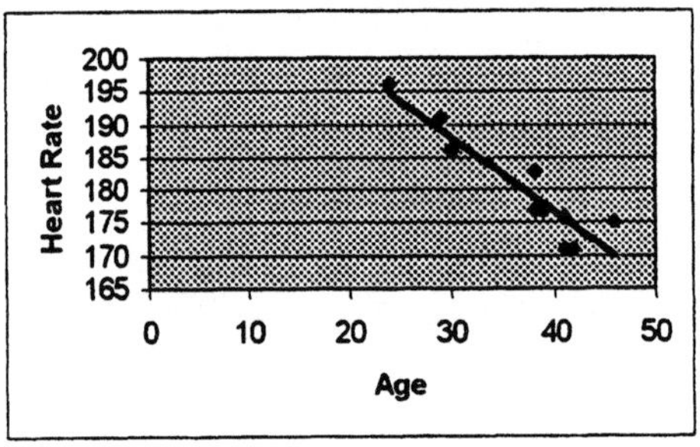

x: age	y: heart rate	Zx	Zy	ZxZy
30	186	-1.1	.75	-.8
38	183	.185	.38	.07
41	171	.65	-1.1	-.73
38	177	.185	-.375	-.07
29	191	-1.2	1.375	-1.65
39	177	.34	-.375	-.13
46	175	1.4	-.625	-.875
41	176	.65	-.5	-.325
42	171	.8	-1.125	-.9
24	196	-1.96	2	-3.92

$\bar{x}=36.8$ $s_x=6.5$	$\bar{y}=180$ $s_y=8$			-9.33

Calculation of Regression Equation:

$$r=\frac{\Sigma Z_x Z_y}{n}=\frac{-9.33}{10}=-.93$$

$$b=\frac{rs_y}{s_x}=\frac{(-.93)(8)}{(6.5)}=-1.15$$

$$a=\bar{y}-(\bar{x}b)=180-(36.8)(-1.15)=222$$

$$Y=222-1.15X$$

(2) The regression equation is: Y= 222-1.15X
(3) Peak heat rate tends to decrease as age increases
(4) The slope represents the fact that the peak heart rate an individual can reach during intensive exercise decreases by 1.14 for each increase in age of 1 year.
(5) Predictor variable = age; response variable = peak heart rate
(6) $r=-.93, r^2=.86$
(7) 86% of the variance in heart rate is due to the regression line.
(8) for a 28-year old, the predicted value is 190.34

3. Cancer
(1) Scatter diagram:

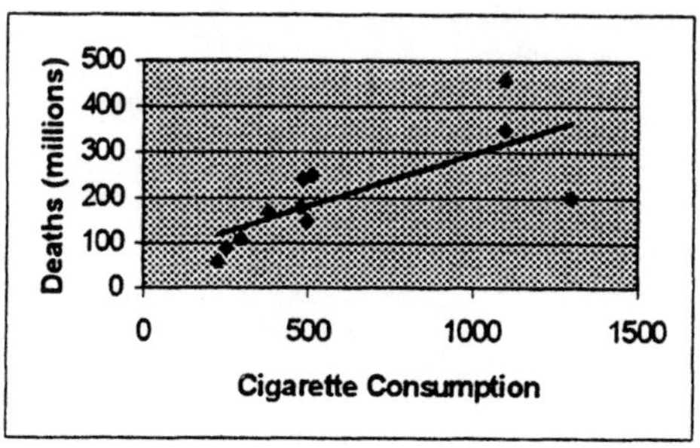

x: cigarettes	y: deaths
480	180
500	150
380	170
1100	350
1100	460
230	60
490	240
250	90
300	110
510	250
1300	200
x=607 s=378.5	Y=205.5 S=117

$r=.74 \quad r^2=.54$

$$b = \frac{rsy}{sx} = \frac{.74(117)}{387.5} = .223$$

a=205.5 –(607)(.223)=71

(2)Y=71 + .22X
(3) Deaths from lung cancer (in millions) increases with per capita cigarette consumption
(4) The slope represents the fact that lung cancer (in millions) increases by .22 (.22 million) for every increase in one cigarette per capita.
(5) Predictor variables: per capita consumption of cigarettes; responses variable: lung cancer death (in millions)
(6) $r=.74$ $r^2=.54$.
(7)The regression line explains 54% of the variance in lung cancer deaths (in millions).
(8) For per capita cigarette consumption of 800 cigarettes, the predicted number of lung cancer deaths (in millions) is: Y=71 + .22(800)=247.

Chapter 10

Exercise 10.1

1. The effect (e) is: immunity from cowpox; the antecedent factors (B, C, ...) here are not specified but presumably they differ from individual

to individual so that none of them is common except the cause, which is "taking cowpox" which we can symbolize as "A." The conclusion is taking cowpox (A) causes immunity (e).

Individuals	Antecedent Factors	Effect
SP	A B - - - - - -	e
MB	A - C - - - - -	e
EW	A - - D - - - -	e
SN	A - - - E - - -	e
JM	A - - - - F - -	e
WR	A - - - - - G -	e
. . .	A - - - - - - H	e

3. The effect(e) is interference with speech. No antecedent factors are mentioned other than fear (A); so one may assume they differ from individual to individual. Conclusion: fear (A) is the cause of interference with speech (e).

Groups	Antecedent Factors	Effect
Combat Anxiety	A B -	e
Stage Fright	A - C	e
Animals	A - -	e

5. Effect e: cataracts. Although a common antecedent factor was being a physicist, in the present case one can ignore it since (presumably) this was not causally related to cataracts. The cause: nuclear radiation (A). Conclusion: nuclear radiation (A) is the cause of cataracts (e).

Individuals	Antecedent Factors	Effect
#1	A B C - - - -	e
#2	A - - D E - -	e
#3	A - - - - F G	e
. . .		

Exercise 10.2

1. This is the *Method of Difference*. Cause (A): removal of brain; effect e: uninhibited movement. Conclusion: removal of brain causes uninhibited movements.

Individuals	Antecedent Factors	Effect
#1	- - B C	-
#2	A B C	e

3. Two studies are reported here. The first one employs the Joint Method of Agreement and Difference:

	A: husked rice	e: disease
Individuals	Antecedents	Effect
#1	A B - -	e
#2	A - C -	e
#3	A - - D	e
#4	- E - -	-
#5	- - F -	-
#6	- - - G	-

Thus, A is the cause of e.

The second one is a comparison between two groups: the diseased group before they ate the husks and the same group after they ate them. This can be represented in either of two ways—by the Joint Method:

Individuals	Antecedents	Effect
#1	A B - -	E
#2	A - C -	E
#3	A - - D	E
#1	- E - -	-
#2	- - F -	-
#3	- - - G	-

or by the Method of Difference involving three pairs:

#1 ABC	e	A is the cause of e
#1 - BC	-	
#2 AEF	e	A is the cause of e
#2 - EF	-	
#3 AIJ	e	A is the cause of e
#3 - IJ	-	

Could one do the same thing with the first employment of the Joint Method? Only if one had matched pairs, i.e., two individuals who had all factors in common except the cause. But note that in this example there are no such matched pairs. Hence, whether one can use the method of difference or not depends upon the issue of whether the individuals in the two groups are matched or not.

Exercise 10.3

1. Let B be alimentary factors (appetite, hunger, need for food), which is known to cause e_1: hoarding of worthless inedible pellets covered with aluminum foil. Let A be sensory and perceptual factors. Let e_2: hoarding of plain food pellets.

AB cause e_1 and e_2
B is known to cause e_1
Thus A causes e_2

Individuals	Antecedent Factors A B	$e_1 + e_2$
	B	e_1
	-	e_2

3. Let G: glass, DW: distilled water, P: perspiration; A: atmosphere. e (the total effect) is divided into e_1, e_2, e_3, e_4, G, DW, P and A. are the likely causes of $e_1 + e_2 + e_3 + e_4$

Antecedent Factors: G, DW, P, A effect= $e_1 + e_2 + e_3 + e_4$

DW	--	--	--	e_1
--	G	--	--	e_2
--	--	P	--	e_3
--	--	--	A	e_4

Thus A is the cause of e_4.

Exercise 10.4

1. Leukemia: A: distance from the bomb, e: incidence of Leukemia

A+: 1 kilometer	e+: few survivors
A++: 1.0-1.5 kilometers	e++ 13,730 survivors
A+++: 1.5 -2 kilometers	e+++: 23,060 survivors
A++++: 2 kilometers	e++++: 156,400 survivors

Group	Antecedent Factors	Effect
#A	A+BC	e+
#B	A++BC	e++
#C	A+++BC	e+++
#D	A++++BC	e++++

Thus, A and e are causally connected (inversely related).

3. Pituitary case: A: size of pituitary e: size of person
A: normal size e: normal
A+: large size e+: giant
A-: small size e-: dwarf

	Antecedent Factors	Effect
Normal	A	e
Giant	A+	e+
Dwarf	A-	e-

Therefore, A is the cause of e.

Exercise 10.5

1. *Method of Residues*: Here ABC are the known particles and e_1, e_2, e_3 is the total energy of the nuclear system.

Antecedent Factors A B C	Effect: $e_1 + e_2 + e_3$
A	e_1
B	e_2
C	e_3

Thus Particle C is the cause of e_3.

3. *Method of Difference*: the relevant antecedent circumstance is the injection (A) versus the absence of the injection; the relevant effect is the presence of spirochetes (e) versus their absence.

Individual	Antecedent Factor	Effect
#1	A B C	e-
#1	- B C	e

Chapter 11

Exercise 11.1

1. Perhaps the most startling discovery made in astronomy this century is that the universe is populated by billions of galaxies and that they are systematically receding from one another like raisins in an expanding pudding.

It is being claimed that the recession of the universe is analogous to the expansion of raisins in a pudding.

3. One of the pleasures of science is to see two distant and apparently unrelated pieces of information suddenly come together. In a flash what one knows doubles or triples in size. It is like working on two large but separate sections of a jigsaw puzzle and, almost without realizing it until it happens, finding that they fit into one.

Here, it is being claimed that scientific discovery is like solving a jigsaw puzzle.

Exercise 11.2

1. [Artistic energy and its periodic release in a burst of creativity can be compared to blowing up a balloon until it pops.] The building up of energy can be compared to the filling of the balloon. The pop of the balloon, like creativity, appears to be sudden but it has really been preceded by a storing up of vital energy.

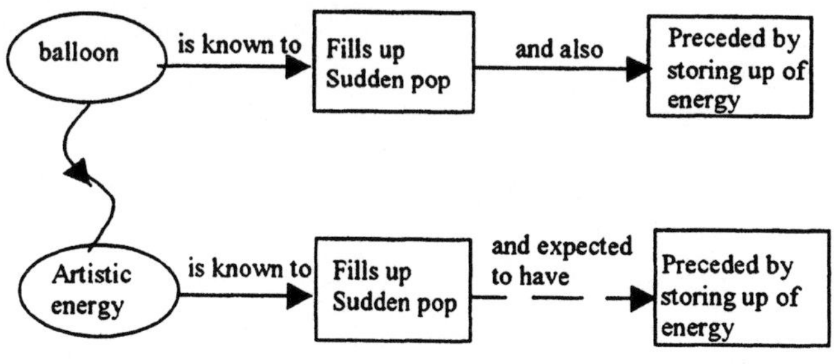

Descriptive analogy: a balloon is like artistic energy.
Basis of Resemblance: both balloons and artistic energy "fill up" over time and if no pressure is released, they pop suddenly.
Linking Generalization: the building up of pressure is preceded by the storing up of energy. The linking generalization appears to be strong.

2. This should be rewritten as: [The universe is not formed by a fortuitous concourse of atoms] any more than the accidental jumbling of the alphabet would fall into a most ingenious treatise of philosophy.

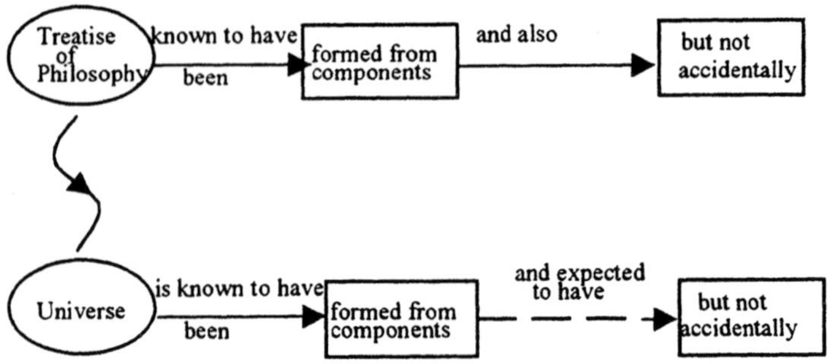

Descriptive Analogy: The universe is like a treatise.
Basis of Resemblance: universe is formed from components (atoms) and treatise of philosophy is formed from components (alphabet).
Linking Generalization: when treatises of philosophy are formed from components, they are not accidentally arranged. The linking generalization appears strong.

5. It seemed to us that the synchronous behavior of malaria parasites, all coming to cell division at the same time every 24 hours or a multiple thereof, was remarkably like the behavior of microfilariae, all entering or leaving the peripheral blood at the same time every 24 hours. Microfilariae are carried from one patient to another by mosquitoes, which suck blood mostly at night, and the swarming of the microfilariae in the peripheral blood is arranged to coincide with this time of sucking blood. The biological purpose of the cycle of the microfilariae is clearly to help them encounter mosquitoes and so get transmitted to new patients. Since malaria is also carried from one person to another by mosquitoes, it seemed to us that [the periodic behavior of the malaria parasites in the blood was similarly designed somehow or other to facilitate transmission by mosquitoes].

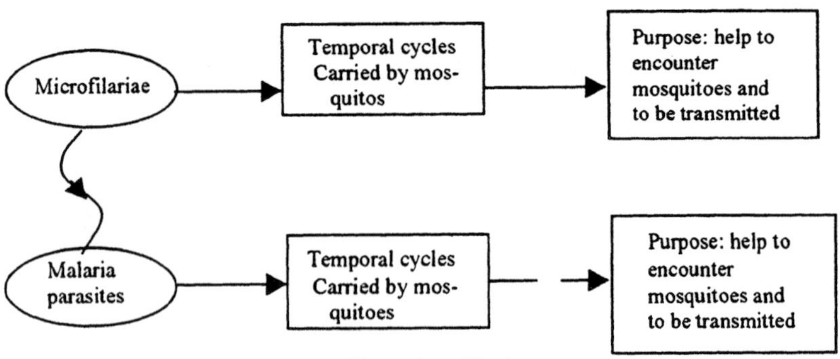

Descriptive Analogy: Malaria is like microfilariae.

Basis of resemblance: 24 hour temporal cycle in both cases, and both are carried by mosquitoes.
Linking Generalization: when microfilariae are carried by mosquitoes, the swarming of mosquitoes coincides with mosquito sucking. It seems to be moderately strong. It does seem to be a biological law-like regularity, but how strong is it? Is there, e.g., an underlying biological mechanism accounting for the law? Furthermore, the linking generalization seems to be very restrictive and particular, applicable only to the behavior of mosquitoes.

Exercise 11.3

A.

1. d e a c b. The probable truth of the statement decreases because the statements are progressively more precise (exact) and hence more likely to be false.

B.

1.a. More probable because of greater similarity in age; young rats are probably more virile.

b. More probable because isolation from all comforts would increase drive.

c. The difference may decrease the probability because the sexual behavior of gray rats may be quite different from albino rats.

d. Privation of food would probably strengthen the drive and thus strengthen the conclusion.

e. It may or may not be relevant depending on how color affects rat perception.

Exercise 11.4

1. Hamlet's Clock.

Structure of Argument:

Dr. Haldane's clock, *Hamlet*, and living organisms are produced by a process involving careful selection, and are alterable in a random way (by shooting a bullet through the clock, by introducing a typographical error into *Hamlet*, and by radiation causing a mutation in an organism).

In Dr. Haldane's clock and *Hamlet*, random alterations would almost invariably be detrimental.

Therefore, in living organisms random alterations (mutations) would almost invariably be detrimental.

Evaluation of Argument:

1. *Strength of Conclusion*: the conclusion seems to be stated in very strong terms—that mutations "almost" invariably are detrimental. Thus, the argument is weak.

2. *Number of Entities*: living organisms are analogous to only *two* entities, which would make it a weak argument. However, if Dr. Haldane's clock is regarded as representing all carefully constructed mechanisms, and if *Hamlet* is regarded as representing all carefully constructed works of art, then living organisms would be held to be analogous to a great number of entities. Hence, it would be strong.

3. *Variety of Premises*: only two kinds of things are mentioned in the premises mechanisms and works of art and these seem to be dissimilar to each other. Hence, the argument is strong according to this criterion.

4. *Similarity*: only two features are mentioned selection and random alterability. Hence, according to this criterion the argument is weak.

5. *Dissimilarity*: there are many dissimilarities between living organisms and mechanisms and works of art, e.g., a clock is mechanical and not a living organism. Hence, a living organism could adapt to a mutation, whereas a clock cannot repair itself. A typographical error can be erased, whereas a bullet or mutation is more permanent. A bullet is mechanical, external and destructive, whereas a mutation is organic, internal and not always detrimental. Furthermore, there is an important difference between one mutation versus many mutations compared to one random change versus many such changes in works of art and mechanisms.

6. *Relevance:* since random mutations are causally relevant to the structure and value of an entity, the relevance of the analogy is strong.

Overall Evaluation: overall, the argument appears to be moderately strong.

3. Insulin:

Structure of Argument:

Drug addiction and insanity are involuntary diseases and both are organically based.

Insulin helps drug addicts.

Thus, insulin should help insane.

Evaluation of Argument:

1. *Strength of Conclusion:* the conclusion is stated in weak terms. Hence, the argument is strong.

2. *Number of entities in the premises*: not mentioned.

3. *Diversity of premises*: weak. They were all addicted to morphine, so there is no variety here. No other kinds of variety seem apparent,

e.g., were they of different brain types (no), personality types (unknown), etc. Hence the argument is weak.

4. *Positive Similarity*: Few similar features are noted; however, drug addicts manifest psychotic symptoms (e.g., morbid fears, anxieties, persecution, jumpy nerves) just as the insane do. In addition, one can suggest other ways in which they are similar (e.g., involuntary affliction, organically or chemically based). The argument appears to be moderately strong on this criterion.

5. *Dissimilarity*: There are important dissimilarities between morphine addicts and psychotic patients: one is drug caused, whereas it is unclear what causes insanity. Many individuals believe insanity, if there is such a thing, is not organically caused at all but rather socially caused. In morphine addiction, there is a chemical dependency which gradually increases but this is not so with insanity. Thus, on this criterion, the argument is weak.

6. *Relevance*: This is unclear. If we use the linking generalization between organic or chemical cause and symptoms, we would seem to be on relatively good grounds.

Overall evaluation: weak. This is really a sloppy use of analogy. Part of the problem is with a careless definition of "insanity" and how, in particular it relates to one type of insanity--psychosis. A better analogical argument would have been between morphine addiction and psychosis. But again the dissimilarity between these two seems to be very great. Thus, overall, the argument seems to be quite weak.

5. *Science and myth*:

Structure of Argument:

Myth is characterized by its metaphorical status

Science is like myth (neither is true)

Thus, science is characterized by its metaphorical status

Evaluation of Argument:

1. *Strength of conclusion*: the conclusion is stated strongly. Hence, the argument is weak.

2. *Number of entities in the premises*: although only one kind of case, (e.g., poetry), is mentioned, presumably there are lots of examples of poetry. Here the argument is strong.

3. *Diversity of premises*: little variety is mentioned between types of myth. Hence, it is weak.

4. *Positive Similarity*: there is some similarity between myth and science: neither are absolutely true; (it is claimed) both use "theory" as "metaphorical linkage between facts," etc. Here the argument seems moderately strong.

5. *Dissimilarity*: there seems to be a strong dissimilarity: science progresses and gets closer to the truth but myths don't; myths aren't temporarily true; theory in science is not a metaphorical linkage between facts. Hence, the argument is weak.

6. *Relevance*: the relevance aspect is particularly important here. The linking generalization seems to be: if something is not unrestrictedly true, and if theory is a metaphorical linkage between facts, then it is not real. But such a linking generalization, it can be argued, is not true.

Overall evaluation: although it would take a long story to tell, this analogy seems weak. Although there are similarities between science and myth, the dissimilarities seem to outweigh the similarities.

Exercise 11.5

1. The Solar system is a model of the atom. There are several ways in which the two seem not to be alike: (l) the revolving electrons, since they are undergoing centripetal acceleration, should emit light, (2) if light is emitted by the orbital electron, this radiation should result in a reduction of the total energy of the electron and its radius should decrease, (3) the model cannot predict that the radiation of light is in the form of discrete frequencies. Later these difficulties were removed by Bohr's model of the atom.

3. A machine is the model of the body. There is a strong positive analogy: both make decisions based on the past, they both decide between two alternatives, they both have a special apparatus for making future decisions based on the past, they both have devices for resisting the increase of entropy. There is also a negative analogy: the nervous system is alive, the machine is not; the nervous system can make creative use of the past, a machine cannot; the nervous system can readapt itself after damage, a machine cannot.

Chapter 12

Exercise 12.1

1 . Pellagra

[1] Background: epidemic of "pellagra" in south, which was endemic; the symptoms of pellagra were: gastric disturbances, skin eruption, nervous derangement thought to be caused by a microbe

[2] Problem: what was the cause of pellagra?

[3] New Observations: O_1: sufferers were in contact with victims

[4] Hypothesis H_1: microbes cause pellagra

[5] Evaluation of H: H_1 would explain O_1

[6] Deduction: if H_1 then I: all persons in close contact with victims should get disease.

[7] Test: if I then Prediction P_1: in a hospital, orderlies, nurses and doctors should catch the disease.

But not-P_1

Thus not-I (by Modus Tollens)

Thus not-H_1 (by Modus Tollens)

[8] Decision: H_1 was rejected

Now the cycle is repeated beginning again with step [3]

[3] New Observations: O_2: only the poor suffered

O_3: victims' diet consisted of mush and grits but no meat, milk

O_4: only children (6-12) had the disease. Children under 6 were given milk; children over 12 were given meat

[4] Hypothesis: H_2: lack of milk and meat is the cause of pellagra

[5] Evaluation: H_2 would explain O_2, O_3, and O_4 [but what about O_1?]

[6] Deduction: if H_2 then I: if these dietary deficiencies are eliminated, then no disease

[7] Test: if I then Prediction P_2: if milk and meat were given to 6-12 year olds [test], then no pellagra should result [result]

P_2 was true

Thus, I was confirmed

Thus H_2 was confirmed

[8] Decision: H_2 was accepted.

3. Columbus

[1] Background: accepted beliefs about the world, e.g., the earth is spherical

[2] Problem: what is the geography of the world?

[3] New Observations: although no new observations were undertaken, several old ones are listed [Note: these "facts" could have been listed under "background"]: 2/3 of earth had been explored; many geographers claimed westward passage to India; Portuguese navigator had sailed 450 leagues west and had picked up carved wood; mariner had reported sighting land to the far west; reeds, unlike any in Europe, floated from west; Westerly winds deposited non-European tree trunks and corpses

[4] Hypothesis: there is land to the west of Europe; it is attainable; it is fertile; it is inhabited.
[5] Evaluation: H could explain the facts mentioned in [3]
[6] Deduction: If H is true, then I: if one sails westward [test], one should find this land [result]
[7] Test: Columbus secures ships and sails westward. Result: he finds land. Thus H is confirmed
[8] Decision: accept hypothesis

Exercise 12.2

1. H_1: Werner's Neptunism (catastrophism) vs. H_2: Hutton's Uniformitarianism. H_1 implied I_1: If molten rock is solidified, it would not become crystalline but glassy and I_2: If limestone is heated, then it will decompose. H_2 implied not-I_1 and not-I_2. Hall showed that a modified version of I_1 is true:

I'_1: if molten rock is cooled slowly, it becomes crystalline, and if molten rock is cooled rapidly, it becomes glassy. Hall also showed that I_2 is true. Thus, since H_2 predicts I'_1 and I_2 and H_1 predicts not-I'_1 and not-I_2, and since I'_1 and I_2 are true, H_2 is confirmed and H_1 is disconfirmed.

3. H_1: Becher & Stahl's phlogiston vs. H_2: Lavoisier's oxygen.
From H_1 one could derive the prediction P: If metal is calcined (i.e., heated so that it results in calx—a powdery residue), then (because it should lose its phlogiston) it should weigh less

So, $H_1 \supset P$ and $H_2 \supset \sim P$

But P was true; thus H_2 was falsified and H_1 was confirmed

Exercise 12.3

1. Data: the data to be explained are the differences between the appearances of objects that are far away and those that are nearby. Hypothesis: the hypothesis proposed is that the distant objects are younger than the nearby ones.

(1) Logically possible: the hypothesis appears to be logically consistent and to be clearly stated.

(2) Relevance: the hypothesis appears to be relevant to the data to be explained since the appearance of things does depend upon their age.

(3) Testable: the hypothesis is testable (in principle) but not directly testable. An indirect test of the hypothesis might involve collecting data on closer galaxies and their respective ages.

(4) Coherence with existing scientific knowledge: given our current knowledge of astronomy, especially cosmology, the hypothesis, if true, would be coherent with this set. We know that galaxies do age over time; our current knowledge of optics is also consistent with this.

(5) Simplicity: it is difficult to assess the overall simplicity of the hypothesis. It appears not to be a very complicated hypothesis since it consists of a relation between two variables.

(6) Scope: its scope is not very broad since it covers just the case of the appearances of galaxies and their age.

(7) Depth: it has minimal depth since the hypothesis is not offering a micro-level hypothesis.

3. Data: the data to be explained consist of slips of the tongue (saying "he" instead of "she"), emotional agitation, and laughter. Hypothesis: the psychoanalytic hypothesis is that Dr. P. unconsciously wished he had obtained a divorce.

(1) Logically possible: the hypothesis appears to be logically consistent (assuming that unconscious desires are logically possible--some would claim they are not). Are the basic theoretical notions--unconscious wish--clearly specified? Not in the present context.

(2) Relevance: the hypothesis is relevant to the data since if it were true, then it would explain why the individual made the verbal mistakes and manifested the behavior he did.

(3) Testable: although the hypothesis is not directly testable--since unconscious wishes cannot be observed and since Dr. P. does not have access to them because they are unconscious, the hypothesis is indirectly testable. Indeed, the passage indicates how corroborating evidence was collected.

(4) Coherence with existing scientific knowledge: this is one of the more difficult questions to answer because psychoanalytic theory is a theory of the nature of the mind. Is this hypothesis consisted with our neurological knowledge? Our knowledge of other psychological processes? Many would argue it is and that the knowledge of these disciplines is consistent with psychoanalysis. Others would deny this.

(5) Simplicity: how simple is the hypothesis? It does not appear to be very simple since (on one interpretation) the notion of an unconscious part of the mind duplicates all of the properties of the conscious but then makes all of these processes unconscious. Hence, it appears to be multiplying entities beyond necessity.

(6) Scope: how much explanatory power does this hypothesis have? It explains a great variety of behavior--slips of the tongue, agitated

behavior and laughter, etc.—that are very diverse. Hence, it appears to have wide scope.

(7) Depth: here the hypothesis appears to have considerable depth since it concerns entities that occur on a level below that of observable phenomena. Although these micro-level processes are not physical, they are psychological.

Chapter 13

Exercise 13.1

1. This is an observation, since Pettit and Nicholson merely measured the temperature. It doesn't say how they measured it but this seems irrelevant: suppose they had actually gone there and stuck a thermometer in the atmosphere. There was no independent variable being manipulated, merely an observation—the measuring of temperature.

3. This is an observation, although it is one used to test two conflicting hypotheses. The observation consisted of examining the age of microorganisms in sediment samples of the earth.

5. Observation. No experiment was performed; instead the percentage of salt in the pond water was measured. Once again, there is no experimental variable being manipulated.

Exercise 13.2

1. [1] indicates there is an observation reported: sufferers were always in contact with other victims. This apparently was not an experiment. [2] indicates another observation he made. [3] obviously reports further observations he made: only the poor suffer. [4] is another observation concerning the diet. [5] is another observation concerning age groups and diet. [6] reports the experiment: he manipulated the independent variable--type of diet—by giving milk and meat to the children (the experimental group). The dependent variable was presence/absence of pellagra. Was there a control group? No.

3. There are two experiments reported here. In the first one, the independent variable was "inoculation" with mutton juice (containing bacteria) into two liquids--mineral solution and turnip solution. The dependent variable was appearance of life. There was no control group here.

5. The independent variable is presentation of sounds from a metronome. The dependent variable is secretion. There is no control group.

7. We have an experimental group and a control group. Independent variable: reinforcing of plural nouns spoken. Dependent variable: percentage of plural nouns spoken.

Exercise 13.3

1. [1] indicates there is an observation reported: sufferers were always in contact with other victims. This apparently was not an experiment. [2] indicates another observation he made. [3] obviously reports further observations he made: only the poor suffer. [4] is another observation concerning the diet. [5] is another observation concerning age groups and diet. [6] reports the experiment: he manipulated the independent variable--type of diet--by giving milk and meat to the children. The dependent variable was presence/absence of pellagra. What type of experiment was this? Was there a control group? No. It appears to be a simple before and after study: observation of a single group followed by treatment followed by observation (O_1 X O_2).

3. There are two experiments reported here. In the first, we have the following situation: weighing sulfur, burning sulfur, then weighing sulfur. Hence, we have O_1XO_2, a before and after experiment. The second experiment is the following: first mercury is weighed, then it is heated, which produces an oxide and disappearance of air. Then the mercury is again weighed and so is the oxide: O_1 X O_2. Then the oxide is burned and weighed, again a case of O_1 X O_2. After inoculating 4 cows, these were divided into 2 groups of 2 each). Then the experimental group received the "cure" whereas the other group did not. This was followed by observations of both groups. This appears to be a pre-test--post-test control group or a post-test only control group.

7. We have an experimental group and a control group. Independent variable: reinforcing of plural nouns spoken. Dependent variable: percentage of plural nouns spoken. In order to determine there was an increase in the percentage of plural nouns spoken, Greenspoon had to have made an initial observation on the frequency of plural nouns. Furthermore, we are not told if the groups are randomized or not. Hence, we have Non-equivalent Control Groups:

Oa X Ob

Oa Ob

If they were randomly assigned, we would have Pre-test—Post-test Control Group:

R Oa X Ob
R Oa Ob

9. Injection of beryllium salt in both experimental and control groups was followed by introduction of the experimental variable (injection of ATA) versus control condition (no injection). The dependent variable was life versus death. With no randomization we have:

X Oa (Static-Group Comparison)
Oa

If randomization occurred, we have Post-test Only Control Group. Assuming that observations were made before the introduction of the experimental variable, namely E observed whether they were alive or not, we would have (with no randomization) Non-equivalent Control Group, or (with randomization), Pre-test Post-test Control Group. Since the initial observation seems trivial and since the important question concerns the difference between experimental and control groups, the better answer seems to be Static Group Comparison or Post-test Only Control Group.

11. This is a time series design. The independent variable was the crackdown on speeding and the dependent variable was number of deaths. The different observations were the various years.

13. This is Nonequivalent Comparison Group Design. The two groups were taken as already constituted without random assignment: how could one randomly assign midshipmen to second year versus third year groups? Here, we are working with "naturally constituted" groups.

Oa (X) M Ob

Oa M Ob

15. This is a retrospective (ex post facto) design. There are two groups: heart attack victims versus non-heart attack victims. Did matching occur?

17. This is a prospective study. The cause was the pill and the observation was death rate. But the pill was never introduced as an experimental variable.

Chapter 14

Exercise 14.1

1. Basketball Players. (1) Because we are not ignoring order and because we are sampling with replacement, we have 25 possible samples of size 2 (N^n=25, where N=total number in the populations, in this case 5, and n = size of the sample, in this case 2).

Sample	$\overline{x}$
76,76	76
76,78	77
76,79	77.5
76,81	78.6
76,86	81
78,76	77
78,78	78
78,79	78.5
78,81	79.5
78,86	82
79,76	77.5

79,78	78.5
79,79	79
79,81	80
79,86	82.5
81,76	78.5
81,78	79.5
81,79	80
81,81	81
81,86	83.5
86,76	81
86,78	82
86,79	82.5
86,81	83.5
86,86	86

$\mu_{\bar{x}} = 80$

$\sigma_{\bar{x}} = 2.4$

Means	Frequency
76	1
76.5	0
77	2
77.5	2
78	1
78.5	4
79	1
79.5	2
80	2
80.5	0
81	3
81.5	0
82	2
82.5	2
83	0
83.5	2
84	0
84.5	0
85	0
85.5	0
86	1

Note: if we have a population of three elements and we take all possible samples of size three, we would have 27 entries; with a population of 500 and a sample size of 5, we would have 31,250,000,000,000. Obviously we can't employ such figures in our calculation. So to simplify matters, we will not assume sampling with replacement, nor will we pay attention to order. In that case, all possible samples of size two from a population of five would just be 10. This is calculated by the formula:

$$\frac{n}{x!(n-x)!}$$

where ! stands for the factorial sign so that 2!=2x1 and 3! = 3x2x1. Hence

$$\frac{5!}{2!(5-2)!} = \frac{5x4x3x2x1}{2x1(3x2x1)} = \frac{5x4}{2x1} = 10$$

where n=the total number in the population and x =equals sample size.

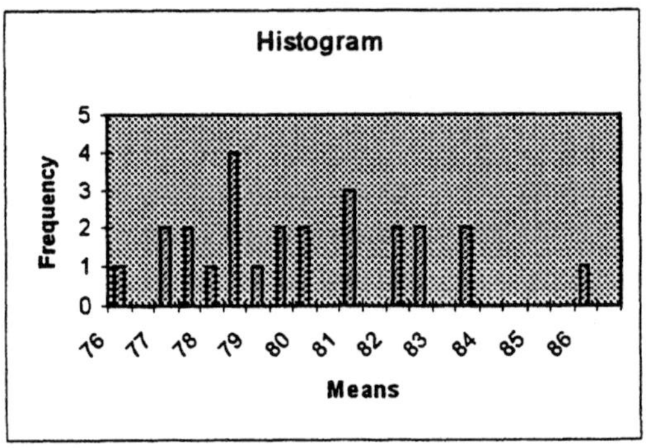

The two histograms illustrate the central limit theorem

(4) $\mu_{\bar{x}} = 80$, $\mu = 80$. $\mu_{\bar{x}} = \mu$. Although several of the values of $\bar{x}$ equal 80 (two of them), most of them do not.

(5) $\sigma = 3.4$. $\sigma_{\bar{x}} = 2.4$. $\sigma_{\bar{x}} = \frac{\sigma}{\sqrt{n}} = \frac{3.4}{\sqrt{.4}} = 2.4$

Exercise 14.2

1.(1a) Since $\bar{x} = \frac{1664}{15} = 110.9$, the estimate of μ is 110.9. (1b) No because it is unlikely that $\bar{x}$ will exactly equal μ. Some sampling error will occur. (2a). A 95% confidence interval would be: 105.9 to 115.9. (2b). It may or it may not, but we can be 95% confident that μ is between 105.9 and 115.9.

3. (1a). 210. (1b). No because it is unlikely that $\bar{x}$ will exactly equal μ. Some sampling error will occur. (2a). A 95% confidence interval would be: 198 to 221.6. (2b). It may or it may not, but we can be 95% confident that μ is between 198 and 221.6.

Glossary

A Priori Evaluation: a preliminary evaluation of a proposed hypothesis occurring prior to the actual empirical testing of the theory (hence 'a priori'). Such evaluation involves rational criteria, indicating how worthy a hypothesis is of being subsequently developed and tested.

A Priori Knowledge: Knowledge obtained independently of (prior to) experience, in the sense that such knowledge is validated independently of sensory experience.

A Posteriori Knowledge: knowledge obtained subsequent to experience, in the sense that such knowledge is validated by reference to sensory experience.

Abductive Inference: a form of inference used by C. S. Peirce (*1839-1914*) in which, on the basis of observed facts, one infers what a likely explanation of it is: facts F are observed; if hypothesis H were true, these facts would be explained; thus, hypothesis H is likely.

Acceptance region: the set of values of the test statistic that leads to the non-rejection of Ho.

Ad Hoc Hypothesis: a hypothesis that is introduced for the sole purpose of explaining an observed phenomenon, usually negative, which casts doubt upon one's hypothesis. Hence the hypo thesis was introduced only for the data in question.

Algorithm: a mechanical rule for generating a solution

Alternative Hypothesis (H_1): The statistical hypothesis that is the "alternative" to the Null Hypotheses, typically the negation of the Null Hypothesis, the hypothesis favored by the scientist.

Ampliative: a term used for inductive inference in which the conclusion "goes beyond" the information contained in the premises.

Analogical Generalization: a form of inductive inference in which, since one individual or class shares one or more properties with a second individual or class, it is concluded that one or more properties of the first is also shared by the second.

Analyzed Data: data that are the outcome of description, summary, measurement, and analysis

Antecedent: in a conditional proposition, the proposition that follows the 'if' (and precedes the 'then').
Argument: a group of propositions consisting of a **conclusion** and one or more **premises** such that the premises are offered as support for the conclusion.
Aristotle *(384-322 B.C.):* A Greek philosopher who formalized logic, wrote several texts on scientific method as well as numerous books on science.
Atomic Proposition: a simple proposition not containing other propositions as parts, e.g., snow is white.
Atomism: theories that describe reality as pluralistic, composed of separate, discrete and irreducible entities.
Auxiliary Assumptions: the additional assumptions required in order to generate an observable prediction from one's theory. These are usually supplementary to the hypothesis under test and include factors involving background conditions, instrumentation, mathematics, etc.
Bacon, Sir Francis *(*1561-1626): famous philosopher of science, who advocated "the new logic" of induction as opposed to medieval Aristotelian logic. Principal work: *The New Organon*
Basis of Resemblance: the set of properties shared by the two components of the analogy.
Bayesian Inference: a form of inductive inference going back to Thomas Bayes (1702-1761) in which one evaluates the probability of a hypothesis on the basis of some evidence in relation to antecedent evidence, the probability of the hypothesis, and the probability of the evidence given the hypothesis.
Bayesianism: a theoretical approach to inductive inference and statistics centering on the use of Bayes' Theorem, requiring consistency with the probability calculus
Blind Study: a study in which the subjects are unaware of the hypothesis being tested or which treatment group they are in.
Carnap, Rudolph (1891-1970): a famous Viennese-American philosopher of science and logician who helped establish the famous Vienna Circle and Logical Empiricism. Well-known for his attempt to construct a quantitative logic of confirmation (*The Logical Foundations of Probability*). Perhaps the most famous philosopher of science in the first half of the twentieth century.
Categorical Syllogism: a deductive argument consisting of three categorical propositions which contain exactly three terms, each of which occurs in exactly two of the constituent propositions
Census: a collection of data from the entire population

Coefficient of Determination: r^2
Coherence Theory of Truth: the truth of a proposition is a property of its logical coherence or consistency with a body of already accepted propositions.
Compound Proposition: those propositions containing other propositions as parts, e.g., either snow is white or grass is green.
Conclusion: the proposition an argument is alleged to establish.
Concomitant Variation: Given two variables (X and Y), if X varies in direct proportion to Y (or inverse proportion to X), then X and Y are causally related.
Conditional Argument: an argument consisting of three propositions (two premises and a conclusion) such that one premise is a conditional proposition, one premise either affirms or denies the antecedent or consequent, and the conclusion either affirms or denies the antecedent or consequent not mentioned in the second premise
Conditional Probability: the probability of one thing conditional on (dependent on) another thing, e.g., the probability of a hypothesis given the evidence.
Conditional Proposition: a compound proposition consisting of two statements (the **antecedent** and the **consequent**) joined by 'if-then.'
Confidence Interval: the intervals of values of which one is confident that a population parameter likely is present.
Confirmation: the degree to which evidence supports or makes a conclusion likely
Confounding: a mixture of the influences of the independent variable and extraneous variables.
Conjunction (Conjunctive Proposition): a proposition containing at least two propositions (called conjuncts) joined by the connector word *and.*
Conjunctive Argument: an argument containing two conjuncts as premises and the conjunction as conclusion.
Connectives : those words, e.g., and, or, etc., which connect propositions together thus forming compound propositions.
Connotation: the set of properties defining a certain concept or word.
Consequent: in a conditional proposition, the proposition that follows the 'then'.
Contrapositive: The result of transposing p and q in a conditional and attaching a negation to each of them. The contrapositive of $(p \supset q)$ is obtained by first transposing p and q to get $(q \supset p)$ and then negating both: $(\sim q \supset \sim p)$.

Contrary-to-Fact (Counter-factual) Conditional: conditionals in which the antecedent refers to a situation that is, in fact, false and the consequent asserts what would follow from this situation, e.g., if the moon were made of green cheese, then it would be high in calories.
Control: the process of preventing the influence of extraneous variables the dependent variable.
Control Group: the group receiving standard or zero treatment.
Controlled Experiment: an experiment employing a control group.
Conventionalism: a version of nominalism which maintains that all classification is conventional.
Correlation Coefficient (r): the numerical degree of strength of the linear correlation between two variables.
Correspondence Theory of Truth: the truth of a proposition consists in its correspondence to the fact(s) that it asserts.
Corroboration**:** a technical term used by Karl Popper to indicate the evaluation of a hypothesis after it has been subjected to test and has withstood attempts to falsify it. Popper contrasts corroboration with confirmation.
Critical Value: the values of the test statistic separating the rejection and the non-rejection regions.
Counterbalancing: a control procedure in which order effects are distributed across all treatment conditions by including all orders of the independent variable or randomly selecting orders of the independent variable
Crucial Experiment: ("experiments of the cross") an experiment that will decide between the truth of two contrary hypotheses.
Cumulative Frequency: a value obtained from summing the frequencies of all classes less than the specified class limit.
Cumulative Relative Frequency: the result of dividing the **cumulative frequency** by the total number of data.
Curve fitting: imposing the best fitting (mathematical) curve on a graphical representation of data.
Decision Rule: A rule, specified by the experimenter, indicating what sample results will lead to the rejection of Ho and what results will not.
Deductive Argument: an argument is deductive if and only if it is claimed that the premises necessarily imply the conclusion.
Deductive Logic: that branch of logic concerned with deductive arguments.
Deductivism: the view that scientific knowledge is obtained by the technique of deducing logical propositions from earlier ones.
Dependent Variable: the variable whose values change as a result of a change in the independent variable; ordinarily thought to be the

effect.

Derived Measurement: measurement presupposing the measurement of other properties

Descartes, René (1596-1650): famous philosopher, mathematician and physicist, who advocated the use of deductive logic as the foundation of scientific method. Principal work: *Discourse on Method*

Descriptive: as contrasted with prescriptive, concerned with characterizing what is (the factual).

Descriptive Analogy: the comparison (analogy) made between the two components of an analogy, e.g., "he's a wolf"; "she's foxy." Often used to draw further inferences in an analogical argument.

Descriptive Statistics: methods for collecting, organizing, summarizing, and analyzing information obtained from data..

Disjunction (Disjunctive Proposition): a proposition containing at least two simpler propositions (called the *disjuncts*) joined by anoperator called disjunction--*or*.

Disjunctive Argument: an argument consisting of three statements (two premises and one conclusion) such that one premise is a disjunction, one premise affirms or denies one of the disjuncts, and the conclusion affirms or denies the other disjunct.

Disjunctive Syllogism: an argument in which one premise denies one disjunct and the conclusion affirms the other disjunct.

Double- Blind Study: an experimental procedure in which both the subjects and the experimenter are unaware of the hypothesis being tested or which treatment group subjects are in.

Duhem-Quine Thesis: the thesis that all theories (hypotheses, etc.) are underdetermined by the scientific evidence.

Eduction: an inductive inference in which the premises and the conclusion are about particulars.

Eliminative Induction: an inductive inference in which the premises contain a disjunction of hypotheses and a statement that all but one hypothesis is false, and the conclusion then affirms that the remaining hypothesis is true, e.g., either H_1 or H_2 or H_3; H_1 is false and H_2 is false; thus, H_3 is true.

Empirical (factual): a general term used to characterize the property of answering questions by observing the natural world (typically through the senses).

Empirical Evidence: evidence obtained through the use of ones senses as in observation or experimentation.

Empiricism: all knowledge of fact must be validated in sense experience. Whatever is knowable is knowable is knowable directly or

by inference from propositions based upon sensations. Stresses the a posteriori and acquired ideas.
Entheme: a syllogism with a missing premise.
Epistemology: the theory of knowledge. That branch of philosophy concerned with the definition, nature and criteria of knowledge.
Evidence: that which is offered as support for some claim.
Ex Post Factor Design: quasi-experimental designs in which the dependent variable has already occurred and one attempts to determine what the independent variable was that caused it. See **retrospective designs.**
Experiment: research in which one or more variables are manipulated and their effects upon other variables noted.
"Experiments in order to see": exploratory experiments performed with no hypothesis in mind.
Experimental Control: techniques employed to insure that variations in the dependent variable are related to manipulation of the independent variable and not to extraneous variables.
Experimental Group: the group receiving the treatment condition.
Experimental Treatment Condition: the specific variation in the independent variable whose effect on the dependent variable is being studied.
Extraneous Variable: a variable, irrelevant to the independent variable being tested, that affects the dependent variable.
Factorial Experiment: an experiment in which the effect of more than one independent variable is evaluated simultaneously.
Fallacy: an inference involving the mistaken judgment that certain premises give support for a conclusion.
Extension: the set of classes falling under a concept or word.
Extensive Measurement: measurement involving extensive properties, i.e., properties in which there is a physical process of addition.
Extrapolation: using a regression equation to predict values outside the actual range of the x-variable, i.e., above or below the actual values.
Fallibilism: The view that knowledge is fallible, i.e., is not infallible, is revisable, and may change, etc. This view was advocated by the 19th century pragmatist philosopher of science, C. S. Peirce and also adopted in the 20th century by Karl Popper.
Falsificationism: Sir Karl Popper's view that, in science, one attempts to falsify a hypothesis, not to confirm it.
Fisher, Ronald (1890-1962): the original founder of statistical hypothesis testing., and one of the co-founders of the modern "synthetic theory of evolution". Fisher's theory of statistical hypothesis testing employed only Ho and significance tests.

Formal Fallacy: an invalid argument form that has a superficial resemblance to a valid argument form.

Formal Logic: that kind of logic dealing with the syntactical and structural aspects of an argument and ignoring the content of an argument.

Formal Science: as contrasted with empirical science, the formal sciences are concerned with studying the formal aspects of a subject

Frequency Distribution: a listing of classes and their frequencies.

Function: in mathematics, a function is a **relation** such that for each member of the domain (the independent variable) there is one and only range element (the dependent variable).

Fundamental Measurement: measurement not presupposing the measurement of other properties.

Grouped data table: a table listing classes, frequency, relative frequency, etc.

Heuristics: unlike **algorithms,** a set or rules that are generally useful for generating a solution.

Hempel, Carl (1905-1996) Famous German-American philosopher of science, associated with the Berlin branch of Logical Empiricism.

Histogram: a graph displaying classes on the horizontal axis and frequencies of the classes on the vertical axis. The height of the vertical bar is equal to the frequency of the data.

Huygens, Christian (1629-1695): famous Dutch mathematical physicist, known for his studies of the pendulum and optics. Author of *Treatise on Light* (1690).

Hypothetico-Deductive Method: a method of inquiry which consists of asking what observable state of affairs would obtain if a given hypothesis were true and then investigating to see whether these states of affairs do obtain. So called because one *deduces* a predicted observation from the *hypothesis*.

Idealism (metaphysical) : one kind of idealism according to which what is real consists of our ideas. Opposed to realism.

Idols of the Human Mind: according to Bacon, these are the tendencies inherent in our biological and psychological nature, and in society, which lead us from the path of truth. There are four: (1) *Idols of the tribe*--our biological nature; (2) idols of the cave--our psychological nature; (3) idols of the market place--our social nature; and (4) idols of the theater--language and ideology.

Implication: one proposition or set of propositions implies a second if and only if it gives logical grounds or evidence for the second.

Imply: to provide a basis from which an inference may be drawn. Words, actions, etc., may have implications .

Independent Variable: the variable under the control of the experimenter (usually thought to be a cause).

Inductive Argument: an argument that is non-deductive or probable; as contrasted with a deductive argument, an inductive argument lends some evidential support to the conclusion but does not entail it.

Inductive Argument: an argument in which the premises do not entail the conclusion but render the conclusion likely (probable) to some degree.

Inductive Logic: that branch of logic dealing with inductive arguments. Inductive logic attempts to formulate correct principles of inductive inference.

Inductive Strength: in inductive logic, the counterpart to validity. Inductive strength is the degree to which the premises of an inductive argument support the conclusion.

Inductivism: the philosophical view that maintains the scientific method consists exclusively of induction and inductive inference. Associated (incorrectly) with the views of Bacon and Mill.

Infer: the activity of drawing a conclusion on the basis of evidence. The result of this is an inference. Contrast ***Imply***. A proposition implies another proposition; it does not infer another proposition.

Inference Indicators: words commonly used to signal premises or conclusions of an argument, e.g., "hence," " because," "since," "therefore."

Inference to the Best Explanation: another term for ***Abductive***

Inferential statistics: methods for drawing and measuring the reliability of conclusions about a population from information in a sample.

Informal Fallacy: a fallacious argument due not to some formal fallacy but erroneous on nonformal grounds, e.g., because of ambiguity or vagueness.

Innate Ideas: ideas/ principles possessed by the mind independently of experience.

Intension: contrasted to extension. See **connotation.**

Intensive Measurement: measurement involving intensive properties, i.e., properties that are ordered but not additive.

Interaction: the effect of the independent variable on the dependent variable varies depending on the value of another independent variable.

Interpolation: using a regression equation to predict missing values of the X-variable in between the upper and lower values.

Interval Scale: a scale in which there are equal intervals between ranks.

Intuition: immediate apprehension of an object that is direct, non-discursive, non-mediated.
Intuitive Induction: a form of induction advocated by Aristotle in which, after having sensed several particulars of a certain class and having remembered them, one then immediately cognizes the underlying universal essence of this class.
Is-ought (fact-norm): the traditional distinction between questions of empirical fact (what is the case) and questions of values and norms (what ought to be the case).
***Invalidity*:** an argument is invalid if and only if it is alleged to be deductive and it is not valid, i.e., it is possible for the premises to be true and for the conclusion to be false.
Joint Method of Agreement and Difference: A conjunction of the positive method of agreement and the negative method of agreement.
Laplace, Pierre Simon de (1749-1827): French mathematician and astronomer, famous for his celestial mechanics and theory of probability. Advocate of strong determinism.
Law of Large Numbers: see **Limit**
Least squares criterion: the straight line that best fits a set of data points is the one for which the sum of the squared errors is smallest.
Likelihood: the probability of the evidence given the hypothesis (E/H).
Linking Generalization: the generalization between a common set of properties and a conclusion drawn from the basis of resemblance, e.g., the lawful connection between the internal combustion engine and gas mileage or the biochemistry of an animal and its susceptibility to cancer.
Limit: a sequence f_n (n = 1,2,3. . .) has a limit L as n goes to infinity iff for every positive number e, no matter how small, there exists a number N such that if n>N, $/f_n$-L/<e
Linnaeus, Carolus (1707-1778): Scandinavian biologist, whose taxonomy of living organisms was the standard account for centuries.
.***Logic*:** that study devoted to analyzing and evaluating the adequacy of arguments.
Logic of discovery: the issue of whether there is a "logic" behind the discovery or creation of hypotheses.
Logical Contradictories: p and q are logical contradictories = not both p and q are true and not both p and q are false, that is, ~(p and q) and ~(~p & ~q). Thus either p or q is true.
Logical Contraries: p and q are logical contraries = not both p and q are true but both might be false, i.e., ~(p & q).
Logical Dependence (**logical relation**): p is logically related to q [q logically depends on p] = the truth (or falsity) of p (q) influences or

determines the truth (or falsity) of q (p).
Logical Equivalence: p is logically equivalent to q = p implies q and q implies p (i.e., if p then q and if q then p).
Logical Incompatibility: p is logically incompatible with q = not both p and q, i.e., ~(p & q).
Matching: a procedure for balancing uncontrolled extraneous variables in which the extraneous variables are evenly distributed across subjects; subjects are paired with other subjects who are similar on the extraneous variables and one member of each pair is then randomly assigned to the control and to the experimental groups
Mathematical Induction: a form of reasoning in mathematical proofs in which one argues, for example, that since the number 1 possesses a property p, and since it is true that if any number n possesses p, then n+1 possesses p, it follows that all natural numbers possess p.
Mean:

Sample Mean $\bar{x}=\frac{\sum x}{n}$

Population Mean $\mu=\frac{\sum x}{N}$

Median: the number dividing the bottom 50% from the top 50% of the data
Mendel, Gregor (1822-1884): Austrian monk, who experimented on peas in the monastery garden. Founder of modern genetics.
Mendeleef, Dmitri (1834-1907): Russian chemist and founder of the modern periodic table of elements
Method of Difference: If a phenomenon is always present when one common antecedent is always present and is always absent when a common antecedent is always absent, the common antecedent is the cause of the phenomenon.
Method of Residues: If C causes E, and C_1 is a part of C, and C_1 causes E_1 (a part of E), then (C-C_1) is the cause of (E-E_1).
Methodology: the theory of method.
Mill, John Stuart (1806-1873): Famous 19th century British logician and philosopher, well known as an advocate of utilitarianism and liberalism. His text book on logic was the standard text book on logic for almost 50 years.
Mode: the most frequently occurring data.
Model: a representation of one system by another (usually more familiar) whose workings are analogous to that of the first.

Multivariate: more than one variable, either independent or dependent.
Negative Analogy: the ways in which the two components of an analogy or model are different, e.g., humans are made of protoplasm but computers are made of transistors, electrical circuits, etc.
Negative Method of Agreement: If the absence of two or more instances of a phenomenon are accompanied by the absence of one antecedent factor, the absent factor is the cause of the phenomenon.
Neutral Analogy: the set of properties not known to be shared by the two components of an analogy and not known not to be shared by them. It is the basis for making predictions about the modeled object.
Newman, Jerzy (1894-1981): one of the founders of modern statistics (together with Pearson). The Newman-Pearson theory of statistics was a successor (and rival) to that of Fisher and became the standard interpretation of statistical hypothesis testing in which there are two hypotheses (Ho, H1) and two types of errors.
Nominal Scale: a scale in which numerals are assigned to qualitatively different categories.
Nominalism: the view that no natural divisions and classes exist in nature since no universals exists in nature; only particular individuals exist in nature. Categories are names (labels) we apply to individuals.
Null Hypothesis: The statistical hypothesis one wants to "nullify", usually that there is no systematic difference due to the independent variable.
Ogive: a graph in which a point is plotted above each class limit at a height equal to the cumulative relative frequency. The points are then formed with connecting lines.
Ordinal Scale: a scale in which there is a ranked order (1st, 2nd, 3rd, etc.).
Pearson, Egon: along with Newman, founder of the Newman-Pearson theory of statistical hypothesis testing.
Occam's Razor: a principle of simplicity names after William of Occam: entities should not be multiplied beyond necessity.
Order Effects: present in before and after (repeated measures) studies when the order or sequence with which subjects are exposed to values of the independent variables confound the effects of the independent variable.
Placebo: a substance without any pharmacological effect
Positive Analogy: the similarities between the two components in an analogy of model, i.e., the basis of resemblance

Positive Method of Agreement: If two or more instances of a phenomenon have one antecedent factor in common, the common factor is the cause of the phenomenon.
Parameter: a descriptive measure for a population.
Peirce, Charles (1839-1914) American philosopher and one of the founders of Pragmatism.
Perfect Induction: a form of inductive inference in which the premises enumerate the entire membership of a finite class and the conclusion asserts that all members of this class have the property.
Personalism: a view associated with Savage as an alternative to 'subjectivism'
Philosophy of science: that branch of philosophy concerned with the metaphysics, epistemology and ethics of science. More narrowly, that branch of philosophy concerned with the methods, concepts and underlying assumptions of science.
Popper, Sir Karl (1902-1995): famous Viennese-British philosopher of science, who advocated falsification of hypotheses instead of their confirmation. Perhaps the most famous philosopher of science of the second half of the twentieth century.
Population: the collection of all individuals under investigation.
Posterior Probability: In Bayes' Theorem, the probability of a hypothesis given the evidence P(H/E).
Practice Effects: due to changes that occur on a measure of the dependent variable attributable to previous exposure to that measure, e.g., taking two tests.
Predictor Variable: the variable on the basis of which a prediction is made.
Premises: the propositions in an argument offered as evidence or proof of the conclusion.
Prescriptive (normative): concerned with prescribing (as opposed to describing), with indicating what ought to be, with what is good or valuable, etc
Principle of Indifference: two events or states of affairs have an equal probability if there are no grounds for expecting one to occur instead of the other.
Prior Probability: the probability of a hypothesis H before the evidence E: P(H).
Probability Calculus: the formal set of rules for inferring (calculating) probabilities.
Probability distribution: a listing of the probability (frequencies) of a class of events.
Proposition: the logical content contained in a statement.

Prospective Study: quasi-experiments n which the independent variable is already present in subjects and the scientist observes the future effects of the independent variable.

Quantitative Induction: an inductive inference to a quantitatively expressed, mathematical law.

Quasi-Experiment: designs in which there is always confounding (or possible confounding) usually because there is no random assignment or control; full experimental control is lacking but some control is present

Random Assignment: placing subjects into groups within an experiment randomly: each subject has an equal chance of being assigned to any group.

Random Selection: choosing subjects from a population in such a way that each member in the population has an equal chance of being selected as a subject in the various treatment conditions.

Random Variable: a numerical quantity whose values depend on chance.

Radical Empiricism: a more radical form of empiricism. The view that science can only deal with those things observable by the senses and that all knowledge is reducible to sense-impressions.

Random sampling: sampling technique in which each sample has an equal chance of being selected.

Random Variable: a variable whose values depend on chance.

Range: the difference between the maximum and the minimum data.

Ratio Scale: interval scales with an absolute zero.

Regression Equation: the equation of a regression line: $Y = a + bX$.

Regression line: the straight line that best fits a set of data points according to the least squares criterion.

Rejection Region: the set of values for the test statistic that leads to the rejection of Ho.

Relation: in mathematics, a relation is any set of ordered pairs.

Relative Frequency: the ratio of the frequency of a class to the total number of pieces of data.

Relative Frequency Distribution: a listing of classes and their relative frequencies.

Relative Frequency Histogram: a histogram with relative frequencies displayed on the vertical axis.

Relative Frequency Polygon: a listing of the midpoint of each class and their relative frequency. The height of the vertical point is equal to the relative frequency. The points are joined together with connecting lines.

Representative sample: a sample that is an accurate reflection of the properties of the population.

Response Variable: a variable, whose value, one is interested in predicting.
Retrospective Design: quasi-experiments which are "backward-looking". See **Ex Post Facto Designs.**
Rationalism: the method of establishing knowledge by reason (deduction), usually involving general propositions. Stresses the a priori and innate ideas. Pure mathematics is the prime example.
Raw Data: the actual physical recording (observation) of a particular experience.
Real definition: the definition of a thing in terms of its underlying nature
Realism: in the theory of classification, the view that classification and categories are real in the sense that the divisions exist in nature, independently of our conceptual schemes.
Reasoning: the activity or process of finding reasons, weighing them, and drawing conclusions from them.
Reichenbach, Hans (1891-1953): Famous 20th century German-American logician and philosopher of science. A member of the Berlin group of logical empiricism and key proponent of the relative frequency interpretation of probability.
Sample: a part of the population from which information is obtained
Scientific Revolution: the period of time from 1450 to 1700 in which modern science replaced the earlier medieval world view
Sentence: a linguistic or written form of words that is well-formed grammatically
Serendipity: a fortunate discovery by chance or accident ("being in the right place at the right time").
Significance Level (α) of a hypothesis test: the probability of making a Type I error, i.e., the probability of rejecting a true Ho.
Simple Enumeration: inductive inference in which the premises mention a particular number of individual cases.
Skepticism: several schools of ancient Greek thought (third century BC–fourth century AD). According to the skeptic, no real knowledge (justification) is possible, or if it is possible, no one can know this.
Standard deviation:

Sample standard deviation: $s = \sqrt{\dfrac{\Sigma(x-\bar{x})2}{n-1}}$

Population standard deviation $\sigma = \sqrt{\dfrac{\Sigma(x-\mu)2}{N}}$

Standard Error of the Mean: the standard deviation of the sample mean, the amount of sampling error to be expected .

Statement: a sentence that is either true or false. Contrasted with questions, commands, and exclamations.
Statistical Generalization: a proposition stating that some proportion of the members of one class are members of another class, e.g., x% of As are Bs.
Statistical Significance: a test result falling into the rejection region (as specified by the α-level of significance).
Statistical Syllogism: a syllogism in which the major premise is a statistical generalization: x% of Fs are Gs, a is an F: it's likely that a is a G.
Statistics: a statistic is a descriptive measure for a sample. By contrast, the science of statistics is concerned with the methodology of collecting, describing, and analyzing data and the drawing of conclusions from them.
Soundness: an argument is sound if and only if (1) it is valid and (2) its premises are true.
Syllogism: an argument consisting of exactly two premises and one conclusion.
Tautology: a proposition that is true under all possible assignments of truth values to its component propositions
Teleology: the theory of goals, ends, or purposes. The present is explained by reference to the future "for the sake of which" something in the present is occurring.
Theoretical Entity: an unobservable entity that is postulated to explain some phenomenon, e. g., electron, atom, gene, black hole, virus.
Truth Values: the values of 'truth' = true, false ***Valid***: an argument is valid: (1) if it is impossible for the premises to be true and at the same time for the conclusion to be false, (2) if the premises were to be true, then the conclusion would also have to be true, (3) if the premises entail the conclusion.
Type I Error: Rejecting the Null Hypothesis when it is true
Type II Error: Not rejecting the Null Hypothesis when it is false.
Universal Generalization: a proposition stating that all members of a class have a certain property, e.g., all copper conducts electricity.
Variable: a property (factor, quality) that can take on different *values*.
Wittgenstein, Ludwig (1889-1951): Viennese-British logician and philosopher. Famous for his early work on "logical atomism" (*Tractatus-Logico-Philosophicus*), largely repudiated in his later work on family resemblance (*Philosophical Investigations*).

z-score: the number of standard deviations a datum value is away from the mean of the data set. E.g., a sample z-score is: $z = \frac{x - \bar{x}}{s}$.

References

Paperback editions are abbreviated as: pb.

American Psychiatric Association (1994). *Diagnostic and Statistical Manual of Mental Disorders* (4th ed.). Washington, DC: A.P.A.

Aristotle (1984). *Posterior Analytics.* In J. Barnes, (Ed.), *The Collected Works of Aristotle* (Vol. 1, pp. 114-166). Princeton, NJ: Princeton University Press.

Aristotle (1984). History of Animals. In J. Barnes, (Ed.), *The Collected Works of Aristotle* (Vol. 1, pp. 774-993). Princeton, NJ: Princeton University Press.

Bacon, F. (1985). *The New Organon and Related Writings* (J. H. Anderson, Ed.). Indianapolis, IN: Bobbs-Merrill. (Original work published 1620) (pb)

Baird, D. (1992). *Inductive Logic: Probability and Statistics*. Englewood Cliffs, NJ: Prentice-Hall.

Baum, R. (1975). *Logic.* New York: Holt, Rinehart & Winston.

Bayes, T. (1958). An essay toward solving a problem in the doctrine of chances. *Biometrika*, 45, 293-315. (Original work published 1763).

Beauchamp, T. L. (Ed.) (1974). *Philosophical Problems of Causation.* Encino, CA: Dickenson. (pb)

Berkeley, G. (1995). *Principles of Human Knowledge and three Dialogues* (H. Robinson, Ed.). New York: Oxford University Press. (original work published 1710) (pb)

Beveridge, W. I. B. (1957). *The Art of Scientific Investigation.* New York: Vintage. (pb)

Black, M. (1952). *Critical Thinking* (2nd ed.). Englewood Cliffs, NJ: Prentice-Hall.

Black, M. (1962). *Models and Metaphors*. Ithaca, NY: Cornell University Press.

Black, M. (1967). Induction. In P. Edwards (Ed.), *The Encyclopedia of Philosophy* (Vol. 4, pp. 169-181). New York: Macmillan.

Black, M. (1970). The Inductive Support of Inductive Rules. In B. Brody

(Ed.), *Readings in the Philosophy of Science* (pp. 578-589). Englewood Cliffs, NJ: Prentice-Hall. (original work published 1954)

Boyd, R., Gasper, P., & Trout, J. D. (Eds.) (1991). *The Philosophy of Science*. Cambridge, MA: M.I.T. Press. (pb)

Broad, W. & Wade, N. (1987). *Betrayers of the Truth: Fraud and Deceit in the Halls of Science*. New York: Simon & Schuster. (pb)

Bunge, M. (1967). *Scientific Research*. New York: Springer-Verlag.

Burtt, E. A. (1932). *The Metaphysical Foundations of Modern Science* (2nd ed.). New York: Doubleday (pb)

Butterfield, H. (1957). *The Origins of Modern Science 1300-1800* (rev. ed.). New York: Free Press. (pb)

Campbell, D. T. & Stanley, J. C. (1963). *Experimental and Quasi-Experimental Designs for Research*. Boston: Houghton-Mifflin. (pb)

Carnap, R. (1960). Statistical and Inductive Probability. In E. H. Madden (Ed.), *The Structure of Science* (pp. 269-278). Boston: Houghton-Mifflin. (Original work published 1955)

Carnap, R. (1962). *The Logical Foundations of Probability* (2nd ed.). Chicago: University of Chicago Press.

Carnap, R. (1972). The Aim of Inductive Logic. In S. A. Luckenbach (Ed.), *Probabilities, Problems, and Paradoxes*_(pp. 104-120). Encino, CA: Dickenson. (Original work published 1962)

Cattell, R. B. (1966). Psychological Theory and Scientific Method. In R. Cattell (Ed.), *Handbook of Multivariate Experimental Psychology* (pp. 1-65). Chicago: Rand McNally.

Chang, R. (1984). *Chemistry* (2nd ed.). New York: Random.

Copi, I. & Cohen, C. (1986). *Introduction to Logic* (8th ed.). New York: Macmillan

Cohen, M. R. & Nagel, E. (1934). *An Introduction to Logic and Scientific Method*. London: Routledge & Kegan Paul.

Conant, J. B. (1957). Pasteur's and Tyndall's Study of Spontaneous Generation. In J. B. Conant & L. K. Nash (Eds.), *Harvard Case Studies in Experimental Science* (Vol. 2, pp. 487-540). Cambridge, MA: Harvard University Press.

Cook, T. D. & Campbell, D. T. (1979). *Quasi-experimentation: Design and Analysis Issues for Field Settings*. Boston: Houghton-Mifflin. (pb)

Cramer, H. (1950). *The Elements of Probability Theory*. New York: John Wiley.

Coyne, A. M. (1984). *Introduction to Inductive Reasoning*. Lanham, MA: University Press of America. (pb)

Descartes, R. (1985). Discourse on Method. In J. Cottingham, et al. (Eds. & Trans.), *The Philosophical Writings of Descartes* (Vol. 1, pp.109-176). (Original work published 1637) (pb)

Dewey, J. (1938). *Logic: The Theory of Inquiry*. New York: H. Holt.

Duhem, P. (1962). *The Aim and Structure of Physical theory* (2nd ed., P. P. Weiner, Trans.). New York: Atheneum. (Original work published 1914) (pb)

Edwards, W., Lindman, H., & Savage, L.J. (1963). Bayesian Statistical Inference for Psychological Research. *Psychological Review,* 70, 193-242.

Ellis, B. (1968). *Basic Concepts of Measurement*. New York: Cambridge University Press. (pb)

Fath, E. A. (1926). *The Elements of Astronomy*. New York: McGraw-Hill.

Feibleman, J. (1972). *Scientific Method*. The Hague: M. Nijhoff.

Feyerabend, P. (1993). *Against Method* (3rd ed.). New York: Verso. (pb)

De Finetti, B. (1964). Foresight: its Logical laws, its Subjective Sources (H. E. Kyburg, Jr., Trans.). In H. E. Kyburg, Jr. & H. E. Smokler (Eds.), *Studies in Subjective Probability* (pp. 93-158). New York: John Wiley. (Original work published 1937)

Fisher, R. A. (1966). Has Mendel's work been rediscovered? In C. Stern & E. R. Sherwood (Eds.), *The Origin of Genetics: a Mental Sourcebook* (pp. 139-172). San Francisco: W. H. Freeman.

Fisher, R. A. (1990). *Statistical Methods, Experimental Design, and Scientific Inference* (J. H. Bennett, Ed.). Oxford: Oxford University Press.(pb)

Foster, M. H. & Martin, M. L. (Eds.) (1966). *Probability, Confirmation and Simplicity*. New York: Odyssey Press.

Frank, P. (1957). *Philosophy of Science*. Englewood Cliffs, NJ: Prentice-Hall. (pb)

Freedman, D., Pisani, R., Purves, R., & Adhikari, A. (1991). *Statistics* (2nd ed.). New York: W. W. Norton.

Freudenthal, H. (Ed.) (1961). *The Ccncept and Role of the Model in Mathematics and Natural and Social Sciences*. Dordrecht: D. Reidel.

Gholson, B., et al. (Eds.) (1989). *Psychology of Science*. Cambridge: Cambridge University Press.

Giere, R. N. (1979a). *Understanding Scientific Reasoning*. New York; Holt, Rinehart. (pb)

Giere, R. N. (1979b). Foundations of Probability and Statistical Inference. In P. D. Asquith & H. Kyburg (Eds.), *Current Research in Philosophy of Science* (pp. 503-533). East Lansing, MI: Philosophy of Science Assoc.

(pb)

Glymour, C. (1980). *Theory and Evidence*. Princeton, NJ: Princeton University Press. (pb)

Gooding, D., Pinch, T., & Schaffer, S. (Eds.) (1989). *The Uses of Experiment: Studies in the Natural Sciences*. New York: Cambridge University Press. (pb)

Goodman, N. (1983). *Fact, Fiction and Forecast* (4th ed.). Cambridge, MA: Harvard University Press.

Gustafson, W. (1994). *Reasoning from Evidence: Inductive Logic*. New York: Macmillan. (pb)

Hacking, I. (1975). *The Emergence of Probability*. New York: Cambridge University Press. (pb)

Hacking, I. (1983). *Representing and Intervening*. Cambridge: Cambridge University Press. (pb)

Haggard, H. W. (1946). *Devils, Drugs, and Doctors*. New York: Pocket.

Harding, S. (Ed.) (1976). *Can Theories be Refuted?* Boston: D. Reidel.

Hanson, N. R. (1958). *Patterns of Discovery*. New York; Cambridge University Press. (pb)

Hanson, N. R. (1970). Is There a Logic of Scientific Discovery? In B. A. Brody (Ed.), *Readings in the Philosophy of Science* (pp. 620-633). Englewood Cliffs, NJ: Prentice-Hall. (Original work published 1961).

Harman, G. (1965). The Inference to the Best Explanation. *Philosophical Review*, 74, 88-95.

Hempel, C. (1945). Studies in the Logic of Confirmation. *Mind*, 54, 1-26; 97-121. Reprinted in C. Hempel: *Aspects of Scientific Explanation* (pp. 3-51). Glencoe, IL: Free Press, 1965. (pb)

Hempel, C. (1966). *Philosophy of Natural Science*. Englewood Cliffs, NJ: Prentice-Hall. (pb)

Hesse, M. (1965). *Models and Analogies in Science*. Notre Dame, IN: University of Notre Dame Press.

Holland, J.., Holyoak, K. J., Nisbett, R. E. & Thagard, P. (1986). *Induction: Processes of Inference, Learning, and Discovery*. Cambridge, MA: M.I.T. Press.

Holton, G. & Roller, D. (1958). *Foundations of Modern Physical Science*. Reading, MA: Addison-Wesley.

Hospers, J. (1988). *An Introduction to Philosophical Analysis* (3rd ed.). Englewood Cliffs, NJ: Prentice-Hall.

Howson, C. & Urbach, P. (1993). *Scientific Reasoning: the Bayesian Approach*. La Salle, IL: Open Court. (pb)

Hume, D. (1896). *A Treatise of Human Nature* (L.A. Selby-Bigge, ed.). Oxford: Oxford University Press. (Original work published 1740) (pb)

Hume, D. (1902). *An Enquiry concerning Human Understanding* (2nd ed., L.A. Selby-Bigge, Ed.). Oxford: Oxford University Press. (Original work published 1748) (pb)

Hung, E. (1997). *The Nature of Science: Problems and Perspectives*. Belmont, CA: Wadsworth. (pb)

Hurd, D. L. & Kipling, J.J. (Eds.) (1964). *The Origins and Growth of Physical Science* (2 vols.). Baltimore, MD: Penguin. (pb)

Huygens, C. (1964). The Wave Theory of Light. In H. Hurd & J.J. Kipling (Eds.), *The Origins and Growth of Physical Science* (Vol. 2, pp. 164-181). Baltimore, MD: Penguin. (Original work published 1690) (pb)

Jeffrey, R. C. (1983). *The Logic of Decision* (2nd ed.). Chicago: University of Chicago Press.

Jevons, W. S. (1905). *The Principles of Science: a Treatise on Logic and Scientific Method* (2nd ed.). London: Macmillan.

Johnson, W. E. (1964). *Logic: Part II. Demonstrative Inference: Deductive and Inductive. New York: Dover. (Original work published 1922).* (pb)

Johnson, W. E. (1964). *Logic: Part V: The Logical Foundations of Science*. New York: Dover (Original work published 1924). (pb)

Joseph, H. W. B. (1916). *An Introduction to Logic* (2nd ed.). Oxford: Clarendon Press.

Kemeny, J. (1959). *A Philosopher Looks at Science*. New York: D. Van Nostrand.

Keynes, J. M. (1973). *A Treatise on Probability*. New York: Cambridge University Press. (Original work published 1921). (pb)

Knorr-Cetina, K. (1981). *The Manufacture of Knowledge: An Essay on the Constructionist and Contextualist Nature of Science*. Oxford: Pergmanon.

Koestler, A. (1971). *The Case of the Midwife toad*. London: Hutchinson.

Kuhn, T. (1968). History of Science. In *International Encyclopedia of the Social Sciences*. (Vol.14, pp. 74-83). New York: Macmillan.

Kuhn, T. (1970). *The Structure of Scientific Revolutions* (2nd ed.). Chicago: University of Chicago Press. (pb)

Kyburg, H. E. & Smokler, H.E. (Eds.)(1964). *Studies in Subjective Probability*. New York: Wiley. (pb)

Lakoff, G., & Johnson, M. (1980). *Metaphors we Live by*. Chicago: University of Chicago Press. (pb)

Langley, P., Simon, H. Bradshaw, G., & Zytkow, J. M. (1987). *Scientific Discovery: Computational Explorations of the Creative Process*.

Cambridge, MA: M.I.T. Press.

Laplace, P. S. de (1951). *A Philosophical Essay on Probabilities* (F. W. Truscott & F. L. Emory, Trans.). New York: Dover. (Original work published 1794) (pb)

Latour, B. & Woolgar, S. (1986). *Laboratory Life: The constuction of Scientific Fact* (2nd ed.). Princeton; Princeton University Press. (pb)

Laudan, L. (1969). Theories of Scientific Method from Plato to Mach: a Bibliographical Review. *History of Science*, 7, 1-63.

Leatherfod, W. H. (1974). *The Role of Analogy, Model and Metaphor in Science*. Amsterdam: Elsevier. (pb)

Lial, M. L., & Miller C. D. (1973). *Precalculus Mathematics*. Glenview, IL: Scott, Foresman & Co.

Linnaeus, C. (1962). Systema Naturae. In D. D. Runes (Ed.), *A Treasury of World Science* (pp.636-641). Patterson, NJ: Littlefield & Adams. (pb)

Lipton, P. (1991). *Inference to the Best Explanation.* London: Routledge & Kegan Paul. (pb)

Locke, J. (1975). *An Essay Concerning Human Understanding* (P.H. Niddith, Ed.). Oxford: Oxford University Press. (Original work published 1690). (pb)

Losee, J. (1993). *A Historical Introduction to the Philosophy of Science* (3rd ed.). New York: Oxford University Press. (pb)

Luckenbach, S. A. (Ed.) (1972). *Probabilities, Problems, and Paradoxes.* Encino, CA: Dickenson.

Mackie, J. (1967). Mill's Methods of Induction. In P. Edwards (Ed.), *The Encyclopedia of Philosophy* (Vol. 5, pp. 324-332). New York: Macmillan.

Mendel, G. (1966). Experiments on Plant Hybrids (E. R. Sherwood, Trans.). In C. Stern & E. R. Sherwood (Eds.), *The Origin of Genetics: a Mental Sourcebook* (pp. 1-48). San Francisco: W. H. Freeman. (Original work published 1865)

Mendeleef, D. I. (1964). The Periodic Table of the Elements. In D. L. Hurd & J. U. Kipling (Eds.), *The Origins and Growth of the Physical Sciences* (Vol. 2, pp. 81-100). Middlesex, England: Penguin Books. (Original work published 1869) (pb)

Mendeleef, D. I. (1962). The Relation between the Properties and Atomic Weights of Elements. In D. D. Runes (Ed.), *Treasury of World Science* (pp. 739-746). Patterson, NJ: Littlefield, Adams. (Original work published 1868) (pb)

Mill, J. S. (1872). *A System of Logic; Ratiocinative and Inductive* (8th ed.). London: Longman.

Mises, von. R. (1957). *Probability, Statistics, and Truth* (3rd ed., J. Neyman, D. Scholl & E. Rabinowitsch, Trans.). New York: Dover. (Original work published 1951) (pb)

Morgan, C.T. & King, R.A. (1975). *Introduction to Psychology*. New York: McGraw-Hill.

Morrison, D. E. & Henkel, R. E. (Eds.) (1970). *The Significance Test Controversy*. New York: Aldine.

Moscovici, S. (1985). Social Influence and Conformity. In G. Lindzey & E. Aronson (Eds.), *Handbook of Social Psychology* (3rd ed., Vol. 2, pp. 347-412). New York: Random House.

Moser, P.K., Mulder, D. H., & Trout, J.D. (1998). *The Theory of Knowledge: a Thematic Introduction*. New York: Oxford University Press. (pb)

Moser, P.K, & van der Nat, A. (1995). *Human Knowledge: Classical and Contemporary Approaches* (2nd ed.). Oxford: Oxford University Press. (pb)

Mulkay, M. (1979). *Science and the Sociology of Knowledge*. London: George Allen & Unwin (pb).

Musgrave, A. (1993). *Common Sense, Science and Scepticism: A Historical Introduction to the Theory of Knowledge*. Cambridge: Cambridge University Press. (pb)

Nagel, E. (Ed.) (1961). *The Philosophy of Scientific Method*. New York: Hafner. (pb)

Newell, A. & Simon, H. (1981). Computer Science as Empirical Inquiry. In J. Haugeland (Ed.), *Mind Design* (pp. 35-66). Cambridge, MA: M.I.T. Press. (Original work published 1976) (pb)

Nickles, T. (Ed.) (1980). *Scientific Discovery, Logic, and Rationality*. Dordrecht, Holland: D. Reidel.

O'Connor, D. J. & Carr, B. (1982). *Introduction to the Theory of Knowledge*. Minneapolis, MN: University of Minnesota Press. (pb)

Oldroyd, D. (1986). *The Arch of Knowledge: An Introductory Study of the Philosophy and Methodology of Science*. New York: Methuen. (paperback)

Pearson, K. (1911). *The Grammar of Science* (3rd ed.). New York: E. P. Dutton.

Peirce, C. S. (1957). The General Theory of Probable Inference. In J. Buchler (Ed.), *Philosophical Writings of Peirce* (pp. 190-217). New York: Dover. (Original work published 1883) (pb)

Pickert, G. & Goerke, L. (1986). Construction of the System of Real nmbers. In H. Behnke, et al (Eds.) 1986). *Fundamental of Mathematics* (Vol. 1, pp. 93-165). Cambridge, MA: MIT Press.

Poincaré, H. (1952). *Science and Hypothesis* (W. J. Greenstreet, Trans.). New York: Dover. (Original work published 1905) (pb)

Popper, K. (1965). *The Logic of Scientific Discovery* (2nd ed.; K. Popper, Trans.). New York: Harper & Row. (Original work published 1935) (pb)

Popper, K. (1965). *Conjectures and Refutations* (2nd ed.). New York; Harper & Row. (pb)

Popper, K. (1979). *Objective Knowledge: An Evolutionary Approach* (Rev. ed.). Oxford: Oxford University Press. (pb)

Price, H. H. (1933). *Perception.* London: Methuen.

Quine, W. V. O. (1961). Two Dogmas of Empiricism. In *From a Logical Point of View* (pp. 20-46). New York; Harper & Row. (Original work published 1951) (pb)

Ramsey, F. P. (1960). Truth and probability. In *The Foundations of Mathematics and other Logical Essays* (R. B. Braithwaite, Ed.; pp. 166-184). Patterson, NJ: Littlefield, Adams. (Original work published 1926) (pb)

Randall, J. H. (1940). *The Making of the Modern Mind.* New York: Columbia University Press (pb)

Reichenbach, H. (1949). *Theory of Probability* (2nd ed., E. H. Hutten & M. Reichenbach, Trans.). Berkeley, CA: University of California Press.

Robinson, R. (1950). *Definition.* Oxford: Clarendon Press.

Rosch, E. E. & Mervis, C. B. (1975). Family Resemblances: Studies in the Internal Structure of Categories. *Cognitive Psychology*, 3, 382-439.

Salmon, M.H. (1989). *Introduction to Logic and Critical Thinking* (2nd ed.). New York: Harcourt, Brace & Jovanovich.

Salmon, M.H., et al (1992). *Introduction to the Philosophy of Science.* Englewood Cliffs, NJ: Prentice Hall. (pb)

Salmon, W. C. (1967). *The Foundations of Scientific Inference*. Pittsburgh: University of Pittsburgh Press. (pb)

Savage, L. J. (1972). *The Foundations of Statistics.* New York: Dover (Original work published 1954)

Sienko, M. J. & Plane, R.A. (1961). Chemistry (2nd ed.). New York: McGraw-Hill.

Simpson, G. G. (1961). *Principles of Animal Taxonomy.* New York: Columbia University Press.

Simpson, G. C. & Beck, W. S. (1965). *Life: An Introduction to Biology* (2nd ed.). New York: Harcourt, Brace & World.

Sinclair, W. J. (1909). *Semmelweis: His Life and His Doctrine*. Manchester, England: Manchester University Press.

Skyrms, B. (1986). *Choice and Chance: an Introduction to Inductive Logic* (3rd ed.). Belmont, CA: Wadsworth. (pb)

Stevens, S. S. (1960). On the Theory of Scales of Measurement. In A. Danto & S. Morgenbesser (Eds.), *Philosohy of Science* (pp. 141-149). New York; Meridian. (Original work published 1946) (pb)

Strawson, P. F. (1953). *Introduction to Logical Theory*. London: Methuen.

Swinburne, R. (Ed.) (1974). The *Justification of Induction*. London: Oxford University Press. (pb)

Taylor, R. (1967). Causation. In P. Edwards (Ed.), *The Encyclopedia of Philosophy* (Vol. 2, pp. 56-66). New York: Macmillan.

Wallas, G. (1924). *The Art of Thought*. New York: Harcourt & Brace.

Wallis, W. A. & Roberts, H. V. (1965). *The Nature of Statistics*. New York: Free Press.

Walter, R.S. (1967). Laws of science and lawlike statements. In P. Edwards (Ed.), *The Encyclopedia of Philosophy* (Vol. 4, pp. 410-414). New York: Macmillan.

Weatherford, R. (1982). *Philosophical Foundations of Probability Theory*. London: Routledge & Kegan Paul.

Weiss, N. A. (1996). *Elementary Statistics* (3rd ed.). Reading, MA: Addison-Wesley.

Weisz, P. B. (1959). *The Science of Biology*. New York: McGraw-Hill.

West, J. F. (1965). *The Great Intellectual Revolution*. New York: Citadel (pb)

Whewell, W. (1989). *Theory of Scientific Method* (2nd ed; R. E. Butts, Ed.). Indianapolis, IN: Hackett. (pb)

Whitehead, A. N. (1967). *Science and the Modern World*. New York: Free Press. (pb)

Wittgenstein, L. (1953). *Philosophical Investigations* (G.E.M. Anscombe, Ed.). New York: Macmillan. (pb)

Wright, G. H. von (1951). *A Treatise on Induction and Probability*. New York: Harcourt & Brace.

Index

Author's Biographical Sketch

*A*fter obtaining an undergraduate degree in philosophy, the author pursued graduate work in psychology, obtaining an M.A. in 1963. Returning the philosophy, he obtained his Ph.D. from the University of Minnesota (1970), specializing in the Philosophy of Science, in particular, the philosophy of psychology. In addition to numerous articles in various journals and anthologies, he is the author *Piaget's Theory of Knowledge: Genetic Epistemology and Scientific Reason* (1986), the editor of *The World View of Contemporary Physics: Does it Need a New Metaphysics?* (1988), the co-editor of *The Philosophy of Psychology* (1996), and *The Encyclopedia of Behaviorism* (1999). He is also editor of the international journal: *New Ideas in Psychology.* He is currently completing a book on naturalistic psychological epistemology (*Genetic Epistemology: Developmental Psychology and Naturalistic Epistemology*), and is working on a book on philosophical issues in developmental psychology (*The Logic of Development*). He is Professor of Philosophy at Colorado State University, married, with two children. He likes skiing, backpacking, biking, and golf. He also sings in a barbershop quartet.